T0136974

Quantum Computing in the Arts and Humanities

Eduardo Reck Miranda
Editor

Quantum Computing in the Arts and Humanities

An Introduction to Core Concepts, Theory and Applications

 Springer

Editor
Prof. Eduardo Reck Miranda ⓘD
Interdisciplinary Centre for Computer
Music Research (ICCMR)
University of Plymouth
Plymouth, UK

ISBN 978-3-030-95540-3   ISBN 978-3-030-95538-0   (eBook)
https://doi.org/10.1007/978-3-030-95538-0

This Springer imprint is published by the registered company Springer Nature Switzerland AG
The registered company address is: Gewerbestrasse 11, 6330 Cham, Switzerland

Preface

Quantum Computing Beyond Science

Publications reporting research into quantum information processing started to emerge around the middle of the 1970s; e.g. Holevo (1973), Poplavskii (1975) and Ingarden (1976), to cite but three. Then, groundbreaking work by Feynman (1982) glimpsed at the possibility of developing a quantum computing device: Feynman proposed a mathematical quantum mechanical system capable of performing computations. This possibility commenced to break free from academic circles into the realm of industry after Deutsch (1985) proposed an abstract Turing-like quantum machine capable of universal computation.

The Turing machine is a widely accepted model of computation, which informed the development of commercial classical computers, as we know them today. With this in mind, the industry and investors gained confidence that it might be possible to harness quantum mechanics to build a somewhat different type of computer. That is, a computer that might be able to efficiently tackle problems that are not easily tractable today, such as predicting how biological molecules interact and making sense of colossal amounts of information. We will need to tackle these problems more efficiently than ever to overcome contemporary and future challenges. These include, for instance, climate monitoring and the development of vaccines to combat new pathogens. Deutsch's machine also paved the way for the development of what is known as the 'circuit model' for programming quantum computers, which is the norm nowadays.

Since then, large consumer electronics companies and many start-up ones have made tremendous progress in the race to build quantum computers, in particular in the last decade. But still, there is some leg to go before these machines become mainstream, if at all. Nevertheless, research and development are progressing fast in universities and companies all over the world.

Despite the hype surrounding this new computing technology, sceptics question the benefits of quantum computers over classical ones. What will a quantum computer be able to do that a standard computer would not be capable of? Today's high-performance supercomputers are rather powerful. And they will continue to improve. Consider the Fugaku supercomputer, developed by Fujitsu and RIKEN in Japan: it performs circa 600 quadrillion operations per second. To put this in perspective, a top of the range desktop computer does approximately 100 billion operations per second. That is, Fugaku is six million times faster than a decent desktop computer.

All the same, quantum computing is here to stay. What started as a desire to build a machine for physicists to study quantum mechanics[1] has been evolving towards the ambition to develop general-purpose quantum computing technology for a wide range of applications, some of which we have not even dreamed of yet. And more, I advocate that processing speed is not the only benefit that such machines will bring to society. They will bring new philosophies, new approaches to design, new algorithms, new creative artefacts, new economies and so on. In the long run, using an actual quantum computer for mundane applications may as well become a matter of choice or luxury rather than necessity. Who knows?

Much ongoing research into quantum computing worldwide focuses on improving hardware and developing solutions for science and technology. Nevertheless, there also is growing activity within areas where quantum computing may impact beyond science and technology, including the arts, humanities and fringe interdisciplinary initiatives. These are exciting because they might influence developments in the quantum computing industry; e.g., create unforeseen new markets.

This book brings a collection of chapters by scholars championing research in their respective areas (such as Philosophy, Linguistics and Music, amongst others) informed by quantum computing. The authors offer thought-provoking discussions about cognition, perception, language, music, games, visualization of the sub-atomic world, brain-machine communication and more.

I am indebted to all contributing authors who enthusiastically supported this book project. They made possible what seemed impossible. We have here what is probably the first book ever on Quantum Computing in the Arts and Humanities.

Plymouth, UK   Eduardo Reck Miranda
October 2021

References

Deutsch, D. (1985). Quantum theory, the church-turing principle and the universal quantum computer. *Proceedings of the Royal Society of London A*, *400*, 97–117.
Feynman, R. P. (1982). Simulating physics with computers. *International Journal of Theoretical Physics*, *21*, 467–88.

---

[1] The impetus to develop a quantum computer stemmed from the fact that some simulations to study quantum mechanics would allegedly require computing power that even state of the art supercomputers might take years to run. The idea then is to harness quantum mechanics to build quantum computers rather than simulate quantum mechanics on standard computers.

Holevo, A. S. (1973). Bounds for the quantity of information transmitted by a quantum commu-
    nication channel, *Problemy Peredachi Informatsii*, *9*, 3–11. English translation in Problems of
    Information Transmission, *9*, 177–83.
Ingarden, R. S. (1976). Quantum information theory. *Reports on Mathematical Physics*, *10*, 43–72.
Poplavskii, R. P. (1975). Thermodynamical models of information processing (in Russian), *Uspekhi
    Fizicheskikh Nauk*, *115*, 465–501.

Contents

From Digital Humanities to Quantum Humanities: Potentials and Applications

Johanna Barzen

Abstract Quantum computers are becoming real. Therefore, it is promising to use their potentials in different application areas, which includes research in the humanities. Due to an increasing amount of data that needs to be processed in the digital humanities the use of quantum computers can contribute to this research area. To give an impression on how beneficial such involvement of quantum computers can be when analysing data from the humanities, a use case from the media science is presented. Therefore, both the theoretical basis and the tooling support for analysing the data from our digital humanities project MUSE is described. This includes a data analysis pipeline, containing e.g. various approaches for data preparation, feature engineering, clustering, and classification where several steps can be realized classically, but also supported by quantum computers.

Keywords Quantum computing · Quantum humanities · Machine learning · Quantum machine learning · Digital humanities · Data analysis · Artificial neural networks · Pattern languages

1 Introduction

As quantum computers are becoming real there are several application areas where the use of quantum computers is expected to be especially promising (National Academies of Sciences, Engineering, and Medicine, 2019): one is simulation, focusing on molecular properties used in material science or in pharma industry, and the other is machine learning e.g. classification, clustering or dimension reduction making heavy use of optimization.

The main difference between classical computers and quantum computers is the different information unit used, namely bits for classical computers and *qubits* for quantum computers (Nielsen & Chuang, 2010). Qubits are not restricted to the two states 0 and 1 but can be in an infinite number of different combinations of these

J. Barzen (✉)

Institute of Architecture of Application Systems (IAAS), University of Stuttgart, Stuttgart, Germany

e-mail: johanna.barzen@iaas.uni-stuttgart.de

© Springer Nature Switzerland AG 2022

E. R. Miranda (ed.), *Quantum Computing in the Arts and Humanities*, https://doi.org/10.1007/978-3-030-95538-0_1

states, referred to as *superposition*. Furthermore, individual qubits can be combined into a quantum register resulting in exponentially growing number of data that such a register can hold. As an example, a quantum register with 50 qubits corresponds to 2^{50} possible combinations of the states of the individual qubits which are again in superposition corresponding to approximately a petabyte of classical data. All these values can be manipulated by a single operation all at once, which is called *quantum parallelism*.

Operations on single qubits or quantum registers, as a combination of qubits, are unitary transformations. They can be used to process the data to achieve everything classical computers can perform (Rieffel & Polak, 2011) but to also use quantum inherent phenomena, e.g. *entanglement*, to—among other potentials—solve a variety of problems that were practically not solvable before, some even with exponential speedup (Nielsen & Chuang, 2010; Preskill, 2018). There are different quantum computing models, e.g., gate-based (Michielsen et al., 2017), measurement-based (Jozsa, 2006), and adiabatic quantum computing (Aharonov et al., 2008), which represent quantum algorithms in various ways. As it can be shown that the different models are formally equivalent (Aharonov et al., 2008; Jozsa, 2006) this chapter restricts the considerations and implementations to the gate-based quantum computing model. Also, quantum computers based on the gate-based quantum computing model are universal quantum computers and most commercially available quantum computers are based on this model (LaRose, 2019).

As different vendors, such as IBM or Rigetti, developed quantum computers in recent years and offer access to them via the cloud (LaRose, 2019; Leymann et al., 2020) quantum computing is becoming a lively research field and first real word applications are developed in industry. When taking a closer look at the research areas quantum computing is applied to, there are mainly applications in the natural sciences that can be found, for example general approaches like (Bhaskar et al., 2015), presenting quantum algorithms and circuits for modules implementing fundamental functions (e.g. the square root, the natural logarithm, and arbitrary fractional powers) or more specific approaches as in molecular simulation in the material sciences (Kandala et al., 2017; McClean et al., 2017) or in artificial intelligence and machine learning (Dunjko et al., 2016; Havlicek et al., 2018; Otterbach et al., 2017). Nevertheless, first applications can be identified in the humanities too, e.g. *quantum social science* as interdisciplinary field which draws parallels between quantum physics and the social sciences (Haven & Khrennikov, 2013), but this is still to be called rudimentary in respect to the potential quantum computers can have in different areas of the humanities.

The chapter is structured as follows: Sect. 2 presents the vision for quantum humanities. Therefore, the potential benefits that can be expected but also the current challenges are outlined. Section 3 introduces the digital humanities project MUSE as use case for quantum humanities. As the use case strongly relies on analysing data Sect. 4 focuses on introducing our data analysis pipeline, how to cope with categorical data, and how pattern languages based on data analysis can be detected. A core concept of data analysis is artificial neural networks. Therefore, Sect. 5 provides a mathematical definition of neurons, neuronal networks, and perceptrons, outlines how they are used for restricted Boltzmann machines and autoencoders, and sketches their realizations on quantum computers. As many quantum algorithms are hybrid, meaning they consist of both, a quantum part and a classical part, Sect. 6 focuses on variational hybrid quantum–classical algorithms. The main idea is outlined as well as an application for clustering, namely a quantum version for solving the maximum cut problem. Therefore, techniques as quantum approximate optimization algorithm (QAOA) and variational quantum eigensolver (VQE) are discussed. To support access to the described techniques, Sect. 7 focuses on our quantum humanities analysis tool (QHAna) that is used for pattern detection. In addition to the substantive goal of the use case from the MUSE project to extract costume patterns in films, QHAna provides easy access for people without quantum computing background to evaluate the benefits of using quantum algorithms and allows them to gain initial application knowledge in the field of quantum humanities. Since both classical analysis algorithms and their quantum counterparts are included, the comparison of their results allows a deeper understanding within quantum humanities. Section 8 concludes the chapter and gives an outlook on future work in quantum humanities.

How to read this chapter: As quantum humanities brings together very heterogenous disciplines (e.g. the respective humanities discipline, physics, mathematics, computer science) and each has its own approaches, terminology, concepts and research culture, a fundamental challenge of quantum humanities is to find a common language to communicate between these disciplines. Thus, the chapter starts with an easy-to-understand introduction to the overall topic, the vision and potentials that does not require any previous knowledge (Sects. 1–4). The underlying ideas and concepts get more and more refined, up to mathematical definitions (Sects. 5–6). We use mathematical definitions as they give little room for interpretation and therefore seems appropriate as long as there is no common language yet. As the mathematical definitions may be challenging to some readers they are accomplished by descriptions and pictures giving a high-level view on the core concepts, so some readers may skip over the mathematical formalisms. Please note that algorithmic details themselves are not the focus. Instead, the underlying concepts of the (classical and quantum) algorithms are provided that are relevant for our approach, as well as the tooling provided by QHAna. QHAna than outlines the practical usage of the introduced concepts in a use case centric manner (Sect. 7).

2 Towards Quantum Humanities

How beneficial the use of computers, their practical support as well as techniques and methods from computer science can contribute to research in the humanities has been proven by the establishment of the digital humanities. With quantum computers becoming available, it is promising to extend the usage of classical computers as done in the digital humanities by the use of quantum computers. As stressed in the introduction, working with quantum computers, programing algorithms and letting them run on quantum hardware is very different to working with classical computers.

Therefore, we coined the term *quantum humanities* to describe addressing research in the humanities with methods and techniques from quantum computing (Barzen & Leymann, 2019; QH, 2021). As the basis for quantum humanities the following section will outline the core components defining digital humanities.

2.1 Digital Humanities

When speaking about *digital humanities* the term combines a broad variety of topics, contents and contexts. This is reflected by the continuously growing amount of literature on divergent approaches in this domain (Berry, 2012; Berry & Fagerjord, 2017; Brown, 2020; Jannidis et al., 2017; Terras et al., 2013). But there are three core elements that can be identified when taking a closer look at approaches that define digital humanities: (i) digital humanities is about bringing together humanities research and information technology, (ii) it is a diverse field, and (iii) it is a still emerging field (DHQ, 2021).

The first of those highlighted three core elements stresses the combination of computer science and the humanities research. Digital humanities therefore are a highly inter- and trans-disciplinary field operating between the traditional humanities and the (in comparison) rather young computer science. The aim is to examine existing and new questions from the humanities by using concepts, methods and tools from computer science. Therefore, digital humanities is understood as a field of methods that unites different topics, approaches and sub-disciplines and as such fosters synergies between different scientific disciplines.

This is rather complex and hints to core element two: digital humanities are a diverse field. 'Humanities' already encompasses a large variety of disciplines, such as philosophy, linguistics, arts, cultural studies, archaeology, musicology and media science, to name just a few. It gets even broader when taking into account that also "computer science" covers quite different subject areas like databases, visualization, or theoretical computer science. In the compilation of so many fields of interest, digital humanities is a very heterogeneous area. It attempts to include the application as well as the further development of tools and methods of computer science in order to possibly obtain new and profound answers to research questions in the humanities while performing a critical self-reflection of the methodological and theoretical approaches.

Even though discussions on digital humanities are around for about 60 years and core areas such as computational linguistics have established themselves through their own courses of study, digital humanities is still a young field of research that has only recently become an independent discipline (Rehbein & Sahle, 2013). This is the third core element of many definitions of digital humanities: Digital humanities is a young field whose own definition is constantly being developed (What is Digital Humanities? 2021).

Nevertheless, a broad variety of research programs and projects, conferences and competence centers, stress the benefits of using computers as tools as well as methods from computer science to support research in the humanities. Techniques like databases, visualization, or distant reading have shown how the use of the classical computer can contribute to access knowledge and to broaden the approaches for addressing already stated problems. Now, with quantum computers there is a new type of computer just becoming reality and it allows to further improve the way in which computers can support research in the humanities. Combining the power quantum computer promises with research done in the humanities—based on the achievements in digital humanities research—is what we call *quantum humanities* (Barzen & Leymann, 2019).

2.2 Potential Benefits of Quantum Humanities

As quantum computers are superior to classical computer in various areas, their use has a high potential contributing to research in the humanities. Especially, since quantum computers are becoming more easily accessible—for example via the cloud (Leymann et al., 2020)—their promising potentials are more and more applicable for first 'proof of concept' application as outlined in Barzen and Leymann (2020) and will be detailed in Sect. 7. In the following, seven potentials (Barzen & Leymann, 2019) will be outlined, which promise to be particularly interesting for supporting digital humanities:

(I): The core benefit of quantum computers is an often stressed potential speedup in several problem areas. Meaning, quantum computers solve certain types of problems much faster (Rønnow et al., 2014) than classical computers (e.g. in the decoding of passwords (Shor, 1995), the determination of global properties of Boolean functions (Deutsch, 1985) or unstructured search (Grover, 1996)). As the amount of data to be processed in the digital humanities is continuously growing, the rapid evaluation of data is becoming increasingly important. Since many algorithms developed for quantum computers can make certain statements about global properties of functions much faster than algorithms developed for classical computers, an enormous time saving can be assumed here. In this context, time saving does not only mean a mere gain in speed of the proper algorithms, but methods of dimensional reduction, for example, that reduce the amount of data to be processed by the proper algorithms, can play a major role here—and the latter can also be performed much more efficiently by quantum computers (e.g. by quantum principal component analysis).

(II): Quantum computers can process large amounts of data in a single step (Nielsen & Chuang, 2010). Due to the superposition of all possible states of the quantum mechanical system, the quantum computer allows true parallelism in the calculation (manipulations), which leads to a significantly higher computing power. This in turn means that complex computational problems, such as those increasingly found in the digital humanities, can be dealt with much more effectively.

(III): The results of quantum computers promise much higher precision in certain areas than those of classical computers (Havlicek et al., 2018; Sierra-Sosa et al., 2020). As the use of quantitative methods more and more leads to the evaluation of data by means of data analytics, for example techniques from the field of machine learning such as clustering and classification, the quality of the results in terms of precision is becoming increasingly important. Various new approaches, for example for support vector machines (SVM) (Havenstein et al., 2018; Havlicek et al., 2018), or for variational quantum classifier (VQC) (Sierra-Sosa et al., 2020), offer different promising fields of application in digital humanities.

(IV): In addition, the use of quantum computers makes it possible to solve problem classes that were previously considered practically unsolvable (e.g. complexity class BQP (Nielsen & Chuang, 2010)). Hereby quantum computers enable approximate solutions for otherwise algorithmically inaccessible problem classes. To what extent and for which fields of applications in the digital humanities this can be beneficially used is still to be explored, but it opens up a whole field of potential questions whose solutions now deem to be possible. A fist promising use case to stress this potential is introduced by (Palmer, 2020). Here natural language processing is performed on a quantum computer using the power of quantum states to code complex linguistic structures and novel models of meaning in quantum circuits to understands both, grammatical structure and the meaning of words.

(V): Finally, there are certain types of problems that can only be solved by a quantum computer, i.e. these types of problems can be proven not to be solvable at all by a classical computer (Raz & Tal, 2018). Being able to solve new types of problems, the digital humanities have the chance to work on corresponding questions that have not yet been tackled at all, perhaps not even identified or considered. To identify and explore these possible new application areas is an exciting and promising task.

(VI): The use of a quantum computer promises to be significantly cheaper than that of a conventional supercomputer: a price of about 200€ per hour compute time on a quantum computer can be assumed (Dickel, 2018). This can be of benefit for the often financially strained humanities, especially for the feasibility of smaller research projects by providing low-cost access to high-performance hardware.

(VII): Furthermore, quantum computers are much more energy efficient than computers with comparable performance (Nielsen & Chuang, 2010). Without having a direct influence on digital humanities, the indirect influence should certainly be mentioned. Due to their energy efficiency quantum computers contain a great potential for change, especially in times of climate change.

The above potentials reveal that it is of great importance to develop application knowledge in the domain of quantum computing at an early stage, especially in a field where the mathematical and physical basics required to use this technology cannot be taken for granted.

2.3 Current Challenges of Quantum Humanities

The description of the potentials outlined above are from an abstract point of view and must be considered much more comprehensively in its complexity, taking into account various factors from related issues of quantum computing as well as from digital humanities.

From a hardware perspective, the potentials stated above apply to an "ideal" quantum computer, which is not yet available. Today's quantum computers are error-prone and have only limited capabilities (Preskill, 2018; Leymann & Barzen, 2020a). They provide only a small number of qubits (LaRose, 2019) allowing only a limited set of input data to be represented within the quantum computer. Besides this, noise affects calculations on quantum computers (Knill, 2007; Preskill, 2018): The states of the qubits are only stable for a certain amount of time—an effect referred to as decoherence—due to unintended interactions between the qubits and their environment (Leymann et al., 2020; Nielsen & Chuang, 2002). In addition, the operations that manipulate the qubits are also error-prone. Nevertheless, so-called NISQ (Noisy Intermediate-Scale Quantum) machines (Preskill, 2018) allow first applications which can be used for prototypical implementations addressing small real-world problems. This can contribute to establish application knowledge—also in the field of quantum humanities. Taking a closer look at the recently published quantum roadmap of IBM (Gambetta, 2020) it emphasizes the importance of starting to build application knowledge now: Here IBM outlines its roadmap to 1 million error-corrected qubits within a decade, providing more than 1,000 qubits in 2023, continuously increasing the number of qubits for the coming years.

Next to the issues related to the hardware of quantum computers, from the software perspective too several questions need to be addressed when working with quantum computers. Even though quantum computers are becoming commercially available (e.g. IBM, D-Wave, Rigetti, Honeywell, IONQ), questions regarding usability and accessibility must be considered for each vendor. Also, more concrete questions like the following need to be taken into account: How to encode your data properly based on the processing demands of the chosen quantum algorithm (Weigold et al., 2021)? How to expand oracles that many algorithms contain to keep the influence of such an expansion on the number of qubits and operations required small? How to cope with readout errors? (Leymann & Barzen, 2020a).

It is also necessary—even if the quantum computer is superior to classical computers in certain areas as stated above—to address several open questions from the application side. In a first step those existing and new problems from the digital humanities need to be determined that are suitable to be considered being solved by quantum computers. Quantum humanities combines all the different disciplines subsumed under the humanities with disciplines like computer science, mathematics, and physics from the quantum computing site. This makes it an even more heterogenous field than the digital humanities is already and demands a high degree of translation capabilities. Defining terms that are used differently in the various disciplines to establish a shared language is one of the fundamental tasks for quantum humanities.

Also, when focusing on the more "pragmatical" benefits by using quantum computers as a tool for answering concrete questing from the digital humanities, e.g. how to run parts of a classification algorithm in a machine learning scenario on quantum hardware, lots of questions still remain and need to be answered. For example: How to choose the best algorithm to address the stated problem? How to cope with categorical data while most algorithms require numerical data?

Based on the following use case we want to contribute reusable knowledge for different applications in quantum humanities and introduce a toolchain supporting data analysis by including quantum computers in the process.

3 Quantum Humanities Use Case: Project MUSE

In recent years, more and more research projects in the field of digital humanities can be identified that are based on data and data analysis to provide new insights into questions stated in the humanities. For example, when taking a closer look at the books of abstracts of the DHd (Digital Humanites im deutschspachigen Raum (DHd, 2021)) conferences—as a well-established conference in the digital humanities—over the last three years (Sahle, 2019; Schöch, 2020; Vogeler, 2018) the term *data* (and its German translation, as well as all compound words containing the term) occurs more than 5,600 times and hint to the significance of data in digital humanities research. Also, in analogy, terms hinting to analysing this data via techniques from machine learning (e.g. searching for: machine learning, artificial intelligence, un-/supervised learning, clustering and classification) has about 1,000 counts, giving an impression of the importance of this approach to analyse data gained during the last years. Since quantum computers are expected to have substantial superiority especially in the field of machine learning (Schuld & Petruccione, 2018), the use case presented in the following is positioned in this area. Here, data from the digital humanities project MUSE is analysed with the of help machine learning and quantum machine learning techniques.

The project MUSE (2021) aims at identifying a pattern language for costumes in films. It contains the method as well as a supporting toolchain to capture all relevant information about costumes, to analyse them and to abstract the significant information into costume patterns. Costume patterns capture those significant elements of a costume (e.g. colour, material, way of wearing) that are used to communicate a certain stereotype or a character trait, for example. When speaking of a "nerd" or a "wild west outlaw" most recipients do have a rather clear idea of what these stereotypes should look like to be recognized as "nerd" or "wild west outlaw" based on media socialization. Costume designers compose their characters by choosing each part of a costume, its specifics and attributes with the explicit intent to support the character, the story, and the actor in achieving a specific effect (Barzen & Leymann, 2014; Schumm et al., 2012). As there are often similar clothes, colours, or materials used to communicate the same stereotype there are conventions that have been developed and knowledge about these conventions is contained in films.

To extract this knowledge the MUSE method consists of five steps (Barzen, 2018; Barzen et al., 2018) which are supported by different tools and are outlined in the following. The method has been proven to be generic by applying it in our parallel project MUSE4Music (Barzen et al., 2016).

3.1 MUSE Ontology

Step 1: Define the domain by means of a comprehensive ontology which might be based on several taxonomies (Barzen, 2013, 2018). These taxonomies structure all relevant parameters that have a potential involvement in the effect a costume might have. The MUSE ontology includes taxonomies of base elements (e.g. pants, shirts, jumpers) and primitives (e.g. sleeve types, zipper, collar types), their materials, designs, colours, way of wearing, conditions, and functions—to name just a few— as well as so-called operators (e.g. worn above, wrapped around, attached) turning base elements into a composed outfit. The hierarchical structure of the taxonomies is important when analysing the data. As various algorithms require numerical data the categorical data of MUSE must be transformed into numerical data based on the structure of the taxonomies, for example (see Sect. 4.2, for more details see Barzen et al. (2021)). With more than 3150 nodes, a comprehensive ontology to describe costumes in a very detailed and structured manner has been developed.

3.2 MUSE Film Corpus

Step 2: Identify—based on well-defined criteria—those films that have a potential "big" impact in terms of costumes on the recipients. Therefore, in a first step genres were determined that promise a reoccurrence of quite similar characters and stereotypes. For MUSE the genres high school comedy, western movies and fairy tales were chosen. In a second step within each of these identified genres the 20 films with the highest box office numbers and scores in rankings were chosen as part of the initial film corpus to be analysed (for more details see Barzen (2018)).

3.3   MUSE Data Set

Step 3: Capturing detailed information about the costumes of the initial film corpus. To support capturing of all relevant parameters, a repository was designed and implemented: The MUSE-repository (an open-source implementation is available under (MUSE GitHub, 2021)) is based on the ontology described in Sect. 3.1 and assists to collect all information about the films of the corpus, the characters and the costumes. Currently (March 2021), the data set contains 5,231 costumes described from 58 films. These costumes contain 28,798 base elements and 63,711 primitives, 159,360 selections of colours and 180,110 selections of materials.

3.4   MUSE Data Analysis

Step 4: Analysing all captured information about costumes to determine those costumes and costume attributes that achieve a similar effect in communicating with the recipient hinting to costume patterns. The analysis consists of two main steps (Falkenthal et al., 2016a): The first step applies data mining techniques, e.g. association rule mining, to determine hypotheses. These hypotheses could, for example, identify those costume elements that are used to communicate a certain character trait like "active person" or "shy person". To refine and verify these hypotheses in the second step, online analytical processing (OLAP) techniques (Falkenthal et al., 2015) are used. As a result, indicators for costume patters are determined.

To improve the process of building hypotheses we are currently extending the analysis of the MUSE data by various techniques from machine learning as well as quantum machine learning (Barzen et al., 2021; Barzen & Leymann, 2020). A more detailed discussion on the currently used methods and techniques is given in the following Sects. 4–7.

3.5   MUSE Costume Patterns

Step 5: Abstracting the results of the analysis step 4 into costume patterns. Patterns in the tradition of Alexander et al. (1977) are documents that follow a pre-defined format to capture knowledge about proven solutions in an abstract way to make this knowledge easily accessible and reusable for other applications. Patterns are related to each other and compose a pattern language based on these relations. As stated above, costume patterns capture the proven solutions about how costume designers address the problem to communicate a certain stereotype like a "wild west sheriff" in terms of all the significant elements of the costume. This contains e.g. base elements, primitives, their relations, colours, ways of wearing, material, if they proved to be significant in the analyse step. As the costume patterns are part of a costume pattern language, they support to solve complex problems by browsing through different solutions related to each other.

To support the accessibility of the costume patterns, tooling support is provided by our generic pattern repository called *Pattern Atlas* ((Leymann & Barzen, 2020b), based on our former pattern repository *PatternPedia* (Fehling et al., 2014)).

4 Analysing Data

Analysing data has two major purposes: discovery and prediction. The collection of techniques focusing primarily on the first purpose is called data mining, and the set of techniques focusing primarily on the second purpose is called machine learning. Despite their different primary purposes, data mining and machine learning have a large overlap in terms of the methods they use: Optimization and statistics are core of both approaches, and data mining is even using selective techniques from machine learning. This is why both disciplines are often subsumed by the term data science.

Another difference is how the suitability or appropriateness of a data mining or a machine learning solution is assessed: The appropriateness of a data mining solution is assessed by its capability to discover previously unknown facts, while the appropriateness of a machine learning solution is assessed by correctly reproducing already known knowledge. Once a machine learning solution reproduced known knowledge the solution is considered to correctly deliver previously unknown knowledge.

Despite these differences, the development of a solution based on either of both disciplines follows the same overall procedure which is described in the next section.

4.1 Data Analysis Pipeline

In the past, we used data mining technologies to analyse data about costumes in films (see Sect. 3.4, for more details; see Falkenthal et al., 2016a; Barzen, 2018). The general procedure shown in Fig. 1 has been used for this analysis, but it is also applicable in analysing data with machine learning techniques (Sect. 5).

Data cleansing (Skiena, 2017) encompasses activities like data acquisition, format conversion and so on. Data preparation (Skiena, 2017) deals with a proper treatment and encoding of data, and feature engineering. Algorithm selection (Kubat, 2017) determines the family of algorithms (e.g. classification, clustering), selects a proper

Fig. 1 Major steps of the data analysis process chain

family member, and choses hyperparameters of the selected algorithm. Result evaluation is often based on visualizing the results (Skiena, 2017) and the creation of hypotheses, which are finally validated. Because these last steps are critical for the success of a data analysis project they are depicted as separate steps in Fig. 1. If no hypothesis can be built or successfully validated, other algorithms may be tried out to finally succeed with at least one proper hypothesis. Successfully validated hypotheses trigger further processing towards the final goal of the project, like finding patterns (see Sect. 4.3).

Note that within a data analysis project, most time is typically spend in data cleansing and data preparation (Skiena, 2017). Also, several algorithms, even several algorithms of the same family of algorithms, are typically applied within an overall data analysis project. And each algorithm applied involves performing several steps of the process shown in Fig. 1. For example, in order to determine patterns of costumes in films in MUSE, the mostly categorial data must be prepared, features must be engineered, and the resulting data must be clustered (see the black shapes in Fig. 2). After clusters have been determined, data about newly captured clothes can be classified to identify which costume the clothes represents.

Each of these tasks can be achieved by a variety of algorithms (grey shapes below black shapes). Data preparation, for example, involves one-hot encoding to turn categorial data into binary vectors, or the set of data points representing clothes may be turned into a distance space based on using Wu-Palmer similarity (see Sect. 4.2). Feature engineering may be performed by training an autoencoder (see Sect. 5.5) or multi-dimensional scaling after the transformation of the data into a distance space.

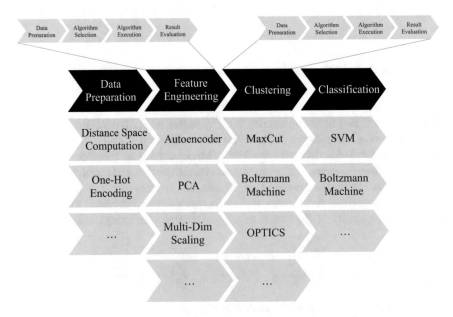

Fig. 2 Data analysis pipeline for determining patterns and succeeding classification

Clustering may be based on solving the maximum cut problem (see Sect. 6.2), training a corresponding Boltzmann machine (see Sect. 5.4), or using the k-means algorithm (Arthur & Vassilvitskii, 2007). Classification can be done by means of a Boltzmann machine too, or by using a SVM.

Furthermore, each of the tasks depicted as black shapes in Fig. 2 is in fact performed as a process chain (see the small process chains refining two of the black shapes). For example, feature engineering requires to select a corresponding algorithm, e.g. multi-dimensional scaling and setting proper hyperparameters (see Sect. 7.3). The corresponding embedding needs the data in a certain format which has to be prepared. After executing the algorithm, the resulting data representation must be evaluated; if it is not appropriate, the hyperparameters must be adapted, and the algorithms must be run again. If the results based on this algorithm is not appropriate at all, another algorithm (e.g. principal component analysis (PCA)) must be tried out. Once the features are properly engineered, clustering is performed, which means that the corresponding process chain is executed. Thus, a *pipeline* results that consists of a sequence of several process chains.

4.2 Categorical Data

Many data in the Humanities are categorical data, i.e. data with a finite set of values, often strings, that cannot be used for calculations. For example, computing the maximum of a set of textures or the average of jobs is meaningless. But most machine learning algorithms presume numerical data or even metrical data. Thus, categorial data must be transformed into metrical data to be able to be processed by such algorithms.

Barzen et al. (2021) discussed a mechanism of how to turn tuples of categorical data the domains of which are tree structured into an approximately equivalent set of vectors in $\mathbb{R}^n$, i.e. into a set of metrical data. Our mechanism is (i) based in the Wu-Palmer similarity (Wu & Palmer, 1994), (ii) the fact that similarities can be turned into distance measures, and (iii) that a finite set with a distance measure can be embedded into an appropriate $\mathbb{R}^n$ by means of multidimensional scaling (MDS) (Cox & Cox, 2001): see Barzen et al. (2021) for all the details.

Data elements the values of which are given by means of a taxonomy are such data with a tree structured domain. In our application area of costumes in films (Barzen, 2018), most data types have domains defined by taxonomies. Thus, based on our mechanism this data can be embedded into a vector space and, consequently, can be processed by machine learning algorithms.

Figure 3 shows a small fraction of one of the taxonomies of our costume data set. To compute the similarity of two nodes in a tree (e.g. a "swimming shorts" and a "cycling shorts" in the figure) the lowest common ancestor of these two nodes must be determined: in our example, this is the node "short pants". Next, the length of the paths (in terms of number of edges) from each of these nodes to their lowest common ancestor must be determined: in our example, these lengths are $L_1 = 2$ and $L_2 = 1$.

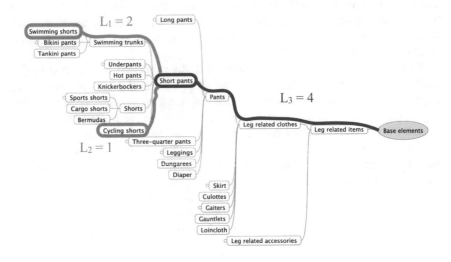

Fig. 3 Example for using the Wu-Palmer similarity measure

Then, the length of the path from their lowest common ancestor to the root of the tree must be determined: in our example, this length is $L_3 = 4$. Finally, the *Wu-Palmer similarity* is defined as

$$\omega(\text{swimming shorts, cycling shorts}) = \frac{2L_3}{L_1 + L_2 + 2L_3} = \frac{8}{2 + 1 + 8} = \frac{8}{11} \approx 0.72.$$

The similarity between two nodes is between 0 and 1, i.e. "swimming shorts" and "cycling shorts" are quite similar. A first quick overview of the similarities between the data elements of a data set can be given by a *similarity matrix* that presents the similarities of pairs of data elements and may render these values by means of a colour code (see Sect. 7.1 for an example).

The method to determine the similarity ω between two values of a tree structured domain (part 1 in Fig. 4) can be extended to determine the similarity of two sets of values of the same tree structured domain: in a nutshell, the similarities ω of all pairs is determined whose first component is from the first set and whose second component is from the second set; a certain mean value of these similarities is computed, resulting in a similarity measure σ between two sets (part 2 of Fig. 4). Based on the similarity measure σ the similarity of tuples of the same dimension can be determined whose components are sets of values of the same tree structured domain (part 3 in Fig. 4): the similarities between each components of the tuples is determined (e.g. the similarity of the colours of two tuples representing costume A and costume B, the similarity of their materials and so on); the mean value of these similarities is computed which results in a similarity measure μ between two tuples. See Barzen et al. (2021) for the details.

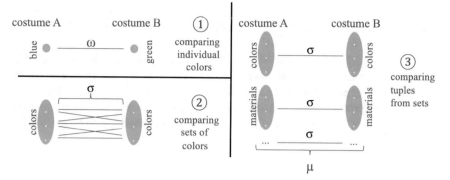

Fig. 4 Computing similarities of sets and tuples

The similarity $\omega(i, j)$ between two nodes i and j (and the similarity $\sigma(i, j)$ of two sets or the similarity $\mu(i, j)$ between two tuples i and j, respectively) can be transformed into the dissimilarity or *distance* $\delta(i, j)$ between i and j, respectively, by means of

$$\delta(i, j) = \sqrt{\omega(i, i) + \omega(j, j) - 2\omega(i, j)}$$

with $\omega(i, i) = \omega(j, j) = 1$ (substituting ω by σ or μ in the formula accordingly). This way, a set of data points the tuples of which have components whose domains are define by taxonomies can be transformed into a set with a distance measure. In analogy to a similarity matrix the distances between the data elements are used to build a *distance matrix*.

Figure 5 depicts this situation: A set M with a distance measure δ is given, and the distances of the elements of M are indicated by the length of the lines between the points. An *embedding* ε maps the set M into $\mathbb{R}^n$, i.e. the elements of M become vectors in $\mathbb{R}^n$ (often referred to as *feature space*). But the vectors are not chosen arbitrarily but the mapping ε tries to keep the distances in M approximately the same than the distances in $\mathbb{R}^n$, i.e. the distances of the corresponding vectors (indicated by the length of the dashed lines between the vectors in $\mathbb{R}^n$) are approximately the same as the distances in M; the distances in $\mathbb{R}^n$ are measured by some norm $\| \cdot \|_p$. Here,

Fig. 5 An embedding maps a set with a distance measure into a Euclidian vector space

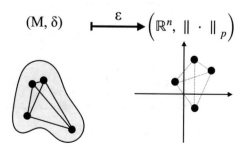

"*n*" and "*p*" are hyperparameters that must be chosen based on former experience. Embeddings can be computed by multidimensional scaling (MDS) (Cox & Cox, 2001).

This way, a set of data points M with components whose domains are defined by taxonomies can be transformed into a set of data points in $\mathbb{R}^n$ such that the distances in M are approximately the distances in $\mathbb{R}^n$. After this transformation, the data points can be further processed by most machine learning algorithms. Our costume data set is transformed in this manner (see Sect. 7.1) and is, thus, available to be analysed by several machine learning algorithms.

4.3   Creating Pattern Languages Based on Data Analysis

The main purpose of our project MUSE is the creation of pattern languages of costumes in films. Pattern languages can be built in several ways. Often, they are built from experience: practitioners recognize that a particular problem occurred repeatedly in the past perhaps in various contexts and that the solutions applied to solve the problem are very similar. Thus, these solutions can be abstracted to identify their underlying principles such that these principles can be applied to solve that particular problem even in new contexts. This way, a new pattern has been identified. Similarly, relations between patterns are established based on experience. Those relations between patterns become links with an associated semantics.

What results is a pattern language: A *pattern language* is a collection of patterns within a particular domain and their relations. From an abstract point of view, a pattern language is a weighted, directed graph. The nodes of this graph are the patterns, the edges are the links between the patterns, and the weights of the edges are the semantics of the links.

In contrast to derive patterns from experience, our project MUSE strives towards deriving patterns by means of analysing data. The method we developed in MUSE is generally applicable, and its use has already been initially verified also in the domain of music (Barzen et al., 2016) and is envisioned by other aspects of films (see Falkenthal et al., 2016b). Applying the MUSE method in other areas of the humanities successfully seems to be possible. After having used data mining techniques in our method first, we are now exploiting machine learning techniques and the initial results we get seem to be more promising than using mining techniques.

To identify a pattern language based on data analysis the process sketched in Fig. 1 in Sect. 4.1 must be extended (see Fig. 6). If the hypothesis validation step resulted

Fig. 6 Creating a pattern language

in a verified hypothesis, this hypothesis must be turned into a pattern document. That is the various sections of the pattern document corresponding to the hypothesis must be authored. Thus, a pattern candidate is created: the resulting document is "only" a candidate but not yet a pattern because its general applicability still needs to be verified. Such a verification is performed in the succeeding pattern verification step. Reiners (2014) proposes a system based on which a community e.g. can discuss the candidate, exchange experiences with it, and can jointly decide to promote the candidate to a pattern. The new pattern must be related to the other patterns of the domain in the pattern relations step. This way a pattern language results. Note that a pattern language is a "living" entity, i.e. over the time patterns will be added to the pattern language. Typically, a first version of a pattern language will be made available or published when "major" problems of the underlying domain can be solved with it.

5   Artificial Neural Networks

An artificial neural network (or neural network for short) is a mathematical model that mimics the biological neural networks of brains of animals. Such a neural network is represented as a directed graph the nodes of which abstract neurons and the edges abstract synapses of a brain. In this section, neurons and neural networks are defined, perceptrons and their use for classification are sketched, and restricted Boltzmann machines, as well as autoencoders are described.

5.1   Neurons

The first model of an artificial neuron roots back to McCulloch and Pitts (1943), nowadays called a McCulloch-Pitts neuron. Such a neuron accepts binary data as input and produces a binary output if a predefined threshold is exceeded. This model was extended and called perceptron (see Sect. 5.3) in Rosenblatt (1958). Today a neuron is abstracted as a mathematical function (Zurada, 1992): it receives multiple values as input and produces a single value as output. What characterizes functions that represent neurons is the way in which the output of a neuron is computed, i.e. the ingredients of such functions and how they interact.

Definition Let $w_1, \ldots, w_n \in \mathbb{R}$ be real numbers called *weights*; the function

$$\Pi : \mathbb{R}^n \to \mathbb{R}, x \mapsto \sum_{i=1}^{n} w_i x_i$$

is called *propagation function*. Furthermore, a function $\alpha : \mathbb{R} \to \mathbb{R}$ is called *activation function*. Then, a *neuron* is a map

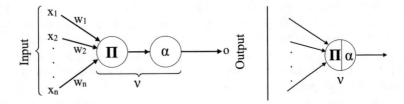

Fig. 7 Ingredients of an artificial neuron

$$\nu : \mathbb{R}^n \to \mathbb{R}, x \mapsto \alpha\left(\sum_{i=1}^{n} w_i x_i\right).$$

∎

Figure 7 shows a graphical representation of an artificial neuron: the left side of the figure represents the details of the definition before, the right side of the figure depicts a more condensed version of this graphical representation in which the two nodes representing Π and α are combined into a single node denoted with $\Pi|\alpha$. The input values $x_1, \ldots, x_n$ of the neuron ν are connected via directed edges with the neuron's propagation function Π; the weight w_i of the input x_i is shown as a number associated with the corresponding edge. The purpose of the propagation function Π is to aggregate the input of the neuron considering the corresponding weights of the input values. The purpose of the activation function α is (i) to decide whether its input suffice to pass on a non-zero output and (ii) what the output value will be.

An important variant defines a neuron as a map $\nu : \{1\} \times \mathbb{R}^n \subseteq \mathbb{R}^{n+1} \to \mathbb{R}$, with

$$\nu(1, x_1, \ldots, x_n) = \alpha\left(w_0 + \sum_{i=1}^{n} w_i x_i\right).$$

Thus, a neuron has one input x_0 of constant value "1" (see Fig. 8). The corresponding weight w_0 is called *bias*, denoted as "b". The bias influences the firing of the neuron: a negative bias results in the neuron firing less frequently, a positive bias results in the neuron firing more frequently.

Fig. 8 A neuron with a bias

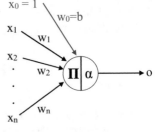

One problem with realizing neurons on a quantum computer is that quantum algo-rithms are unitary transformation, thus, linear, but activation functions of neurons, are typically non-linear. Cao et al. (2017) and Wan et al. (2017) present quantum neurons with different activation functions. A special neuron (a so-called percep-tron—see Sect. 5.3) on a quantum computer of the IBM Quantum Experience family has been described in Tacchino et al. (2018). This implementation proves the expo-nential advantage of using quantum computers in terms of requiring only N qubits in order to represent input bit-strings of 2^N bits. But in order to prepare this input an exponential number of 1-qubit gates and CNOTs are needed, in general (Leymann & Barzen, 2020a).

5.2 Neural Networks

Neural networks are directed graphs the nodes of which are neurons (Kubat, 2017; Skiena, 2017). The edges of the graph specify which neuron passes its output as input to which other neuron. The neurons are partitioned into disjoint sets $L_1,\dots, L_n$ (so-called layers) such that neurons of layer L_i are connected with neurons of layer L_{i+1}. The neurons of layer L_1 get their input from the outside, i.e. not from other neurons; this layer is called the input layer of the neural net. Neurons of layer L_n pass their output to the outside, i.e. not to other neurons; this layer is called the output layer of the neural network. All other layers are referred to as hidden layers. A neural network with at least one hidden layer is sometimes called a deep neural network.

Definition A *neural network* is a tuple (N, E, $\mathscr{L}$, X, Y) with the following properties:

- G = (N, E) is a directed graph,
- $\mathscr{L} = \{L_1,\dots, L_n\} \subseteq \mathcal{P}(N)$ is a partition of the set of nodes, i.e. $\forall$i: Li $\neq \emptyset$, $L_i \cap L_j$ = $\emptyset$ for i $\neq$ j, and $\cup$ Li = N; L_i is called *layer*, L_1 *input layer*, L_n *output layer*, and L_i (1 < i < n) *hidden layer*,
- each node $v_i = \left(\{w_{ji}\}, \Pi_i, \alpha_i\right) \in N$ is a neuron with a set of weights $\{w_{ji}\}$, a propagation function Π_i, and an activation function α_i,
- X = $\{x_1,\dots, x_m\}$ is the set of *input values*,
- Y = $\{y_1,\dots, y_k\}$ is the set of *output values*,
- the set of edges E $\subseteq$ (N $\times$ N) $\cup$ (X $\times$ N) $\cup$ (N $\times$ Y) connects two neurons, or an input value and a neuron, or a neuron and an output value,
- for $v \in L_i$ and (v, v') $\in$ E $\cap$ (N $\times$ N) it is v' $\in L_{i+1}$,
- for $v \in L_i$, x $\in$ X, and (x, v) $\in$ E it is i = 1,
- for $v \in L_i$, y $\in$ Y, and (v, y) $\in$ E it is i = n, and y is the output of v,
- for $(v_j, v_i) \in$ E $\cap$ (N $\times$ N) the output of v_j is passed to v_i where it is processed by v_i's propagation function Π_i weighted by w_{ji},
- for $v \in L_1$ and (x, v) $\in$ E, v processes x by its propagation function Π weighted by w_{ji}.

■

Effectively, a neural network that consumes m input values and that produces k output values is a function $\mathbb{R}^m \rightarrow \mathbb{R}^k$. The computational power of neural networks stems from the fact that any continuous function on a compact set $K \supseteq \mathbb{R}^m \rightarrow \mathbb{R}^k$ can be approximated with arbitrary precision by a neural network (see Kidger and Lyons (2020) for details).

Figure 9 shows the principle structure of a neural network according to our definition before. Note that there are some variants of our definition: sometimes, edges between neurons with the same layer are supported, sometimes edges are allowed between neurons of any layer, sometimes a neuron may be connected to itself, and so an. However, such variations are for use in special cases.

The structure of a neural network, i.e. its set of layers and its edges, is referred to as its *topology*. Defining the topology of a neural network, specifying the activation functions and propagation functions of its neurons is called *modelling* a neural network. Note that most often the propagation function from our definition in Sect. 5.1 is used, but activation functions change across problem domains; even different activation functions for different neurons of a given neural network may be defined.

Training a neural network consists of choosing appropriate weights and biases. For this purpose, the neural network is considered as a function $F : \mathbb{R}^m \rightarrow \mathbb{R}^k$. Furthermore, a set of input vectors $\{x_j\} \in \mathbb{R}^m$ is given together with the corresponding known correct output vectors $\{y_j\} \in \mathbb{R}^k$; note that the set of pairs $\{(x_j, y_j)\}$ is called *training data*. Ideally, the neural network will output y_j for each input x_j; but the neural network is not (yet) realizing the ideal function, i.e. it is just approximating

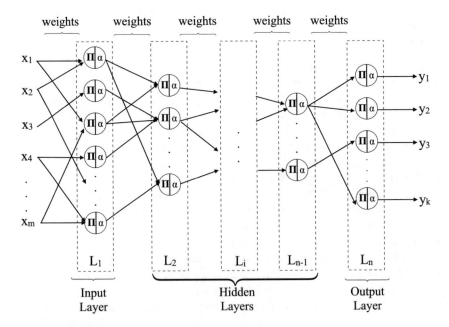

Fig. 9 Principle structure of a neural network

it: it will output $Y_j = F(x_j)$. Thus, the neural network will produce an error $|y_j - Y_j|$ in computing y_j. The goal of training the neural network is to minimize these errors considering all training data by adapting the weights and biases.

For this purpose, the so-called *loss function*

$$L(w, b) = \sum_j \left(y_j - Y_j\right)^2 = \sum_j \left(y_j - F(x_j)\right)^2$$

is minimized (as usual, the square of errors is used instead of their modulus). Here, "w" denotes all weights of the neural network and "b" all its biases. *Training* now means to choose w and b such that $L(w, b)$ is minimized. Lots of mathematical procedures are known to minimize L, e.g. gradient-based methods (e.g. stochastic gradient descent) or derivative-free methods (e.g. Nelder-Mead) can be chosen (Nocedal & Wright, 2006). Note that gradient-based methods require differentiability of the loss function which in turn requires differentiability of F which in turn requires differentiability of the activation functions of the neurons composing the neural network implementing F (see Sect. 5.1). After this training phase, the neural network is ready for use. It is said that the neural network has *learned*. The use of training data, i.e. input values with corresponding known output values, is referred to as *supervised learning*: the processing of the input data by the neural network is supervised by comparing its output with the known given results associated with the input. Learning without supervision is referred to as *unsupervised learning*: this kind of learning is discussed in Sect. 5.5 in the context of autoencoders.

An implementation of quantum neural network that requires a single qubit for each neuron (plus additional ancillae) has been proposed by Cao et al. (2017). The training of this quantum neural network is performed in a hybrid quantum–classical manner, i.e. the optimization is executed by classical software using the Nelder-Mead algorithm. In contrast to training classical neural networks where the individual training data is processed sequentially, the training of this quantum neural network can be done based on a superposition of the input/output pairs of the training data—something impossible for classical neural networks. A quantum neural network has been realized on a near-term quantum processor as described by (Farhi & Neven, 2018); their neural network has been successfully used as a binary classifier on the MNIST data set to distinguish two different handwritten digits. A set of requirements on implementations of quantum algorithms that represent "meaningful" quantum neural networks has been posed in Schuld et al. (2014).

5.3 Perceptrons

A perceptron can decide whether a given data point represented as input is left or right of a given straight line in the plane (or in higher dimensions, right or left of a hyperplane). Thus, a perceptron can be used as a binary classifier of linear separable

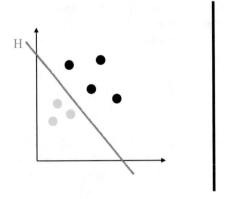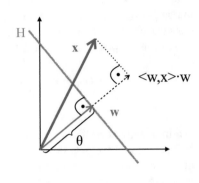

Fig. 10 Binary classifier for linear separable sets

data sets: the left side of Fig. 10 shows two kinds of data points (grey points and black points) in the plane. They are linear separable because a straight line H can be found that splits these two data sets such that on one side of H only the black data points are located, on the other side only grey data points reside.

Any straight line H (or hyperplane in higher dimensions) can be defined by means of a vector w that is orthogonal to H, and whose length $\|w\| = \theta$ is the distance of H from the origin (so-called Hesse normal form—see the right side of the figure). For any point x, the scalar product $\langle w, x \rangle$ is the length of the projection of x onto w, and $\langle w, x \rangle \cdot x$ itself is the projection of x onto the straight line defined by w. Thus, a point x is on H if and only if $\langle w, x \rangle - \theta = 0$. Furthermore, x is "on the right side" of H if the length $\langle w, x \rangle$ of the projection of x onto w is greater than θ (i.e. $\langle w, x \rangle - \theta > 0$), and x is "on the left side" of H if the length $\langle w, x \rangle$ of the projection of x onto w is less than θ (i.e. $\langle w, x \rangle - \theta < 0$). Consequently, the value of $\langle w, x \rangle - \theta$ determines whether a point x is right or left of the hyperplane H defined by w, i.e. it serves as is a binary classifier: a positive value classifies x as a black point, a negative value classifies x as a gray point.

Definition A *perceptron* is a neuron v with Heaviside-function Φ_θ as activation function, where $\theta := \|(w_1, \ldots, w_n)\|$ for the weights $w_1, \ldots, w_n$ of the neuron. ∎

The vector $w = (w_1, \ldots, w_n)$ of the weights of the perceptron v determines a hyperplane H (with $\|w\| = \theta$). When a point $x = (x_1, \ldots, x_n)$ is passed as input to the propagation function Π of v, the scalar product of w and x results:

$$\Pi(x) = \sum_{i=1}^{n} w_i x_i = \langle w, x \rangle.$$

With

$$\nu(x) = \Phi_\theta\left(\sum_{i=1}^{n} w_i x_i\right) \in \{0, 1\},$$

it is $\nu(x) = 1 \iff \Phi_\theta(\langle w, x\rangle) = 1 \iff \langle w, x\rangle \geq \theta \iff \langle w, x\rangle - \theta \geq 0$. Thus, $\nu(x) = 1$ classifies x as "right of H" (considering x on H, i.e. $\langle w, x\rangle - \theta = 0$, as "right of H"), and $\nu(x) = 0$ classifies x as "left of H". This reveals that a perceptron is a binary classifier.

Given a linear separable set of data, how are the weights $w = (w_1, \dots, w_n)$ determined the corresponding hyperplane of which splits the data set into two disjoint parts? In other words: how does a perceptron learn to split the data set correctly? For this purpose, the following procedure is used (e.g. da Silva et al. (2017)):

- Assume a set $T = \{(t_1, c_1), \cdots, (t_m, c_m)\}$ of training data is given, i.e. $t_j = \left(1, x_{1j}, \cdots, x_{nj}, \right)$ is a data point (with its first component set to "1" to introduce its weight as a bias) and $c_j \in \{0, 1\}$ classifies the data point as member of the class "0" or member of the class "1". For example the class "0" may indicate a grey point and the class "1" a black point in Fig. 10. Since the training data specifies the correct result of the classification, the algorithm sketched is a supervised learning algorithm.
- The algorithm is performed in several steps τ; the tuple $w(\tau) = (w_0(\tau), w_1(\tau), \cdots, w_n(\tau))$ denotes the bias $b(\tau) = w_0(\tau)$ and the weights of the perceptron at step τ.
- At step $\tau = 0$, $w(0)$ is chosen randomly ("small" values have been shown to be a good first choice).
- At any step $\tau > 0$ the following is computed for $1 \leq j \leq m$, i.e. for the whole set of training data:

$$y_j(\tau) = \Phi_{b(\tau)}\left(\sum_{i=1}^{n} w_i(\tau)x_{ij} + b(\tau)\right) = \Phi_{b(\tau)}(\langle w(\tau), t_j\rangle) \in \{0, 1\},$$

i.e. $y_j(\tau)$ is the classification of the data point t_j at step τ.

- The value $|c_j - y_j(\tau)|$ measures the error of classifying the data point t_j at step τ as $y_j(\tau)$, while c_j is the correct known classification. The goal is to minimize the overall classification error

$$e(\tau) = \frac{1}{m}\sum_{k=1}^{m}|d_k - y_k(\tau)|,$$

i.e. $e(\tau) < \gamma$, where γ is a predetermined error threshold. If the error is below this threshold the algorithms stops.
- Otherwise, a new bias and new weights are determined: chose an arbitrary data point t_j (i.e. $1 \leq j \leq m$) and compute

$$w_i(\tau + 1) = w_i(\tau) + r \cdot \big(c_j - y_j(\tau)\big) \cdot x_{ij},$$

for $0 \le i \le n$. Thus, if the computed classification $y_j(\tau)$ for data point t_j is correct, i.e. it is the known classification c_j, $w(\tau)$ is not changed "in direction of t_j", i.e. $w(\tau + 1) = w(\tau)$. Otherwise, $w(\tau)$ is modified "in direction of t_j", i.e. $w(\tau + 1) = w(\tau) \pm r \cdot t_j$. Here, r is a predetermined constant, the so-called "learning rate".

Assuming linear separable training data, the above algorithm converges and separates the training data correctly (Novikoff, 1962). Typically, the algorithm converges fast, i.e. the condition $e(\tau) < \gamma$ is met fast. But there are situations in which the convergence is slow and the algorithm should be stopped even if the result is not quite precise. This is covered by the additional condition "$\tau = N$" limiting the number of iterations even if $e(\tau) \ge \gamma$, i.e. the result of the learning process is above the error threshold. Note that the variables γ, r, and N are so-called *hyperparameters* and must be properly set prior to the learning process, e.g. based on former experience.

The separating hyperplane learned by the perceptron depends on the initial values of the bias and the weights $w(0)$ chosen during the initial step $\tau = 0$. This is depicted on the left side of Fig. 11: different initial values $w(0)$ result in different hyperplanes H_1, H_2, H_3. These hyperplanes are differently suited in correctly classifying new data points: two new data points are shown as a triangle and a square. The triangle is classified as a black point by H_1 and H_3 (being "right" of the hyperplanes), and as a grey point by H_2 (being left of it). The square is classified by H_3 as a black point, but as a grey point by H_1 and H_2. This non-determinism of a perceptron learning a separating hyperplane is in contrast to another machine learning technique (which is not based on neural networks at all): a support vector machine (SVM) (Burges, 1998). A support vector machine determines a unique separating hyperplane based on linear separable training data: this is achieved by computing the hyperplane with the maximum margin that does not contain any of the test data (see the right side of Fig. 11—the grey shaded rectangle represents the margin of the hyperplane). A support vector machine would classify the square as a grey point, and the triangle as a

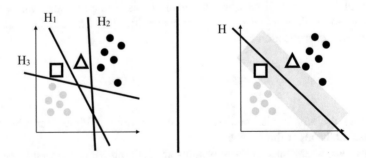

Fig. 11 Computing separating hyperplanes via perceptrons and support vector machines, and using a computed hyperplane as classifier for new data

black point. Furthermore, support vector machines can even be used to classify non-linear separable data sets (Bennett & Campbell, 2000) by embedding the data into a high-dimensional vector space (which already hints that quantum implementations of support vector machines are advantageous because the state spaces of quantum computers are extremely huge).

Wiebe et al. (2016) proposed two quantum perceptrons: the first one can be trained with a quadratic speedup in size of the training data, and the second one quadratically reduces the training time required to achieve a given precision when compared with the algorithm presented in this section. A quantum perceptron has been suggested in Tacchino et al. (2018) that requires exponentially less storage resources for training; it has been implemented and validated on a near-term quantum computer of the IBM Q family of systems.

5.4 Restricted Boltzmann Machines

Restricted Boltzmann machines (RBMs) are attributed to Smolensky (1986). While for general Boltzmann machines learning is not feasible under practical conditions, restricted Boltzmann machines can learn efficiently (see Hinton (2012) for an overview). A restricted Boltzmann machine is a neural network with two layers: an input layer and a hidden layer (see Fig. 12). Each neuron of the input layer has a single input value, and the weight associated with such an input is "1". Input values as well as the output of each neuron are Boolean values; especially the output of a restricted Boltzmann machine, i.e. the output of the neurons of the hidden layer, are Boolean values. Each neuron has a bias (not shown in Fig. 12).

Fig. 12 Structure of a restricted Boltzmann machine; biases are not shown

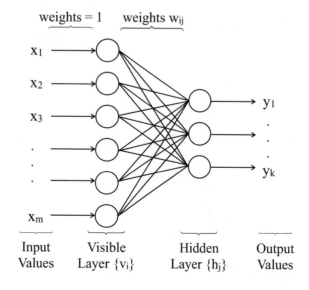

Definition A *restricted Boltzmann machine* is a two layer neural network (N, E, $\mathscr{L}$, X, Y) with following properties:

- the output of each neuron is a Boolean value (especially, $y \in \{0, 1\}$ for $y \in Y$),
- the input values are Boolean values, i.e. $x \in \{0, 1\}$ for $x \in X$,
- each neuron $\nu_i \in N$ has a bias; for $\nu_i \in L_1$ the bias is denoted by a_i, for $\nu_i \in L_2$ the bias is denoted by b_i,
- for $\nu_i \in L_1$ there exists exactly one $x_i \in X$ with $(x_i, \nu_i) \in E$ (ignoring the input "1" for the bias),
- for $\nu_i \in L_1$ the weight w_{ii} of $(x_i, \nu_i) \in E$ is "1",
- $L_1 \times L_2 \subseteq E$.

L_1 is called *visible layer*, L_2 is called *hidden layer*, i.e. the last property enforces that each neuron of the visible layer is connected to each neuron of the hidden layer. ∎

Training of a restricted Boltzmann machine (see Hinton (2012) for all details) is different from the general mechanism for training a neural network (see Sect. 5.2). The input values of a restricted Boltzmann machine are Boolean values. But data to be processed is typically not Boolean. For this purpose, each neuron ν of the input layer represents an observed value o: the value o is observed if and only if the input of ν is "1". Especially, o maybe of any type of data, i.e. not only numerical data but also categorical data. This way, restricted Boltzmann machines can be used with any type of data as input; and similar for output data.

The application areas of restricted Boltzmann machines include, for example, classification (e.g. Chen and Srihari (2015), Larochelle et al. (2012)) and feature learning (e.g. Zheng et al. (2013), Tomczak (2016)). For classification purposes, the training data $T = \{(t, c)\}$ are one-hot encoded, i.e. the class indicator c becomes a vector $c = (k_1, ..., k_r) \in \{0, 1\}^r$ (where r is the number of classes) with $k_j = 1$ if t is of class j, and $k_i = 0$ otherwise (see the left side of Fig. 13). Note that this is multi-class classification, i.e. data can be associated with more than two classes. For feature learning (right side of Fig. 13), the properties of the data correspond to

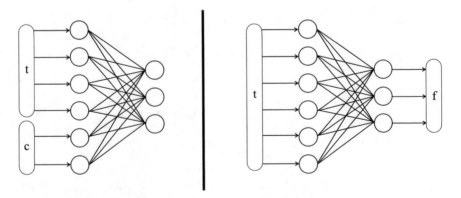

Fig. 13 Using restricted Boltzmann machines for classification and feature learning

the input and the output are the features learned. Typically, these features are used by another algorithm for further processing, e.g. by another restricted Boltzmann machine (so-called "stacked" restricted Boltzmann machines).

A special kind of quantum restricted Boltzmann machines that can approximate any quantum circuit has been specified in Wu et al. (2020). Also, an implementation of such kind of a quantum restricted Boltzmann machine on a NISQ device of the IBM Q family has been provided. Quantum Boltzmann machines (i.e. a general Boltzmann machine which supports also connections between neurons of the same layer) as well as quantum restricted Boltzmann machines have been investigated by Amin et al. (2018) with the result that their quantum Boltzmann machine outperforms a classical Boltzmann machine on a specific learning task. A circuit realizing a quantum restricted Boltzmann machine is given in Zhang and Zhou (2015). No implementation on a real quantum computer has been evaluated but a simulation: it turned out that the quantum restricted Boltzmann machine performed a classification task faster and with higher precision than a classical machine. An implementation of a restricted Boltzmann machine to be realized on a D-Wave quantum annealer is described in Denil and Freitas (2011); obstructions for a real implementation are discussed.

5.5 Autoencoders

Today's concept of an autoencoder goes back to Hinton and Zemel (1994). At a first glance, it can be considered as two restricted Boltzmann machines merged together the purpose of which is to reconstruct their input as output. After training, the autoencoder may be split again into two restricted Boltzmann machines, each of which may be used in combination with further machine learning algorithms.

Definition An *autoencoder* is a three layer neural network (N, E, $\mathscr{L}$, X, Y) with following properties:

- let F be the function represented by the autoencoder, then: $F \approx id$ (i.e. the autoencoder's output approximates its input),
- card L_1 = card L_3 (i.e. input layer and output layer contain the same number of neurons).

The hidden layer L_2 is called *code (layer)*. Input layer and code layer are collectively referred to as *encoder*, code layer and output layer are collectively referred to as *decoder*.

∎

Note that there are variants of autoencoders that support more than one hidden layer, which is not discussed here. Figure 14 depicts on its left side the principle structure of an autoencoder consisting of an encoder and a decoder. The right side of the figure shows its structure as a neural network with three layers, the hidden layer named "code". Roughly, it can be considered as two restricted Boltzmann machines with the code layer being the hidden layer of the encoder and the input layer of the decoder ("stacked machines").

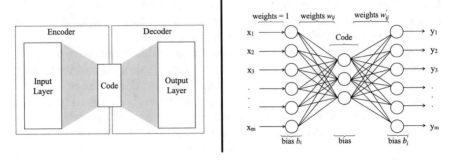

Fig. 14 Principle structure of an autoencoder and its structure as a neural network

The encoder transforms its input into a code (also called a *representation* of the input). If the input is a data point in $\mathbb{R}^m$ and the code is a data point in $\mathbb{R}^k$, the encoder is a map $\varphi : \mathbb{R}^m \to \mathbb{R}^k$. Because the output layer must have the same number of neurons as the input layer, the output is again a data point in $\mathbb{R}^m$, i.e. the decoder is a map $\psi : \mathbb{R}^k \to \mathbb{R}^m$. By definition, it is $\psi \circ \varphi \approx id$. That is, for any input x, the output is $\psi \circ \varphi(x) \approx id(x) = x$; thus, the difference (or error) between x and $\psi \circ \varphi(x)$ should be as small as possible (as usual the square of the error $(x - \psi \circ \varphi(x))^2$ is taken). For a set of training data $T = \{x^k\}$, the sum of these squared errors should be minimal. Now, φ and ψ perform their mapping based on their weights w_{ij} and w'_{ij} as well as their biases b_i and b'_i, respectively. Consequently, the sum of the squared errors is a function L ("loss function") of these weights and biases:

$$L(w, b) = \sum_k \left(x^k - \psi \circ \varphi\left(x^k\right)\right)^2.$$

Training an autoencoder means minimizing this loss function based on some optimization algorithm, i.e. to determine $\left\{w_{ij}, b_i, w'_{ij}, b'_i\right\}$ such that $L\left(\left\{w_{ij}, b_i, w'_{ij}, b'_i\right\}\right) \approx 0$. Note that the training of an autoencoder is unsupervised because no set of training data is required that is labeled to indicate the output for a given input.

After training, the encoder may be split from the decoder. The encoder will then transform a data point $x \in \mathbb{R}^m$ into a data point $\varphi(x) \in \mathbb{R}^k$, and it is known that this $\varphi(x)$ represents x faithful, i.e. it contains "the essence of x" because it includes all information to reconstruct x (by means of ψ): $\varphi(x)$ represents the features of x. This is the reason for using the term "feature learning" for training an encoder. If $k < m$, i.e. if the code has less components than the input, a dimension reduction is achieved. Reducing the dimension of data is key because many algorithms are very sensitive to it, i.e. they are much more efficient if a data point to be processed has less components.

Note the similarity to embeddings as discussed in Sect. 4.2 which also reduce dimensions of data. In contrast to encoders, embeddings try to keep distances between data points intact. While some other algorithms that might be used to process data after dimension reduction require similar distances between the high-dimensional data points and the low-dimensional data points, some other do not consider these distances. Thus, the next algorithm used to process the data influences the decision whether an embedding has to be used or whether an encoder suffice.

Based on the work of Lamata et al. (2019), a quantum autoencoder requiring for each neuron a separate qubit has been worked out in Bondarenko and Feldmann (2020). Each neuron has a unitary operation associated with it that operates on the neuron and all neurons of the preceding layer. The number of such operations needed to apply the autoencoder grows exponentially with the maximum number of neurons per layer. Training of their autoencoder has been simulated in MATLAB. An implementation of a (small) autoencoder on a Rigetti Quantum Computer has been described in Ding et al. (2019); the corresponding circuits are provided. This autoencoder was successfully used for dimension reduction. Similar has been achieved in Pepper et al. (2019) based on a photonic quantum computer. Romero et al. (2017) described a variational quantum–classical way to implement a quantum autoencoder. Figure 15 depicts (for illustration only, not for reading) our implementation of the quantum part of the algorithm based on the simulator of IBM Q und Qiskit (2021). The input data is prepared in four qubits that are transformed by the encoder. Once the encoder finished its processing, the first two qubits are set to $|0\rangle$ while the last two qubits are kept in the state the encoder produced. Effectively, the code of the autoencoder consists of these two last qubits, i.e. a dimension reduction by 50% is achieved. Next, the decoder tries to reconstruct the original input from the code. The decoding result is measured and analysed by a classical part (not shown in the figure); based on this analysis another iteration may be required. Once the autoencoder is successfully trained, the encoder can be used for dimension reduction.

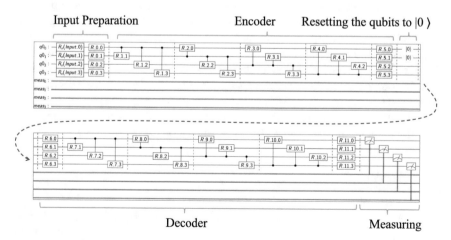

Fig. 15 Example circuit of a quantum autoencoder

6 Variational Hybrid Quantum–Classical Algorithms

As described in Sect. 2.3, near-term quantum computers support "short" computations only due to decoherence of their qubits and lack of fidelity of their operations. However, many problems require more complex computations than supported on such NISQ devices. In this section, the principle mechanisms of how to support such complex computations on NISQ devices is described.

6.1 The Main Idea

Some quantum algorithms consist of both, a pure quantum part and a pure classical part (e.g. Shor's algorithm for factorization): its quantum part is performed on a quantum computer, and its classical part is performed on a classical computer. That is, such an algorithm inherently requires classical pre- or post-processing to achieve its goal, it would not succeed without its classical part. Such a split algorithm is referred to as a *hybrid quantum–classical algorithm*.

Instead of such algorithms that are inherently hybrid, algorithms might be designed to be hybrid from the outset to limit their processing on a quantum computer. For example, this is enforced by today's near-term quantum devices that restrict the amount of processing that can be performed with sufficient precision. Such an algorithm requires additional classical processing to compensate for the restricted amount of work performed by the quantum computer: the algorithm is not inherently hybrid quantum–classical but it is so by design.

Two main problem domains are addressed by such kind of hybrid algorithms by design: problems that at their core can be solved based on the Raleigh-Ritz principle (Yuan et al., 2019), and problems that can be reduced to combinatorial optimization problems. Both kind of algorithms make use of a quantum part that consists of parameterized quantum circuits that prepare a state on a quantum computer and measure it, and a classical part that optimizes the corresponding parameters in iterations in dependence of the measured results (see Fig. 16). The parameterized unitary operator $U(p_1, ..., p_k)$ in Fig. 16 is called an "ansatz": its goal is to prepare states that deliver measurement results m that are good candidates for arguments optimizing a given function F. Thus, the value F(m) is dependent on the parameters $p_1, ..., p_k$. By

Fig. 16 Principle structure of variational hybrid quantum–classical algorithms

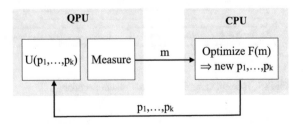

varying these parameters based on classical optimization algorithms a corresponding m may be found (Shaydulin et al., 2019). Finding an appropriate ansatz is hard. Note that a cloud environment is particularly suited to perform hybrid quantum–classical algorithms (Karalekas et al., 2020; Leymann et al., 2020).

Varying the parameters iteratively gives rise to the name *variational* hybrid quantum–classical algorithm (or variational algorithms for short) for this kind of approach. The application areas of variational algorithms span a broad spectrum including factoring of numbers (Anschuetz et al., 2018), solving linear equations (Bravo-Prieto et al., 2020), training autoencoders (Romero et al., 2017), or solving non-linear partial differential equations (Lubasch et al., 2020). A general approach towards a theory of variational hybrid quantum–classical algorithm is proposed in McClean et al. (2016).

6.2 Maximum Cut: A Combinatorial Optimization Problem

The maximum cut problem is to partition the node set of a graph in two sets such that the number of edges between nodes of the different sets is maximal. In Fig. 17, part 1, a graph with four edges {A,B,C,D} is shown where each node is connected to each other node. Part 2 and part 3 of the figure depict two different maximum cuts: the members of one node set are coloured black while the members of the other node set are left white. Dashed edges connect nodes of the same set, i.e. they do not

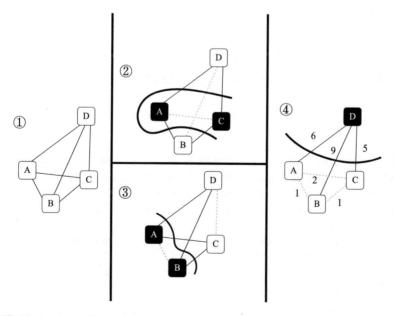

Fig. 17 Maximum cut of a graph: examples

contribute to the cut, while solid lines connect nodes of the two different set, i.e. they contribute to the cut. The number of edges of each cut is four.

Part 4 of the figure adds a weight to each edge, e.g. the weight of an edge indicates the distance between the nodes connected by the edge; such a distance, for example, might be determined by means of a distance measure (see Sect. 4.2). The weighted maximum cut problem strives towards finding a partition of the node set such that the sum of the weights of the edges between nodes of different sets is maximal. Note that by setting each weight to "1", the maximum cut problem is seen to be a special case of the weighted maximum cut problem. Part (4) shows a weighted maximum cut: the node set is partitioned into {A,B,C} and {D}, and the sum of the weights of the corresponding edges is 20. If the weights are distances, the weighted maximum cut is nicely interpreted as determining clusters: the two node sets consist of nodes that are close to each other within each set, but nodes of different sets are far apart. For example, D has distance 6, 9, 5 to A, B, C, respectively, but A, B, C are quite close to each other (distances 1, 1, 2). Thus, an algorithm for determining a weighted maximum cut can be used for clustering.

Let $G = (N, E)$ be a graph with n nodes (i.e. card(N) = n), and let S, T be a cut of N (i.e. $S \cap T = \emptyset$, $S \cup T = N$, $S \neq \emptyset$, and $T \neq \emptyset$). The function $z : N \rightarrow \{-1, +1\}$ indicates membership to S or T, i.e. $z(u) = +1 \iff u \in S$ and $z(u) = -1 \iff u \in T$ (elements of S are labelled $+1$, elements of T are labelled -1—or "black" and "white" in the figure). For brevity, $z_u := z(u)$.

For i, j $\in$ S, it is $z_i = z_j = +1$, thus $z_i \cdot z_j = 1$; and for i, j $\in$ T, it is $z_i = z_j - 1$, thus $z_i \cdot z_j = 1$. Consequently, it is i, j $\in$ S $\vee$ i, j $\in$ T $\Leftrightarrow$ $z_i \cdot z_j = 1$, i.e. (i, j) is not a cut edge. Similarly, $(i \in S \wedge j \in T) \vee (j \in S \wedge i \in T) \Leftrightarrow z_i z_j = -1$, i.e. (i, j) is a cut edge. Reformulation results in:

- (i, j) $\in$ E is not a cut edge $\Leftrightarrow$ $1 - z_i z_j = 0 \Leftrightarrow \frac{1}{2} \cdot (1 - z_i z_j) = 0$
- (i, j) $\in$ E is a cut edge $\Leftrightarrow$ $1 - z_i z_j = 2 \Leftrightarrow \frac{1}{2} \cdot (1 - z_i z_j) = 1$

Thus, a maximum cut maximizes the following cost function:

$$C(z) = \frac{1}{2} \sum_{(i,j) \in E} \left(1 - z_i z_j\right).$$

This is because each cut edge contributes a "1" to the sum, and a non-cut edge contributes a "0": the more cut edges exist, the higher is $C(z)$. Such kind of a formula, i.e. a formula with parameters with values from $\{-1, +1\}$ is called an *Ising formula*, which is important for solving quantum problems on so-called quantum annealers.

In this chapter it is assumed that algorithms are formulated in the gate model. For this purpose, parameters from $\{0, 1\}$ are better suited, i.e. the above cost formula has been transformed. To achieve this, a function $x : N \rightarrow \{0, 1\}$ indicates membership to S or T, i.e. $x(u) = 1 \iff u \in S$ and $x(u) = 0 \iff u \in T$. For brevity, $x_u := x(u)$. This implies:

- (i, j) $\in$ E is not a cut edge $\Leftrightarrow$ i, j $\in$ S $\vee$ i, j $\in$ T $\Leftrightarrow$ $x_i \cdot (1 - x_j) + x_j \cdot (1 - x_i) = 0$

- $(i, j) \in E$ is a cut edge $\Leftrightarrow (i \in S \wedge j \in T) \vee (j \in S \wedge i \in T) \Leftrightarrow x_i \cdot (1 - x_j) + x_j \cdot (1 - x_i) = 1$

The cost formula with binary parameters for a maximum cut is then:

$$C(x) = \frac{1}{2} \sum_{(i,j) \in E} \left(x_i \cdot (1 - x_j) + x_j \cdot (1 - x_i) \right).$$

The term $x_i \cdot (1 - x_j) + x_j \cdot (1 - x_i)$ is a map $\{0, 1\}^n \to \{0, 1\}$, which is also called a *clause*. A combinatorial optimization problem is to determine a $z \in \{0, 1\}^n$ (i.e. binary parameters) that maximizes (or minimizes) a cost function

$$C(z) = \sum_{\alpha=1}^{m} C_\alpha,$$

where each $C_i : \{0, 1\}^n \to \{0, 1\}$ $(1 \leq i \leq m)$ is a clause with n binary parameters. This reveals that the maximum cut problem is a (binary) combinatorial optimization problem. Many other practically relevant problems like the travelling salesman problem are such problems. Next, it is shown how such problems can be solved with quantum computers.

6.3 QAOA

The concept of the quantum approximate optimization algorithm (QAOA) has been introduced by Farhi et al. (2014). It is used to approximately solve combinatorial optimization problems on near-term quantum devices and has been first applied to the maximum cut problem in Farhi et al. (2014). Crooks (2018) provided a QAOA-based implementation of a maximum cut algorithm on a near-term quantum computer of Rigetti and showed that this implementation is faster than the well-known classical Goemans–Williamson algorithm. The results can even be improved following the warm starting idea for QAOA (Egger et al., 2020).

The basic idea is as follows: The cost function C of a combinatorial optimization problem induces a map $C : \mathbb{H}^n \to \mathbb{H}^n$ by interpreting $z \in \{0, 1\}^n$ as the computational basis vector $|z\rangle \in \mathbb{H}^n$ and defining

$$C|z\rangle := \sum_{\alpha=1}^{m} C_\alpha(z) \cdot |z\rangle = f(z) \cdot |z\rangle.$$

Thus, the matrix of C in the computational basis is a diagonal matrix, i.e. the map C is Hermitian and can be used to measure the states of a quantum computer (C is an observable). Let $z' \in \{0, 1\}^n$ such that $f(z') = \max\{f(z)\} = C_{\max}$. The expectation value $\langle C \rangle_\psi$ of measuring C in any state $|\psi\rangle = \Sigma x_z |z\rangle$ is $\langle C \rangle_\psi = \langle \psi | C \psi \rangle =$

$\langle \Sigma x_z | z \rangle | \Sigma x_z f(z) | z \rangle \rangle = \Sigma |x_z|^2 f(z) \le \Sigma |x_z|^2 f(z') = f(z') \Sigma |x_z|^2 = f(z') = C_{max}$, i.e. the expectation value is less than or equal C_{max}. If $|\psi\rangle$ is close to $|z'\rangle$, $\langle C \rangle_\psi$ will be close to C_{max}. This means that with

$$|\psi\rangle = \sum_{z \in \{0,1\}^n} x_z |z\rangle = \sum_{z \ne z'} x_z |z\rangle + x_{z'} |z'\rangle,$$

it must be achieved that $|x_{z'}|^2$ gets close to 1 (so-called amplitude amplification), i.e. a measurement will result with high probability in z'. This is realized by a unitary transformation (the "ansatz"—see Sect. 6.1) the construction of which involves the cost function and rotations that emphasize components x_z of the state ψ the more z contributes to the cost; these angles are the parameters of the ansatz that are classically optimized. For all the details see Farhi et al. (2014).

6.4 Computing Maximum Cuts via Eigenvalues

Let $G = (N, E)$ be a graph with n nodes (i.e. card(N) = n), and let S, T be a partition of N. Such a partition can be described by a bit vector $x \in \{0, 1\}^n$ as follows: $x_i = 1 \iff i \in S$ and $x_i = 0 \iff i \in T$. There are 2^n different bit vectors, i.e. partitions. For each partition the number w_d of edges between nodes of different sets of the partition as well as the number w_s of edges between nodes of the same set of the partition is determined. Next, $x \in \{0, 1\}^n$ is considered as the binary representation of a natural number $x \in \mathbb{N}$, and $w_x := \frac{1}{2}(w_s - w_d)$ is associated with this number x. Finally, the matrix $M \in \mathbb{C}^{2^n \times 2^n}$ is defined where $m_{ij} = 0 \iff i \ne j$ and $m_{ii} = w_i$, i.e. the i-th row of M has as diagonal element the number w_i corresponding to the partition $i \in \{0, 1\}^n$, and all other elements of the row are zero.

M is a diagonal matrix and, thus, Hermitian, i.e. all eigenvalues are real numbers. Each vector of the computational basis is an eigenvector of M. It can be proven that each vector of the computational basis that is an eigenvector of the lowest eigenvalue defines a maximum cut. Thus, determining the lowest eigenvalue of the matrix M and one of its eigenvectors means to determine a maximum cut of the corresponding graph. Note that an analogue construction can be made for arbitrary weighted graphs. Using eigenvalues to solve the maximum cut problem is based on Poljak and Rendl (1995).

6.5 VQE

A variational hybrid algorithm (called variational quantum eigensolver VQE) to determine the lowest eigenvalue of an operator has been first developed by Peruzzo et al. (2014). This work has been extended towards a variational hybrid algorithm to approximate all eigenvalues of an operator (Higgott et al., 2019). As a consequence of Sect. 6.4, the maximum cut problem can be solved by using VQE. In addition, dimension reduction based on principal component analysis (PCA) requires the computation of all eigenvalues of a certain matrix, i.e. this algorithm is key for our work.

The basic idea of VQE is as follows: according to the Raleigh-Ritz principle (Yuan et al., 2019), the lowest eigenvalue $\lambda_{\min}$ of a Hermitian operator O satisfies the following equation:

$$\lambda_{\min} \leq \frac{\langle \psi | O | \psi \rangle |}{\| \langle \psi, \psi \rangle \|} \text{ for any } |\psi\rangle \neq 0.$$

With $\langle \psi | O | \psi \rangle = \langle O \rangle_{|\psi\rangle}$ for each state $|\psi\rangle$ (i.e. $\| \psi \| = 1$), the expectation value of O provides an upper bound of the lowest eigenvalue of O: $\lambda_{\min} \leq \langle O \rangle_{|\psi\rangle}$. Thus, if a state $|\psi\rangle$ can be found that minimizes the expectation value of O, $\langle O \rangle_{|\psi\rangle}$ is close to the lowest eigenvalue of O. (Note that $\langle O \rangle_{|\psi'\rangle} = \lambda_{\min}$ holds for an eigenvector $|\psi'\rangle$ of $\lambda_{\min}$).

In order to find such a $\left| \hat{\psi}' \right\rangle$, series of states $|\psi(p_1, ..., p_k)\rangle$ are iteratively constructed that depend on parameters $p_1, ..., p_k$; in Fig. 16 the unitary operator $U(p_1, ..., p_k)$ prepares this state. The expectation value $\langle O \rangle_{|\psi(p_1,...,p_k)\rangle}$ is measured for each parameterized state, and the measurement result is subject to a classical optimization algorithm that determines new parameters $p_1, ..., p_k$ that reduce the expectation value further; in Fig. 16 the function F(m) is this expectation value. These iterations approximate $|\psi'\rangle$, i.e. the final set of parameters determine $\langle O \rangle_{|\psi(p_1,...,p_k)\rangle} \approx \lambda_{min}$ and $|\psi(p_1, ..., p_k)\rangle$ is an approximation of the corresponding eigenvector.

An efficient measurement of the expectation values is done as follows: Each Hermitian operator O can be written as a linear combination of "simpler" Hermitian operators O_i, i.e. $O = \Sigma x_i O_i$. Here, "simpler" means that O_i is a combination of Pauli operators. The expectation value $\langle O_i \rangle_v$ of such a combination can be efficiently measured in a quantum computer (Hamamura & Imamichi, 2019), and it is $\langle O \rangle_v = \Sigma x_i \langle O_i \rangle_v$, i.e. the expectation value of O can be easily computed classically based on the measured $\langle O_i \rangle_v$—i.e. in Fig. 16 the function F computes $\Sigma x_i O_i$. For all the details see Peruzzo et al. (2014).

7   QHAna: A Quantum Humanities Analysis Tool

As described in Sects. 5 and 6 there are several machine learning techniques for which first implementations are available on quantum computers. Some can be used to just gain initial experiences, others already show improvements in terms of speedup or precision. Bringing together the aspects and concepts outlined in the above sections, QHAna (2021) was developed. QHAna is our quantum humanities data analysis tool that aims for several goals: (i) There is the content-related goal of supporting the identification of patterns by data analysis. A feasibility study is provided by the use case (Sect. 3) that focuses on analysing costume data to improve the understanding of vestimentary communication in films. (ii) By performing this analysis by classical and hybrid quantum–classical algorithms a comparison of both methods is supported, allowing to assess the benefits of using quantum algorithms. (iii) This comparison supports the goal of improving the understanding of potential benefits quantum computing may provide for digital humanities research (see Sect. 2.2). (iv) Additionally, QHAna allows the integration of heterogeneous tools from different quantum computing vendors in a single analysis pipeline. (v) Thus, one of the overarching goals of QHAna is to provide easy access to people without quantum computing background to gain first application knowledge in the domain of quantum humanities. In the following QHAna will be introduced by this core objectives and functions.

7.1 Support the Identification of (Costume) Patterns

The primary goal of QHAna is to support analysing data when aiming at the identification of patterns based on the data analysis pipeline defined in Sect. 4.1. By providing easy access to the analysis methods and techniques outlined above, QHAna is designed to e.g., improve the understanding of vestimentary communication in films by supporting the identification of costume patterns in the MUSE dataset (see Sect. 3.3). Note that QHAna is data independent and therefore, not limited to the data of our use case. Data from other application areas like from the MUSE4Music project (Barzen et al., 2016) is planned to be analysed.

Based on initial question like (i) "which base elements are used to communicate specific stereotypes?", (ii) "do certain age impression and attributes like colour or condition communicate certain character traits within one genre?", (iii) "and if this is true, how to group new costumes to the suitable category when being captured?" several analysis steps need to be performed. The upper part of Fig. 18 gives an impression of the main menu that guides through the pipeline for analysing data. To approach an answer to these sample questions, for example, question (ii) is addressed by clustering algorithms, focusing on finding those costumes that achieve the same effect, while the question (iii) is addressed by classification algorithms. For both, the first step is about preparing the data depending on the requirements of the algorithms. As this is the basis for all the other following steps, it will be described in more detail how QHAna supports this step.

As Fig. 18 depicts, the tab "Data Preparation" allows to prepare the data depending on the requirements of specific analysis algorithms. As described in Sect. 4.2, several algorithms require numerical data and the MUSE data is mostly categorical data. Therefore, the subtabs "Distance Space Calculation" and "One-Hot Encoding" as shown in Fig. 18 provide different options to transform categorical data into numerical ones. For our example we chose the "Distance Space Calculations" that supports the approach described in Sect. 4.2: Here, a dropdown menu allows to choose the

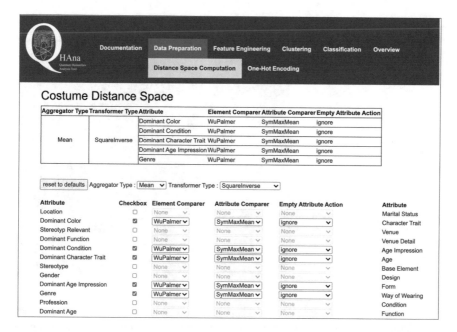

Fig. 18 Screenshot of QHAna "Data Preparation"

domain of the distance space. In our MUSE use case, the "Costume" data set is chosen. Figure 18 depicts how to specify a view for analysing a defined subset of the attributes of the MUSE data set. For this purpose, the attributes of interest of our example question are chosen, e.g., "Dominant Color", "Dominant Condition", "Dominant Character Trait", "Dominant Age Impression", and "Genre" are selected. Furthermore, the user can define which element comparer (that allow to compare two elements within an attribute category) and attribute comparer (that allow to compare sets of elements within an attribute), or which aggregator type (that define how data points are aggregated) and transformer type (that describe the function used to transform similarity measures into distance measures), are most suitable for the use case, as well as how empty attributes are to be treated (e.g. if a costume has no value for the attribute category "Color" the missing element should be "ignored"). The table at the top provides an overview of all selections made.

The tap "Documentation" (right of "Data Preparation") allows to retrieve general information about the currently processed data and its structure (e.g. attribute lists, taxonomies), explanations about the components already implemented in QHAna (e.g. different cluster or classification algorithms), references to software dependencies as well as all relevant related papers used to implement the tool.

After computing the distance space, QHAna supports calculating corresponding distance matrices (see Sect. 4.2). Here, the user can choose how many costumes should be analysed and whether these costumes should be selected randomly from the whole data set or custom specific. Figure 19 gives an example of such a distance

Fig. 19 Screenshot of a distance matrix

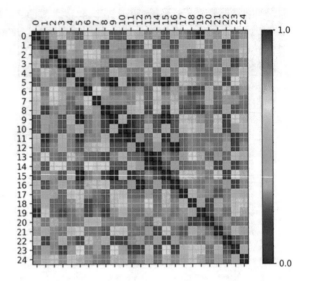

matrix of 25 costumes. Note that the visualization of the distance may become very complex with an increasing number of input data and the resulting visualization may become difficult to comprehend.

Distances between pairs of data elements (e.g. costumes) are represented by a colour code: The distance value of two costumes becomes a coefficient of the distance matrix and this value is turned into a colour: dark blue corresponds to no distance at all, while dark red indicates maximum distance. Each costume is represented by a number (0–24 in Fig. 19). As the distance matrix depicts, some of the costumes are very close to each other based on the selected attributes and others are very far away (i.e. they are not really similar). For example, costume 5 is very close to costume 9 and costume 15, while it has hardly any similarity with (i.e. it is far away from) costume 11 and 22. For a more detailed comparison of these costumes the "Entity Table" subtab of the "Overview" tab (see Fig. 20) allows to compare the costumes of interest in detail by their attributes chosen. Figure 20 depicts costumes 5, 9, 11, 15 and 22 stressing that 5, 9 and 15 have the same attribute values, while costume 11 and 15 differ in their attribute values. As most costumes have more than just 5 attributes, changing the input parameters in the costume distance space definition may have a deep impact on the similarity measures of these costumes. Also, experts sometimes need to verify the concrete costumes manually. For this purpose, the entity table provides for each costume links to the MUSE-repository (the film, the character, the concrete costume) allowing further investigation of the costumes with all the detailed information, in particular it provides the visual representations of the costumes from the MUSE-repository.

After the data is prepared the next step allows to—if needed—perform "Feature Engineering" to reduce the number of relevant attributes. The tab "Feature Engineering" provides several techniques for dimension reduction and feature extraction such as autoencoders, embeddings, or PCA. This provides the basis to use clustering

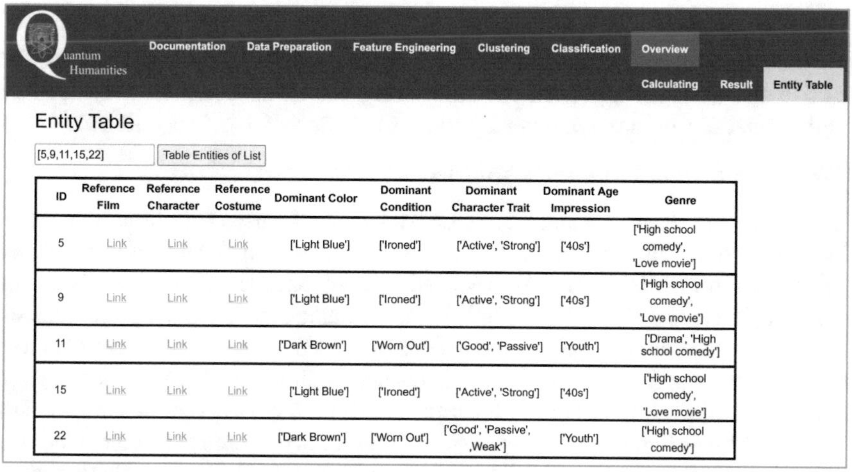

ID	Reference Film	Reference Character	Reference Costume	Dominant Color	Dominant Condition	Dominant Character Trait	Dominant Age Impression	Genre
5	Link	Link	Link	['Light Blue']	['Ironed']	['Active', 'Strong']	['40s']	['High school comedy', 'Love movie']
9	Link	Link	Link	['Light Blue']	['Ironed']	['Active', 'Strong']	['40s']	['High school comedy', 'Love movie']
11	Link	Link	Link	['Dark Brown']	['Worn Out']	['Good', 'Passive']	['Youth']	['Drama', 'High school comedy']
15	Link	Link	Link	['Light Blue']	['Ironed']	['Active', 'Strong']	['40s']	['High school comedy', 'Love movie']
22	Link	Link	Link	['Dark Brown']	['Worn Out']	['Good', 'Passive', ,Weak']	['Youth']	['High school comedy']

Fig. 20 Screenshot of "Entity Table"

techniques that allow to group those costumes together that have the same effect, e.g. communicating a Sheriff by certain base elements like a Sheriff star, Cowboy boots and so on (Barzen, 2018). As there are new costumes captured on a daily basis and they need to be classified in terms of being mapped to the costume pattern they contribute to, running classification algorithms is very promising for the MUSE use case. All those steps supported by QHAna aim at improving the understanding of vestimentary communication in films. How feature engineering, clustering, and classification are supported is unfolded in what follows.

7.2 Comparing Classical and Quantum Machine Learning Algorithms

To get the optimal results when approaching an answer to a stated question the different analysis steps are performed based on different algorithms, often in parallel by different classical implementations as well as by different quantum–classical implementations. This allows comparing the different results of classicals machine learning algorithms amongst each other, comparing different results of quantum machine learning algorithms amongst each other, and comparing results from classicals machine learning algorithms with results from quantum machine learning algorithms. This enables finally selecting the most suitable algorithm for the problem at hand as well as improving the application of an algorithms when comparing the results and optimizing iteratively the hyperparameters. Please note that due to the limitations of today's quantum computers only small sets of data can be processed when using quantum machine learning algorithms.

To outline how such a comparison is enabled by QHAna, two examples are provided, one from "Feature Engineering" where a different technique is used to get so a distant matrix and one from "Clustering" where the comparison of different implementations of maximum cut algorithms (see Sect. 6.2–6.5) is outlined.

Classical and Quantum-based Approaches to Distance Matrices

As described in Sects. 5.4 and 5.5, autoencoders can be used for dimension reduction. Therefore, four implementations of autoencoders have been used to reduce the dimension of MUSE data. We realized two autoencoders classically (based on PyTorch (2021) and TensorFlow (2021)) and two autoencoders in a hybrid manner: in the latter case, a classical autoencoder is first used to reduce all input data to a three-dimensional feature space, while in the succeeding step (i) a quantum inspired autoencoders (using TensorFlow Quantum (2021)) and (ii) a quantum autoencoders (using Qiskit, see Fig. 15 for an impression of the circuit) are used to further reduce the dimension to two (for more details see Barzen et al. (2021). Please note that those autoencoders are currently being integrated into QHAna). By pairwise measuring the fidelity of the quantum states and applying the Fubini-Study metric (Biamonte, 2020) to the measured data, the distances between the resulting quantum states can be computed.

The left side of Fig. 21 gives an example of the distance values of 10 costumes (with the same 5 attributes selected as in Fig. 18) as result of a hybrid quantum inspired autoencoder. As before, 0 (dark blue) corresponds to the smallest distance and 1 (bright red) to the largest possible distance (note that based on the actual data, the largest distance is about 0.30). What can be seen is that the costumes 0 to 4 are highly similar to each other but rather different from costumes 5 to 9 and that the costumes 5 to 9 are again highly similar to each other. The right side of Fig. 21 shows the distances of the same 10 costumes determined by using the Wu and Palmer similarity measure (see 4.2 and 7.1). What can be seen is that quite similar

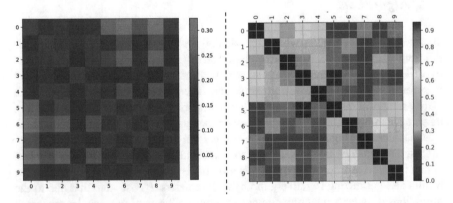

Fig. 21 Distance matrix as result of a hybrid autoencoder (left side) and computed via the Wu and Palmer similarity (right side)

results have been achieved by using these two different techniques: The statistical approach (autoencoder) on the one hand and the approach using taxonomies (Wu-Palmer similarities) on the other hand, identified the same costumes as being close to each other. This hints to a potential costume pattern. Also, it provides first insights on comparing a classical and a quantum-based approach to compute distance values.

Classical and Quantum-based Approaches to Clustering

Implementations of both, classical clustering algorithms and first quantum–classical clustering algorithms are available under the tab "Clustering" in QHAna. Currently, this includes an implementation of the OPTICS cluster algorithm (Ankerst et al., 1999), four different implementations of the maximum cut (a.k.a. maxcut) algorithm (see Sect. 6.2) and four different implementations of the k-means algorithm (Arthur & Vassilvitskii, 2007; Khan et al., 2019). The implementations of the maximum cut range from a classical naive implementation ("classicNaiveMax-Cut"), to two approximative implementations ("sdpMaxCut" based on semidefinite programming (Boumal, 2015) and "bmMaxCut" using a Bureir-Monteiro solver (Boumal et al., 2016)), to a quantum–classical implementation ("qaoaMaxCut") that is based on QAOA (see Sect. 6.3) and implemented in Qiskit (see Max Cut, 2021).

Figure 22 allows a comparison of the results of using different maximum cut implementations. Input are the same 10 costumes as in Fig. 21, using the default values of the hyperparameters per algorithm provided by QHAna. Diagram 1 of the figure shows the costumes and their distances (upper left corner) in an MDS-based embedding. The subtab "Embedding" of "Feature Engineering" supports to use MDS as one implemented approach to map the distance measures of the chosen costumes with their chosen attributes to a metrical feature space. As a result of the embedding via MDS the data is mapped into $\mathbb{R}^n$ in such a way that the original distances of the data elements are nearly the distances in the feature space (see Sect. 4.2). The other three diagrams present clustering results achieved by different maximum cut implementations namely "qaoaMaxCut" (Diagram 2), "sdpMaxCut" (Diagram 3), and "bmMaxCut" (Diagram 4). It can be seen that the two approximative implementations (Diagrams 3 and 4) have identical results: The first cluster (red circles) contains the costumes 0–4, while the second cluster (blue circles) contains the costumes 5–9. The result of the quantum–classical implementation (part 2) is rather different: cluster one (red circles) contains the costumes 3, 4, 7, and 9 while the second cluster (blue circles) contains the costumes 0, 1, 2, 5, 6, and 8. In providing easy access to several implementation the comparison of such results is supported by QHAna. In addition, the results can be manually verified using the entity table of QHAna (see Fig. 20) as outlined in Sect. 7.2. The entity table allows the evaluation of the attributes of the costumes that are part of the clusters and provides all the details of the costumes by linking them to the MUSE-repository. This can be used to improve the understanding of the benefits quantum computing can have, e.g. for quantum humanities research.

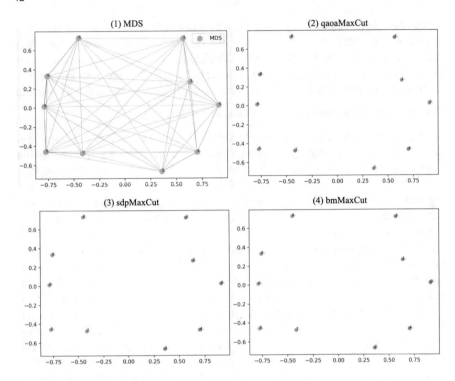

Fig. 22 Values of 10 costumes embedded via MDS (1) and their clustering results of different maximum cut implementations (2–4)

7.3 Improve the Understanding of the Benefits of Quantum Computing

In a fist attempt when taking a closer look at the comparison of Fig. 22 one might decide for the classical implemented clustering for further analysis as it seems more precise. But QHAna also allows to improve the results by adjusting the hyperparameters of the algorithms. For the results of Fig. 22 this would include to potentially improve the results of the quantum–classical implementation of the algorithm as shown in Fig. 22 to equal the results of the classical implementation. Thus, QHAna lists the selectable parameters (a.k.a. hyperparameters) specific to the chosen algorithms, together with descriptions that support selecting the right parameters. Figure 23 gives an impression of the hyperparameters to be selected for the quantum–classical implementation of a maximum cut algorithm including the user token required to run the algorithm on the IBM Q systems.

Figure 24 outlines the influence of the maximum number of iterations performed by the algorithm, while all other parameters keep the default settings of QHAna. As can be seen, increasing the number of iterations seems to improve the results: Diagram 1 is performed with only one iteration, while the number of maximum

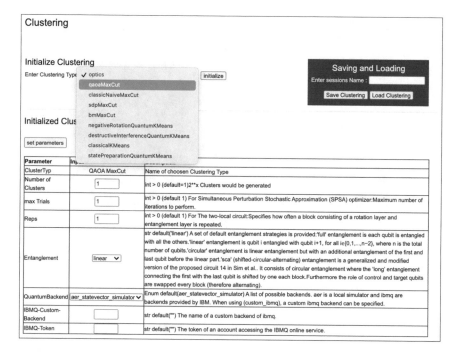

Fig. 23 Screenshot of "Clustering"

iterations is increased by 50 each time (Diagram 2: 50 trials, Diagram 3: 100 trials, and Diagram 4: 150 trials). Thus, the results approach those achieved by the classical maximum cut implementations up to the identification of the same clusters when performed with a maximum of 150 trials (Diagram 4).

By enabling easy comparison of different techniques and implementations of classical and quantum–classical algorithms during the data analysis process, QHAna contributes to a better understanding of the potential benefits (Sect. 2.2) that quantum computing can provide for quantum humanities research. This includes as described above identifying optimal hyperparameters, e.g., approximating the results of the quantum–classical implementation and the classical ones, which then can be applied to larger use cases. But it also includes identifying already available advantages of applying quantum machine learning. An example of these potential advantages of quantum machine learning, is the precision that the quantum kernel estimation (QKE) method for SVM (Havlicek et al., 2018) can achieve.

Currently the tab "Classification" of QHAna provides three implementation of SVM (see Sect. 5.3) methods ("classicSklearnSVM" an implementation of classical support vector machines, "qkeQiskitSVM" an implementation using the quantum kernel estimation (QKE) variant of the SVM, and an implementation of an variational quantum SVM (Havlicek et al., 2018)) and two classifiers based on neural networks (an implementation based on a classical neural network, an implementation based on a hybrid quantum–classical neural network). Figure 25 gives an example of the

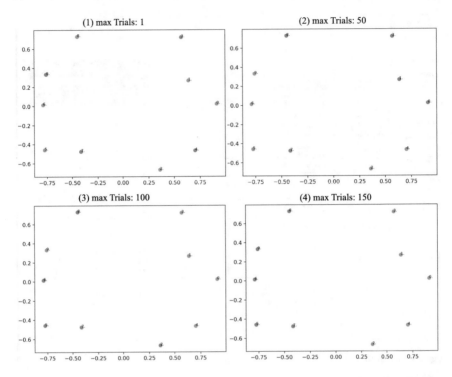

Fig. 24 Clustering results of the quantum–classical maximum cut implementations when changing the number of maximum iterations performed

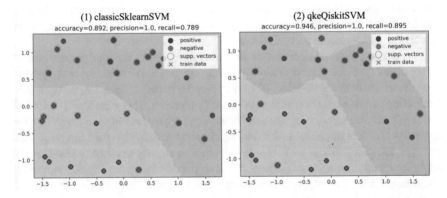

Fig. 25 Classification results of the classically implemented SVM (left) and of the QKE based implemented SVM (right)

results of the classically implemented SVM (Diagram 1) and of the QKE based SVM (Diagram 2). Input for this example are 30 costumes and their distance values based on the attributes stereotype and role relevance embedded by MDS, and classified as "positive" or "negative" based on another training set for the classifier. As can be seen in Fig. 25 these classes are not linear separable. Thus, finding the right classifier is a challenging task. Figure 25 depicts that (by using the default parameters QHAna provides) the result of the QKE variant (Diagram 2) is much more precise (accuracy of 94% versus 89%) than the result of the classical variant (Diagram 1).

7.4 Integration of Heterogeneous Tools

The prototype of QHAna has a modular structure. Thus, it can easily be extended by further algorithm implementations. Various implementations are already available (see Sect. 7.1–7.3). Because QHAna can also serve REST APIs the effective integration of components accessible via such APIs is straightforward; this way, further implementations of techniques described in the theoretical sections of this contribution into QHAna is achieved: This includes—but is not restricted to—extending the feature engineering step by quantum principal component analysis, for example, or add quantum restricted Boltzmann machines, using more clustering algorithms and integrating more classification algorithms. Also, algorithms used in other domains (like HHL, VQE and so on) can be integrated into QHAna.

Another significant advantage of QHAna is that it provides comfortable access to different backends to run the algorithms on. Figure 23 shows, for example, that there are different simulators and IBM quantum computing backends integrated and selectable already. Currently we are integrating more backends like PennyLane (PennyLane, 2021) and TensorFlow.

7.5 Provide Easy Access to Quantum Computing

Working with quantum computers still requires advanced knowledge in physics and mathematics and is therefore a great challenge for research in the field of digital humanities. As there are promising potentials in using the quantum computer in this area (see Sect. 2.2), it is of interest to provide a simple, straightforward access to this new technology without having to deal with all the mathematical, physical, and algorithmic details. This is what QHAna aims for. As Fig. 23 depicts, even very complex quantum–classical algorithms are provided in an abstract manner allowing to adjust their hyperparameters, but default values of these parameters are set. Thus, these algorithms and their benefits can be applied by someone not familiar with qubits, QPU connectivity, quantum circuits, and so on. In Sects. 5 and 6 we provided the core concepts to enable an understanding of the underlying ideas of the (classical and quantum) algorithms used in QHAna, but no details about classical algorithms nor quantum algorithms are given. These algorithms and their implementations evolve rapidly and potentially change quite frequently. Their details are not the focus of QHAna. Instead, providing access to a complex new technology to learn about and participate in the advantages given for further research is at the heart of QHAna.

8   Conclusion and Outlook

As outlined in this contribution there are a lot of potentials in applying quantum computers to research done in the humanities but there is also a long way to go to fully benefit from them. Thus, the benefits and challenges of quantum computers were stressed and—based on our use case MUSE—first application knowledge for further quantum humanities is provided.

As the use case is focused on machine learning, a data analysis pipeline for machine learning tasks was introduced (that is especially useful for detecting patterns in a domain) and core concepts that are promising to be applied in the fields of the humanities were introduced and discussed. Therefore, artificial neural networks were described by providing a mathematical definition of neurons, neural networks, and perceptrons. Also, their use for restricted Boltzmann machines and autoencoders as well as first realizations of these on quantum computers were described. Quantum algorithms are often hybrid; therefore, the main idea of variational hybrid quantum–classical algorithms was sketched as well as an application for clustering based on hybrid techniques like quantum approximate optimization algorithm (QAOA) and variational quantum eigensolver (VQE) were discussed.

To provide a straightforward access to the described techniques an analysis tool called QHAna has been introduced. The presented prototype has a modular architecture and implementation; thus, it can be extended easily. We plan to realize an import and export functionality such that—after key steps of the supported pipeline—the intermediate results can be exported, and results from other steps can be imported. Especially, this will allow to consider results from external algorithms (i.e. algorithms not available via QHAna) in the pipeline as well as to pass results from QHAna for further processing to external algorithms. We also plan to integrate workflow features such that users may define their own sequence of invoking algorithm implementations (those provided by QHAna as well as external ones) to support custom pipelines of data analysis in a multitude of domains: this will significantly improve the flexibility of QHAna.

As quantum computers are continuously improving and the number of qubits is expected to increase constantly and significantly (Gambetta, 2020), it is of great importance to develop application knowledge in the domain of quantum computing at an early stage. Therefore, we envision to generalize the application knowledge that can be extracted from our approach to *quantum humanities patterns* to make it reusable for different use cases in other application domains of the humanities. Providing solution knowledge on such a promising and innovative topic as quantum computing is already relevant in itself. First samples for pattern languages in this domain can be seen (Leymann, 2019; Weigold et al., 2020, 2021). It is of particular relevance to provide application knowledge in a field where the mathematical and physical basics required to use this technology cannot be taken for granted, but fields, which are essential for the critical reflection of digital methods, such as the digital humanities. Especially, if a quantum computer is not only to be used as a tool like for quantum machine learning but for thinking about totally new questions that have not yet been tackled at all, perhaps not even identified or considered, knowledge about the corresponding concepts, methods, and application potentials is key. To identify and explore these possible new application areas is an exciting task and needs to be started by building knowledge on how quantum computers can be applied to problems stated in the humanities.

Acknowledgements I am very grateful to Frank Lemann for discussing several subjects of this chapter. Also, I would like to thank Felix Truger, Philipp Wundrack, Marcel Messer, Daniel Fink and Fabian Bühler for their valuable input and implementing several aspects of our use case.

This work was partially funded by the BMWi project PlanQK (01MK20005N) and the Terra Incognita project Quantum Humanities funded by the University of Stuttgart.

References

Aharonov, D., van Dam, W., Kempe, J., Landau, Z., Lloyd, S., & Regev, O. (2008). Adiabatic quantum computation is equivalent to standard quantum computation. *SIAM Review, 50*(4), 755–787.

Alexander, C., Ishikawa, S., Silverstein, M., Jacobson, M., Fiksdahl-King, I., & Angel, S. (1977). *A pattern language: Towns, buildings, constructions.* Oxford University Press.

Amin, M. A., Andriyash, E., Rolfe, J., Kulchytskyy., B., & Melko, R. (2018). Quantum Boltzmann machine. *Physical Review X, 8,* 021050.

Ankerst, M., Breunig, M. M., Kriegel, H., & Sander, J. (1999). OPTICS: ordering points to identify the clustering structure. In *ACM SIGMOD international conference on Management of data.* ACM Press.

Anschuetz, E. R., Olson, J. P., Aspuru-Guzik, A., & Cao, Y. (2018). Variational quantum factoring. arXiv:1808.08927

Arthur, D., & Vassilvitskii, S. (2007). "k-means++: The advantages of careful seeding". In *Proceedings of the eighteenth annual ACM-SIAM symposium on Discrete algorithms* (pp. 1027–1035). Society for Industrial and Applied Mathematics.

Barzen, J. (2013). Taxonomien kostümrelevanter Parameter: Annäherung an eine Ontologisierung der Domäne des Filmkostüms. Technical Report No. 2013/04, Universität Stuttgart. https://www.iaas.uni-stuttgart.de/publications/TR-2013-04_Taxonomien-_kostuemrelevanter_Parameter.pdf

Barzen, J. (2018). Wenn Kostüme sprechen—Musterforschung in den Digital Humanities am Beispiel vestimentärer Kommunikation im Film. Dissertation, University of Cologne. https://kups.ub.uni-koeln.de/9134/

Barzen, J., Breitenbücher, U., Eusterbrock, L., Falkenthal, M., Hentschel, F., & Leymann, F. (2016). The vision for MUSE4Music. Applying the MUSE method in musicology. In: *Computer science—Research and development. Advancements of service computing: Proceedings of summer SoC 2016* (Vol. 32, pp. 3–4).

Barzen, J., Falkenthal, M., & Leymann, F. (2018). Wenn Kostüme sprechen könnten: MUSE—Ein musterbasierter Ansatz an die vestimentäre Kommunikation im Film. In P. Bockwinkel, B. Nickel, G. Viehhauser (Eds.), *Digital humanities. Perspektiven der Praxis, Frank & Timme* (pp. 223–241).

Barzen, J., & Leymann, F. (2014). Costume languages as pattern languages. In *Proceedings of Pursuit of Pattern Languages for Societal Change—Preparatory Workshop*.

Barzen, J., & Leymann, F. (2019). Quantum humanities: A vision for quantum computing in digital humanities. In *SICS software-intensive cyber-physical systems* (pp. 1–6). https://doi.org/10.1007/s00450-019-00419-4

Barzen, J., & Leymann, F. (2020). Quantum humanities: A first use case for quantum-ML in media science. In *ISAAI'19 Proceedings—Artificial Intelligence, Digitale Welt* (Vol. 4(1), pp. 102–103).

Barzen, J., Leymann, F., Falkenthal, M., Vietz, D., Weder, B., & Wild, K. (2021). Relevance of near-term quantum computing in the cloud: A humanities perspective. In *Cloud Computing and Services Science, Communications in Computer and Information Science*. Springer (to appear).

Bennett, K. P., & Campbell, C. (2000). Support vector machines: Hype or Hallelujah? *ACM SIGKDD, 2*(Issue 2).

Berry, D. M. (2012). *Understanding digital humanities*. Palgrave Macmillan.

Berry, D. M., & Fagerjord, A. (2017). *Digital humanities: Knowledge and critique in a digital age*. Polity Press.

Bhaskar, M. K., Hadfield, S., Papageorgiou, A., & Petras, I. (2015). Quantum algorithms and circuits for scientific computing. arXiv:1511.08253

Biamonte, J. (2020). Lectures on quantum tensor networks. arXiv:1912.10049

Bondarenko, D., & Feldmann, P. (2020). Quantum autoencoders to denoise quantum data. *Physical Review Letters, 124*, 130502.

Boumal, N. (2015). A Riemannian low-rank method for optimization over semidefinite matrices with block-diagonal constraints. arXiv:1506.00575

Boumal, N., Voroninski, V., & Bandeira, A. S. (2016). The non-convex Burer-Monteiro approach works on smooth semidefinite programs. In *Advances in Neural Information Processing Systems 29 (NIPS 2016)*.

Bravo-Prieto, C., LaRose, R., Cerezo, M., Subasi, Y., Cincio, L., & Coles, P. J. (2020). Variational quantum linear solver. arXiv:1909.05820

Brown, K. (Ed.) (2020). *The Routledge companion to digital humanities and art history*. Routledge.

Burges, Ch. J. C. (1998). A tutorial on support vector machines for pattern recognition. *Data Mining and Knowledge Discovery, 2*, 121–167.

Cao, Y., Guerreschi, G. G., & Aspuru-Guzik, A. (2017). Quantum neuron: an elementary building block for machine learning on quantum computers. arXiv:1711.11240

Chen, G., & Srihari, S. H. (2015). Restricted Boltzmann machine for classification with hierarchical correlated prior. arXiv:1406.3407v2

Cox, T. F., & Cox, M. A. A. (2001). *Multidimensional scaling*. Chapman and Hall.

Crooks, G. E. (2018). Performance of the quantum approximate optimization algorithm on the maximum cut problem. arXiv:1811.08419v1

da Silva, I. N., Spatti, D. H., Flauzino, R. A., Liboni, L. H. B., & dos Reis Alves, S. F. (2017). *Artificial neural networks: A practical course*. Springer International Publishing.

Denil, M., & de Freitas, N. (2011). *Toward the implementation of a quantum RBM*. Oxford University Research Archive.

Deutsch, D. (1985). Quantum theory, the Church-Turing principle, and the universal quantum Turing machine. *Proceedings of the Royal Society of London, A400*, 97117.

DHd. (2021). Digital Humanites im deutschspachingen Raum. https://dig-hum.de/

DHQ. (2021). Digital humanities quarterly. http://www.digitalhumanities.org/dhq/.

Dickel, C. (2018). A cloud quantum computer business plan. In: QuTech. http://blog.qutech.nl/index.php/2018/07/18/a-cloud-quantum-computer-business-plan/

Ding, Y., Lamata, L., Sanz, M., Chen, X., & Solano, E. (2019). Experimental implementation of a quantum autoencoder via quantum adders. *Advanced Quantum Technologies, 1800065*, 2019.

Dunjko, V., Taylor, J. M., & Briegel H. J. (2016). Quantum-enhanced machine learning. arXiv:1610.08251

Egger, D. J., Marecek, J. & Woerner, S. (2020). Warm-starting quantum optimization. arXiv:2009.10095

Falkenthal, M., Barzen, J., Breitenbücher, U., Brügmann, S., Joos, D., Leymann, F., & Wurster, M. (2016a). Pattern research in the digital humanities: How data mining techniques support the identification of costume patterns. In *Computer Science—Research and Development. Advancements of Service Computing: Proceedings of SummerSoC 2016* (Vol. 32, pp. 3–4).

Falkenthal, M., Barzen, J., Breitenbücher, U., Fehling, C., Leymann, F., Hadjakos, A., Hentschel, F., & Schulze, H. (2016b). Leveraging pattern applications via pattern refinement. In *Proceedings of the International Conference on Pursuit of Pattern Languages for Societal Change (PURPLSOC)*.

Falkenthal, M., Barzen, J., Dörner, S., Elkind, V., Fauser, J., Leymann, F, & Strehl, T. (2015). Datenanalyse in den Digital Humanities—Eine Annäherung an Kostümmuster mittels OLAP Cubes. In *Datenbanksysteme für Business, Technologie und Web (BTW)* (Vol. 16). *Fachtagung des GI-Fachbereichs "Datenbanken und Informationssysteme" (DBIS)*.

Farhi, E., Goldstone, J., & Gutmann, S. (2014). A quantum approximate optimization algorithm. In MIT-CTP/4610.

Farhi, E., & Neven, H. (2018). Classification with quantum neural networks on near term processors. arXiv:1802.06002v2

Fehling, C., Barzen, J., Falkenthal, M., & Leymann, F. (2014). PatternPedia—Collaborative Pattern identification and authoring. In *Proceedings of Pursuit of Pattern Languages for Societal Change—Preparatory Workshop*.

Gambetta, J. (2020). IBM's roadmap for scaling quantum technology. In IBM Research Blog. Retrieved September 15, 2020, from https://www.ibm.com/blogs/research/2020/09/ibm-quantum-roadmap/

Grover, L. K. (1996). A fast quantum mechanical algorithm for database search. In *Proceedings of the twenty-eighth annual ACM symposium on Theory of computing* (pp. 212–219).

Hamamura, I., & Imamichi, T. (2019). Efficient evaluation of quantum observables using entangled measurements. arXiv:1909.09119v2

Haven, E., & Khrennikov, A. (2013). *A Quantum social science*. Cambridge University Press.

Havenstein, C., Thomas, D., & Chandrasekaran, S. (2018). Comparisons of performance between quantum and classical machine learning. *SMU Data Science Review, 1*(4), Article 11.

Havlicek, V., Córcoles, A. D., Temme, K., Harrow, A. W., Kandala, A., Chow, J. M., & Gambetta, J. M. (2018). Supervised learning with quantum enhanced feature spaces. arXiv:1804.11326

Higgott, O., Wang, D., & Brierley, S. (2019). Variational quantum computation of excited states. *Quantum, 3*, 156.

Hinton, G. E., (2012). A practical guide to training restricted Boltzmann machines. In G. Montavon, G. B. Orr, K. R. Müller (Eds.), *Neural networks: Tricks of the trade. Lecture Notes in Computer Science* (Vol. 7700). Springer.

Hinton, G. E., & Zemel, R. S. (1994). Autoencoders, minimum description length and Helmholtz free energy. In *Advances in neural information processing systems* (Vol. 6, pp. 3–10).

Jannidis, F., Kohle, H., & Rehbein, M. (Eds.) (2017). *Digital humanities: Eine Einführung*. J.B. Metzler.

Jozsa, R. (2006). An introduction to measurement based quantum computation. NATO Science Series, III: Computer and Systems Sciences. *Quantum Information Processing-From Theory to Experiment, 199*, 137–158.

Kandala, A., Mezzacapo, A., Temme, K., Takita, M., Brink, M., Chow, J. M., & Gambetta, J. M. (2017). Hardware-efficient variational quantum Eigensolver for small molecules and quantum magnets. arXiv:1704.05018

Karalekas, P. J., Tezak, N. A., Peterson, E. C., Ryan, C. A., da Silva, M. P., & Smith, R. S. (2020). A quantum-classical cloud platform optimized for variational hybrid algorithms. *Quantum Science and Technology, 5*, 024003.

Khan, S. U., Awan, A. J., & Vall-Llosera, G. (2019). K-means clustering on noisy intermediate scale quantum computers. arXiv:1909.12183

Kidger, P., & Lyons, T. (2020). Universal approximation with deep narrow networks. In *Machine learning research* (Vol. 125, pp. 1–22). *Proc. 33rd Annual Conference on Learning Theory (COLT 2020)*.

Knill, E. (2007). Quantum computing with very noisy devices. arXiv:quant-ph/0410199 (preprint)

Kubat, M. (2017). *An introduction to machine learning*. Springer International Publishing AG.

Lamata, L., Alvarez-Rodriguez, U., Martın-Guerrero, J. D., Sanz, M., & Solano, E. (2019). Quantum autoencoders via quantum adders with genetic algorithms. *Quantum Science and Technology, 4*, 014007.

Larochelle, H., Mandel, M., Pascanu, R., & Bengio, Y. (2012). Learning algorithms for the classification restricted Boltzmann machine. *Journal of Machine Learning Research, 13*, 643–669.

LaRose, R. (2019). Overview and comparison of gate level quantum software platforms. *Quantum, 3*, 130. arXiv:1807.02500

Leymann, F. (2019). Towards a pattern language for quantum algorithms. In *First international workshop (QTOP 2019)*.

Leymann, F., & Barzen, J. (2020a). The bitter truth about gate-based quantum algorithms in the NISQ era. *Quantum Science and Technology, 5*, 044007.

Leymann, F., & Barzen, J. (2020b). Pattern atlas. arXiv:2006.05120

Leymann, F., Barzen, J., Falkenthal, M., Vietz, D., Weder, B., & Wild, K. (2020). Quantum in the cloud: Application potentials and research opportunities. In *Proceedings of the 10th International Conference on Cloud Computing and Services Science* (pp. 9–24). SciTePress.

Lubasch, M., Joo, J., Moinier, P., Kiffner, M., & Jaksch, D. (2020). Variational quantum algorithms for nonlinear problems. *Physical Review A, 101*, 010301.

McClean, J. R., et al. (2017). OpenFermion: The electronic structure package for quantum computers. arXiv:1710.07629

McClean, J. R., Romero, J., Babbush, R., & Aspuru-Guzik, A. (2016). The theory of variational hybrid quantum-classical algorithms. *New Journal of Physics, 18*, 023023.

McCulloch, W. S., & Pitts, W. (1943). A logical calculus of the ideas immanent in nervous activity. *Bulletin of Mathematical Biophysics, 5*, 115–133.

Michielsen, K., Nocon, M., Willsch, D., Jin, F., Lippert, T., & De Raedt, H. (2017). Benchmarking gate-based quantum computers. *Computer Physics Communications, 220*, 44–55.

MUSE. (2021). Muster Suchen und Erkennen. https://www.iaas.uni-stuttgart.de/en/projects/muse/

MUSE GitHub. (2021). https://github.com/Muster-Suchen-und-Erkennen/muse-docker

National Academies of Sciences, Engineering, and Medicine. (2019). *Quantum computing: progress and prospects*. The National Academies Press. https://doi.org/10.17226/25196

Nielsen, M., & Chuang, I. L. (2010). *Quantum computation and quantum information*. Cambridge University Press.

Nocedal, J., & Wright, S. J. (2006). *Numerical optimization*. Springer Science + Business Media, LLC.

Novikoff, A. (1962). On convergence proofs for perceptrons. In *Proc. Mathematical Theory of Automata*, Vol. 12, pp. 615–622.

Otterbach, J. S., et al. (2017). Unsupervised machine learning on a hybrid quantum computer. arXiv: 1712.05771v1

Palmer, M. (2020). Quantum computers will create better versions of Alexa and Siri. In: Sifted. Retrieved December 10, 2020, from https://sifted.eu/articles/quantum-computers-nlp/

PennyLane. (2021). https://pennylane.ai/

Pepper, A., Tischler, N., & Pryde. G. J. (2019). Experimental realization of a quantum autoencoder: The compression of qutrits via machine learning. *Physical Review Letters, 122*, 060501.

Peruzzo, A., McClean, J., Shadbolt, P., Yung, M.-H., Zhou, X.-Q., Love, P. J., AspuruGuzik, A., & O'Brien, J. L. (2014). A variational eigenvalue solver on a quantum processor. *Nature Communications, 5*, 4213.

Poljak, S., & Rendl, F. (1995). Solving the max-cut problem using eigenvalues. *Discrete Applied Mathematics, 62*, 249–278.

Preskill, J. (2018). Quantum computing in the NISQ era and beyond. *Quantum, 2,* 79.

PyTorch. (2021). https://pytorch.org/

QH. (2021). Quantum humanities. https://www.iaas.uni-stuttgart.de/forschung/projekte/quantum-humanities/

QHAna. (2021). Quantum humanities data analysis tool. https://github.com/UST-QuAntiL/qhana

Qiskit. (2021). https://qiskit.org/

Qiskit Max Cut. (2021). https://qiskit.org/documentation/locale/de_DE/tutorials/optimization/6_e xamples_max_cut_and_tsp.html

Raz, R., & Tal, A. (2018). Oracle separation of BQP and PH. Iin: Electronic colloquium on computational complexity, Report No. 107.

Rehbein, M., & Sahle, P. (2013). Digital Humanities lehren und lernen. Modelle, Strategien, Erwartungen. In: H. Neuroth, N. Lossau, A. Rapp (Eds.), *Evolution der Informationsinfrastruktur Kooperation zwischen Bibliothek und Wissenschaft.* Verlag Werner Hülsbusch.

Reiners, R. (2014). An evolving pattern library for collaborative project documentation. Dissertation, University of Aachen.

Rieffel, E., & Polak, W. (2011). *Quantum computing: A gentle introduction.* The MIT Press.

Romero, J., Olson, J. P., & Aspuru-Guzik, A. (2017). Quantum autoencoders for efficient compression of quantum data. *Quantum Science Technology, 2,* 045001.

Rønnow, T., Wang, Z., Job, J., Boixo, S., Isakov, S. V., Wecker, D., Martinis, J. M., Lidar, D. A., & Troyer, M. (2014). Defining and detecting quantum speedup. *Science, 345*(6195).

Rosenblatt, F. (1958). The perceptron: A probabilistic model for information storage and organization. *Psychological Review, 65*(6).

Sahle, P. (Ed.) (2019). DHd 2019. Digital humanities: Multimedial & multimodal. Konferenzabstracts. Mainz/Frankfurt a M.

Schöch, C. (Ed.) (2020). DHd 2020. Spielräume Digital Humanities zwischen Modellierung und Interpretation. Konferenzabstracts. https://doi.org/10.5281/zenodo.3666690

Schuld, M., & Petruccione, F. (2018). *Supervised learning with quantum computers.* Springer.

Schuld, M., Sinayskiy, I., & Petruccione, F. (2014). The quest for a Quantum Neural Network. arXiv:1408.7005v1

Schumm, D., Barzen, J., Leymann, F., & Ellrich, L. (2012). A pattern language for costumes in films. In *Proceedings of the 17th European Conference on Pattern Languages of Programs (EuroPLoP 2012).*

Shaydulin, R., Safro, I., & Larson, J. (2019). Multistart methods for quantum approximate optimization. arXiv:1905.08768

Shor, P. W. (1995). Polynomial-time algorithms for prime factorization and discrete logarithms on a quantum computer. arXiv:quant-ph/9508027

Sierra-Sosa, D., Arcila-Moreno, J., Garcia-Zapirain, B., Castillo-Olea, C., & Elmaghraby, A. (2020). Dementia prediction applying variational quantum classifier. arXiv:2007.08653

Skiena, S. S. (2017). *The data science design manual.* Springer International Publishing AG.

Smolensky, P. (1986). Information processing in dynamical systems: Foundations of harmony theory. In D. E. Rumelhart & J. L. McLelland (Eds.), *Parallel distributed processing: explorations in the microstructure of cognition, Volume 1: Foundations* (pp. 194–281). MIT Press.

Tacchino, F., Macchiavello, C., Gerace, D., & Bajoni, D. (2018). An artificial neuron implemented on an actual quantum processor. arXiv:1811.02266v1

TensorFlow. (2021). https://www.tensorflow.org/

TensorFlow Quantum. (2021). https://www.tensorflow.org/quantum

Terras, M., Nyhan, J., & Vanhoutte, E. (2013). *Defining digital humanities. A reader.*

Tomczak, J. M. (2016). Learning informative features from restricted Boltzmann machines. *Neural Processing Letters, 2016*(44), 735–750.

Vogeler, G. (Ed.) (2018). DHd 2018. Kritik der digitalen Vernunft. Konferenzabstracts.

Wan, K. H., Dahlsten, O., Kristjánsson, H., Gardner, R., & Kim, M. S. (2017). Quantum generalisation of feedforward neural networks. *npj Quantum Information, 3,* 36.

Weigold, M., Barzen, J., Breitenbücher, U., Falkenthal, F., Leymann, F., & Wild, K. (2020). Pattern views: Concept and tooling for interconnected pattern languages. In S. Dustdar (Ed.), *Service-oriented computing. SummerSOC 2020. Communications in computer and information science* (Vol. 1310). Springer, Cham. https://doi.org/10.1007/978-3-030-64846-6_6

Weigold, M., Barzen, J., Leymann, F., & Salm, M. (2021). Data encoding patterns for quantum algorithms. In The Hillside Group (Ed.), *Proceedings of the 27th Conference on Pattern Languages of Programs (PLoP'20)* (to appear).

What is Digital Humanities? (2021). http://whatisdigitalhumanities.com/

Wiebe, N., Kapoor, A., & Svore, K. M. (2016). Quantum perceptron models. arXiv:1602.04799

Wu, Y., Wei, C., Qin, S., Wen, Q., & Gao, F. (2020). Quantum restricted Boltzmann machine universal for quantum computation. arXiv:2005.11970v3

Wu, Z., & Palmer, M. (1994). Verb semantics and lexical selection. In *ACL'94 Proceedings of the 32nd annual meeting on Association for Computational Linguistics*.

Yuan, X., Endo, S., Zhao, Q., Li, Y., & Benjamin, S. (2019). Theory of variational quantum simulation. *Quantum, 3*, 191.

Zhang, P., Li, S., & Zhou, Y. (2015). An algorithm of quantum restricted Boltzmann machine network based on quantum gates and its application. Shock and Vibration Volume. Article ID 756969.

Zheng, X., Wu, Z., Meng, H., Li, W., & Cai, L. (2013). Feature learning with Gaussian restricted Boltzmann machine for robust speech recognition. arXiv:1309.6176

Zurada, J. M. (1992). *Introduction to artificial neural systems*. West Publishing Company.

(All links have been last followed March 11, 2021)

Quantum Computing and Cognitive Simulation

Martha Lewis

Abstract Cognitive Science is the study of the mind and how it relates to human behaviour. As a field, it is highly interdisciplinary, bringing together theories from Psychology, Neuroscience, Biology, and others. The field of cognitive science operates at a number of levels. On the one hand, some aspects of cognitive science look at the behaviour of neurons, whereas other aspects look to explain human behaviour at a more abstract level, seeking to explain human decision making or reasoning. A key area of research in cognitive science is how to formalise human behaviours around judgements of similarity, categorization, and decision making. In the field of Physics, Quantum Mechanics has fundamentally altered our understanding of the way in which particles behave. Quantum mechanics has a number of unintuitive phenomena, some of which can be used model unusual aspects of human behaviour. The application of quantum theory to model human behaviour is wide-ranging. In this chapter we will look at three main areas in which it has been applied. One key area is in how similarity judgements can be modelled. There are a number of phenomena around similarity judgements that are not well modelled using a view of concepts that does not take the state of the observer into account. These include the asymmetry of similarity judgements and the fact that similarity can change depending on other exemplars that are present. Another key area is in judgement and decision-making. Again, puzzling phenomena have been observed regarding the fact that judgements do not follow classical probability theory, known as the conjunction fallacy, can depend crucially on the order in which questions are presented or on the amount of knowledge about the world. Finally, approaches to modelling cognitive phenomena at the neural level will also be discussed within a quantum-theoretic framework.

Keywords Quantum computing · Quantum humanities · Cognitive modelling · Similarity judgements · Decision-making

M. Lewis (✉)
Department of Engineering Mathematics, University of Bristol, Bristol, UK
e-mail: martha.lewis@bristol.ac.uk

© Springer Nature Switzerland AG 2022
E. R. Miranda (ed.), *Quantum Computing in the Arts and Humanities*,
https://doi.org/10.1007/978-3-030-95538-0_2

1   Introduction

Cognitive science is the study of the mind and how it relates to human behaviour. As a field, it is highly interdisciplinary, bringing together theories from Psychology, Neuroscience, Biology, and others. The field of cognitive science operates at a number of levels. On the one hand, some aspects of cognitive science look at the behaviour of neurons, whereas other aspects look to explain human behaviour at a more abstract level, seeking to explain human decision-making or reasoning. There is a long-standing history of vector space models in cognitive science. Theories of categorization such as those developed by Ashby and Gott (1988), Nosofsky (1986), Rosch (1975) utilize notions of distance and similarity that can readily be incorporated in vector space models of meaning. Hampton (1987), Smith and Osherson (1984), Tversky (1977) encode meanings as feature vectors, and models of high-level cognitive reasoning have been implemented within vector symbolic architectures (Gayler, 2003; Plate, 1995; Smolensky, 1990).

Another field in which vector space models play an important role is Physics, and especially quantum theory. Though seemingly unrelated to language, intriguing connections have recently been uncovered. The link between Physics and natural language semantics that vector space models provide has been successfully exploited, providing novel solutions and a fresh perspective for a number of problems related to cognitive science, such as modelling logical aspects in vector spaces (Widdows & Peters, 2003). Methods from quantum logic have also been applied to cognitive processes related to the human mental lexicon, such as word association (Bruza et al., 2009), decision-making (Pothos & Busemeyer, 2013), and human probability judgements (Busemeyer et al., 2011). Furthermore, the categorical model of Coecke et al. (2010), inspired by Quantum Mechanics, has provided a convincing account of compositionality in vector space models and an extensible framework for linguistically motivated research on sentential semantics. More recently, the link between Physics and text meaning was made more concrete by a number of proposals that aim at replacing the traditional notion of a word vector with that of a density matrix—a concept borrowed from Quantum Mechanics which can be seen as a probability distribution over vectors (Bankova et al., 2019; Piedeleu et al., 2015; Sadrzadeh et al., 2018).

A key area of research in cognitive science is how to formalize human behaviours around judgements of similarity, categorization, and decision-making. However, formalizing concepts within a vector space, and relying on distance within vector space to give a notion of similarity, can lead to some puzzling features, at least if we assume that distance and similarity behave classically. For example, it has been shown (Tversky, 1977) that judgements of similarity are not always symmetric, as you would expect if they are just based on distance in a vector space.

In the field of Physics, Quantum Mechanics has fundamentally altered our understanding of the way in which particles behave. Quantum mechanics has a number of unintuitive phenomena, some of which can be used to model unusual aspects of human behaviour. This is not to say that a description of the brain and mind is provided in terms of quantum phenomena at a small scale, although this sort of modelling has been proposed (Hameroff, 2014; Penrose, 1990). Instead, the idea is that the formalism arising from quantum theory provides the right sort of mathematical tools to model these puzzling aspects of human behaviour.

The application of quantum theory to model human behaviour is wide-ranging. In this chapter, we will look at three main areas in which it has been applied. One key area is how similarity judgements can be modelled. There are a number of phenomena around similarity judgements that are not well modelled using a view of concepts that does not take the state of the observer into account. These include the asymmetry of similarity judgements and the fact that similarity can change depending on other exemplars that are present (Tversky, 1977). These phenomena have been addressed in Pothos et al. (2013). Another key area is in judgement and decision-making. Again, puzzling phenomena have been observed regarding the fact that judgements do not follow classical probability theory, known as the conjunction fallacy (Tversky & Kahneman, 1983), and can depend crucially on the order in which questions are presented (Moore, 2002) or on the amount of knowledge about the world (Tversky & Shafir, 1992). These phenomena are modelled within a quantum framework in Busemeyer et al. (2011), Franco (2009), Khrennikov and Haven (2009), Pothos and Busemeyer (2009), Wang and Busemeyer (2013). Related to this area is the phenomenon of contextuality. This can be summarized as the idea that there may be sets of random variables (in Psychology, results of an experiment) that have pairwise joint distributions but for which no joint distribution across the whole set of random variables can be found. This is one of the key aspects of quantum theory. Amazingly, the same theory was also developed in Psychology, and has been formalized in Dzhafarov and Kujala (2014, 2016). Finally, quantum theory has been applied to describe categorization and concept combination. Again, the ways in which humans use concepts has been shown not to be well modelled by classical views of combination like fuzzy set theory, where problems known as *over-* and *underextension* are observed (Hampton, 1987, 1988a, 1988b, 1997; Smith & Osherson, 1984). Approaches to answer these phenomena have been proposed in Aerts (2009), Aerts and Gabora (2005a, 2005b), Aerts et al. (2015), Sozzo (2014, 2015). The problem of modelling how concepts compose can also be addressed via a linguistic route. The problem of modelling concept composition by means of the grammatical structure in language has been addressed in Coecke et al. (2010), and applications of this to the problem of overextension was proposed in Coecke and Lewis (2015).

Approaches to modelling cognitive phenomena at the neural level have also been considered within a quantum-theoretic framework. Fuss and Navarro (2013) show that a quantum random walk approach to modelling choice tasks better simulates human reaction times. Consideration of neuronal activation within a quantum framework is also discussed in Pothos and Truebloodm (2015). The question of modelling how concepts can 'bind' together (for example, the combination of an adjective and a noun) was addressed at a vector-based level in Smolensky (1990) and has been investigated within a cognitive science context in Martin and Doumas (2020). The compositional distributional framework of Coecke et al. (2010) has the potential to model these aspects of neuronal activity well, and combining this theory with the tensor product binding of Smolensky (1990) is an area for future research.

In the remainder of the chapter, we cover the following. We provide a short section on the mathematical notation we will use. In Sect. 2, we will summarize the cognitive phenomena that have been described by quantum theory. In Sect. 3, we describe how quantum theory has been used to address each of these phenomena, as well as discussing quantum-theoretic approaches to modelling neuronal-level phenomena. In Sect. 4, we discuss the implications for using quantum computing to model cognitive phenomena and Artificial Intelligence more generally.

1.1 Mathematical Notation

We assume that the reader has a general understanding of linear algebra. We use Dirac's bra-ket notation to represent vectors, their duals, and inner products.

- A ket $|v\rangle$ is a column vector in a Hilbert space $\mathcal{H}$. We will always consider the space to be finite-dimensional, so $\mathcal{H} = \mathbb{C}^n$.
- A bra $\langle v|$ is the vector dual to $|v\rangle$. It is the conjugate transpose of $|v\rangle$ and can be thought of as a row vector whose elements are the complex conjugates of the elements in $|v\rangle$.
- The inner product of two vectors $|v\rangle$, $|w\rangle$ is represented by $\langle v|w\rangle$.
- If a matrix is represented by M, its multiplication with a vector $|v\rangle$ is given by $M|v\rangle$ and its multiplication with a bra $\langle v|$ is given by $\langle v|M$. Often, these will be combined to give a scalar $\langle v|M|v\rangle$.
- The absolute value of a complex number $\alpha = a + ib$ is represented as $|\alpha| = \sqrt{a^2 + b^2}$.
- The Euclidean norm of a vector is written as $||v|| = \sqrt{\langle v|v\rangle}$.

2 Cognitive Phenomena

One of the cornerstones of human cognition is the ability to categorize. From an evolutionary perspective, it is essential to be able to categorize objects into edible/inedible, or animals into dangerous/safe. Central to the notion of categorization is the notion of similarity. An animal can be classified as safe based on how similar it is to another animal known to be safe. The study of categorization and similarity is therefore central to cognitive science, and has been widely researched (Ashby & Gott, 1988; Nosofsky, 1986; Rosch, 1975). How can we determine whether one item is similar to another or not? One of the approaches to studying the notion of similarity and categorization is to view objects as represented by points in a feature space, and then determining similarity within that space. Categories can then be viewed as forming regions within the feature space. However, under classical assumptions this kind of model does not necessarily model human behaviour well. We describe here some of these key phenomena.

2.1 Asymmetry of Similarity Judgements

In Tversky (1977), the question is addressed on whether objects can be represented as points in a geometric space together with the distance between the points being measured via a metric on the space. The similarity between objects is then viewed as a function of the distance between the objects in the space.

Within this model, a key assumption is that distance is viewed as a metric, meaning that the following hold: given objects x, y, and z, and distance metric d,

$$d(x, x) = 0 \tag{1}$$
$$d(x, y) = d(y, x) \text{ (symmetry)} \tag{2}$$
$$d(x, z) \leq d(x, y) + d(y, z) \text{ (triangle inequality)} \tag{3}$$

Tversky (1977) show that these assumptions do not necessarily hold. In a series of experiments, Tversky showed that when one object is considered more prominent than another, the less prominent object is considered more similar to the more prominent object and vice versa. Firstly, pairs of countries are assessed (by Israeli college students) for which country is most prominent. The pairs of countries are, for example, China and Vietnam,[1] the USA and Mexico, or Belgium and Luxembourg. A separate group of students were then asked to judge which phrase they preferred to use when describing the similarity of the two countries out of "country a is similar to country b" and "country b is similar to country a". Across all pairs, the majority of students chose the ordering in which the more prominent country was given second, for example "Vietnam is similar to China" rather than "China is similar to Vietnam". This kind of effect was seen across a range of topics and modalities, for example, when judging the similarity of shapes, letters, and sequences of sounds.

2.2 Diagnosticity

Tversky (1977) further show that judgements of similarity can be altered by the presence of comparison objects. According to a geometric model of similarity, the similarity of two objects should not be affected by the presence of other objects.

---

[1] In fact, the countries assessed are Red China and North Vietnam.

However, this is shown not to hold for human similarity judgements in the following manner. Pairs of quadruples of countries were designed that differ in only one entry, i.e. there are pairs of sets $\{a, b, c, p\}$ and $\{a, b, c, q\}$. Participants are asked to say which of b, c and q or p country a is most similar to. For example, one such pair is the sets of countries {Austria, Sweden, Poland, Hungary} and {Austria, Sweden, Norway, Hungary}. Participants see only one quadruple and are asked to judge which country Austria is most similar to. Tversky (1977) find that in the set {Austria, Sweden, Poland, Hungary}, most participants judge Austria to be most similar to Sweden, but in the quadruple {Austria, Sweden, Norway, Hungary} participants judge Austria to be most similar to Hungary.

2.3   Conjunction and Disjunction Fallacies

In modelling human decision-making, one assumption is that human behaviour can be expressed by means of probabilities. A key facet of classical probability is that the probability of a conjunction of events will always be less than or equal to the probability of one of its constituents: $P(A\&B) \leq P(A)$. Similarly, the probability of a disjunction of events will always be greater than or equal to the probability of one of its constituents: $P(A) \leq P(A \text{ or } B)$. However, experiments in Tversky and Kahneman (1983) showed that these rules are not adhered to when humans make decisions about category membership. The experiments run as follows. Human participants are given a story about a person, and asked to make certain judgements about them. The most frequently cited is the following:

> Linda is 31 years old, single, outspoken and very bright. She majored in philosophy. As a student, she was deeply concerned with issues of discrimination and social justice, and also participated in anti-nuclear demonstrations.

The description is followed by eight statements about the person's career or personality:

- Linda is a teacher in elementary school.
- Linda works in a bookstore and takes Yoga classes.
- Linda is active in the feminist movement (F).
- Linda is a psychiatric social worker.
- Linda is a member of the League of Women Voters.
- Linda is a bank teller (T).
- Linda is an insurance salesperson.
- Linda is a bank teller and is active in the feminist movement ($T\&F$).

Participants are asked to rank the eight statements associated with each description by how probable each statement is. A large majority (between 85 and 90%, depending on the participant pool) rank the statements in the order $T\&F > T$, that is, they judge that the probability that Linda is both a feminist and a bank teller is higher than the probability that Linda is a bank teller. This phenomenon still holds (to a lesser extent) in other contexts where the participants are asked to place a bet on the two statements T and $T \& F$, in a domain of medical experts (where the story that is told is about symptoms and the task is to diagnose the patient), and in a range of other situations designed to reduce the incidence of the conjunction fallacy. Experiments where fallacy does not hold to such a great extent are when the participant pool is with a range of statistically sophisticated participants, and when the experiment is phrased in terms of numbers of people.

2.4 Question Order Models

In Moore (2002), the effect of order on answers to questions is addressed. If you are asked "Do you try to eat healthily?" and "Do you like McDonalds?", your answer to each question may change depending on which question is asked first. There are a number of ways in which the ordering could affect the answers. Moore (2002) identifies four types of such effect, termed *contrast, consistency, additive,* and *sub-tractive*. The consistency effect runs as follows. A group of participants is asked a pair of questions. For a concrete example, consider the questions "Do you think Bill Clinton is trustworthy" and "Do you think Al Gore is trustworthy". For context, these questions were asked in 1997. If the question is asked first, it is considered to be asked in a *non-comparative* context, meaning that there is nothing to immediately compare the question to. If the question is asked second, it is considered to be asked in a *comparative* context, because it can be considered in comparison to the first. The answers to the questions were distributed as in Table 1. The effect of the question ordering is that the answer to the question in the comparative context is altered to make it more consistent with the answer in the non-comparative context. So, if a participant is first asked whether Gore is honest and trustworthy, and they answer positively, then they are more likely to answer positively to the question of whether Clinton is honest and trustworthy. On the other hand, if a participant is first asked whether Clinton is honest and trustworthy, and answers negatively, then they are more likely to answer negatively to the question of whether Gore is honest and trustworthy. The effect is to make the answers to the two questions more consistent.

Table 1 Example of the consistency effect. * indicates significant difference. Figures from Moore (2002)

Context	Per cent saying yes		
	Clinton	Gore	Gap
First	50	68	+18*
Second	57	60	+3
Difference	+7*	−8*	−15*

Table 2 Example of the contrast effect. * indicates significant difference. Figures from Moore (2002)

Context	Per cent saying yes		
	Gingrich	Dole	Gap
First	41	60	+19*
Second	33	64	+31*
Difference	−8*	+4*	+12*

Table 3 Example of the additive effect. * indicates significant difference. Figures from Moore (2002)

Context	Per cent saying 'All' or 'Many'		
	White	Black	Gap
First	41	46	+5
Second	53	56	+3
Difference	+12*	+10*	−2

The contrast effect occurs when people alter their answers to make the answers contrast with each other. For example, in a Gallup poll in 1995 when asked the question of whether the label 'honest and trustworthy' applies to either Bob Dole or Newt Gingrich, people's answers in the comparative context altered to increase the gap between them, emphasizing the differences between the two. Figures are shown in Table 2.

Under the additive effect, considering either question first has the same effect. In a series of racial hostility polls in 1996, people were asked the questions "Do you think that only a few white people dislike blacks, many white people dislike blacks, or almost all white people dislike blacks?" and "Do you think that only a few black people dislike whites, many black people dislike whites, or almost all black people dislike whites?". In both cases, the percentage responding 'All' or 'Many' increased in the comparative context; see Table 3.

Lastly, in the subtractive context considering either question first again has the same effect, but in this case the effect is to decrease the number of positive answers. The example given in Moore (2002) is the question of whether baseball players Pete Rose and, separately, Shoeless Joe Jackson should be eligible for admission to the Baseball Hall of Fame. In each case, the players are ineligible, and the participants are told why. In each case, the number of people answering favourably to the question of whether they should be allowed decreased in the comparative context (Table 4).

These four examples show clearly the different kinds of effects that question order can have on responses, and that the context in which a question is asked is crucial for its response. These effects are addressed in Wang and Busemeyer (2013), Wang et al. (2013).

Table 4 Example of the subtractive effect. * indicates significant difference. Figures from Moore (2002)

	Per cent saying 'Favourable'		
Context	Rose	Jackson	Gap
First	64	45	+19*
Second	52	33	+19*
Difference	−12*	−12*	0

2.5 The 'Sure Thing' Principle

A counter-intuitive aspect of human decision-making is the violation of the 'sure thing' principle. The sure thing principle was introduced by Savage (1954). This says that if we prefer x to y given any possible state of the world, then we should prefer x to y even when the exact state of the world is unknown. However, people do not always make decisions according to the sure thing principle. Tversky and Shafir (1992) present the following experiment:

> Imagine that you have just taken a tough qualifying examination. It is the end of the fall quarter, you feel tired and run-down, and you are not sure that you passed the exam. In case you failed you have to take the exam again in a couple of months-after the Christmas holidays. You now have an opportunity to buy a very attractive 5-day Christmas vacation package to Hawaii at an exceptionally low price. The special offer expires tomorrow, while the exam grade will not be available until the following day. Would you:
>
> 1. buy the vacation package
> 2. not buy the vacation package
> 3. pay a $5 non-refundable fee in order to retain the rights to buy the vacation package at the same exceptional prices the day after tomorrow—after you find out whether or not you have passed the exam.

31% of respondents said that they would buy the vacation package, 7% said that they would not buy the vacation package, and 60% said that they would reserve the low price for the small fee-paying to wait until the uncertainty was resolved.

However, when the participants were asked to imagine they knew the outcome of the exam, a majority said that they would buy the package:

> Imagine that you have just taken a tough qualifying examination. It is the end of the semester, you feel tired and run-down, and you find out that you [passed the exam/failed the exam. You will have to take it again in a couple of months-after the Christmas holidays]. You now have an opportunity to buy a very attractive 5-day Christmas vacation package to Hawaii at an exceptionally low price. The special offer expires tomorrow. Would you
>
> 1. buy the vacation package
> 2. not buy the vacation package
> 3. pay a $5 non-refundable fee in order to retain the rights to buy the vacation package at the same exceptional prices the day after tomorrow.

Table 5 Matrix of payoffs for the prisoner's dilemma situation

	Other D	Other C
You D	You 10, other 10	You 25, other 5
You C	You 5, other 25	You 20, other 20

In the context where the participant had passed the exam, 54% of respondents said that they would buy the vacation package, 16% of respondents said that they would not buy the vacation package, and 30% of students said that they would pay the $5 waiting fee. Very similarly, in the case where the participant had failed the exam, 57% of respondents said that they would buy the vacation package, 12% of respondents said that they would not buy the vacation package, and 31% of students said that they would pay the $5 waiting fee. Essentially, in the context where the outcome of the exam is known, a majority of participants decide to book the holiday—whether or not it is pass or fail. According to the sure thing principle, those participants should decide to book the holiday in that case where the outcome of the exam is not known. But, as can be seen, this does not happen.

A similar experiment involves gambling:

> Imagine that you have just played a game of chance that gave you a 50% chance to win $200 and a 50% chance to lose $100. The coin was tossed and you have [won $200/lost $100].
>
> You are now offered a second identical gamble
>
> 50% chance to win $200 and 50% chance to lose $100. Would you:
>
> 1. Accept the second gamble
> 2. Reject the second gamble

In the context that the participants had won the first gamble, 69% accepted and 31% rejected the second. In the context that participants had lost the first gamble, the split was 59% accept, 41% reject. Either way, a majority of participants accepted the second gamble.

However, in a context where they did not know the outcome of the first gamble, a majority of the *same set* of participants rejected the second gamble. The split was 36% accept, 64% reject.

The explanation given in Tversky and Shafir (1992) is that the presence of uncertainty makes it more difficult for people to focus on the implications of each outcome.

A further example is given in a prisoner's dilemma situation. In the prisoner's dilemma, a hypothetical prisoner has the choice either to cooperate with the other prisoner, i.e. keep quiet about their activities, or defect, i.e. tell the authorities what they know. The possible outcomes of cooperating or defecting are presented in the following payoff matrix (Table 5).

In an experimental situation, if the participant is told that the other has cooperated, the majority of participants choose to defect. Also, if the participant is told that the other has defected, the majority of participants choose to defect. However, if the participant is not told what the choice of the other is, the majority of participants choose to cooperate.

2.6 Categorization

One of the most important things that humans can do is to categorize their experiences. Is this fruit good to eat, or poisonous? Is that large striped animal dangerous? Categorization can be thought of in terms of similarity to other experiences, and to help us do this we use concepts. There have been a number of theories of concepts. The classical view of concepts dates back to Plato and Aristotle. Under the classical view of concepts, a concept is viewed as a set of necessary and sufficient conditions, formulated as a list of properties. Such a view of concepts is inadequate. Firstly, for many natural concepts it is impossible to give a set of necessary and sufficient conditions (Wittgenstein, 1953). Instead, instances of a category can be thought of as sharing a set of 'family resemblances'. For example: what makes the images in Fig. 1 pictures of chairs, and the images in Fig. 2 not pictures of chairs? It is difficult to decide on a set of necessary and sufficient conditions that rules the kneeling chair to be a chair, but rules out a bar stool.[2,3,4,5]

Furthermore, the classical view does not take into account the graded and context-dependent nature of concepts. Consider the much simpler concept *tall*. Whether or not to describe something as *tall* will firstly be dependent on the set of things that form its context, for example, 4-year-old children, or Dutch women, or giraffes, or mountains. Secondly, even if we have fixed a context, it is not a crisp question of

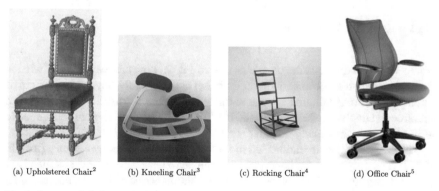

(a) Upholstered Chair[2]   (b) Kneeling Chair[3]   (c) Rocking Chair[4]   (d) Office Chair[5]

Fig. 1 Images of chairs

[2] https://commons.wikimedia.org/wiki/File:Philipp_Rumpf_Studienblatt_eines_Stuhls.jpg.

[3] https://commons.wikimedia.org/wiki/File:Sleekform_Kneeling_Chair.jpg.

[4] https://commons.wikimedia.org/wiki/File:Shaker_no._7_rocking_chair_Rocking_Chair,_1878 %E2%80%931910_(CH_18460985).jpg.

[5] https://commons.wikimedia.org/wiki/File:LibertyChair.jpg.

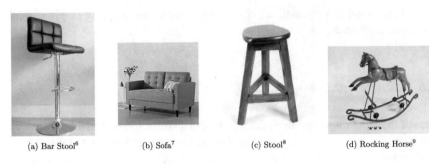

| (a) Bar Stool[6] | (b) Sofa[7] | (c) Stool[8] | (d) Rocking Horse[9] |

Fig. 2 Non-chairs

Fig. 3 Heights of male celebrities

whether we can divide some people as tall. In Fig. 3, how are we to decide what is the necessary and sufficient condition for being tall? Certainly, anyone to the right of Jeremy Clarkson is tall, and anyone to the left of Nikita Khrushchev is not tall, but should we state the condition to be 180 cm? 184 cm? If we say 184 cm and someone is 183.5 cm, are they really not tall?[6,7,8,9]

These sorts of considerations led to the formulation of *prototype theory* (Rosch, 1975). In prototype theory, concepts are assumed to have a prototype, and members of the concept are judged as being more or less similar to the prototype. For example, in the concept *bird*, a robin is judged as a very prototypical bird, whereas an ostrich or a penguin would be judged as much less typical. This allows the consideration

---

[6] https://commons.wikimedia.org/wiki/File:Bar_Stools_(49907001456).jpg.

[7] https://commons.wikimedia.org/wiki/File:52.8_Inch_Sofa_Couch_Loveseat,_Grey.jpg.

[8] https://commons.wikimedia.org/wiki/File:TabouretAFDB.jpg.

[9] https://commons.wikimedia.org/wiki/File:Wooden_Rocking_Horse_with_Weels.jpg.

of concepts in terms of family resemblances, and accounts for the graded nature of concepts.

Further, the *exemplar theory* of concepts views a category as based on specific instances, or exemplars, that are stored in memory. Categorization judgements are based on judgements of similarity to the exemplars, which are easily called to mind.

Exemplar theory and prototype theory are both backed by experimental evidence (Nosofsky, 1986; Rosch, 1975). Key characteristics of these theories are that they can account for the graded nature of concepts, and in both cases, we represent concept instances within some kind of feature space, equipped with a distance metric.

Another key feature of human concept use is that we are able to combine concepts to create new ones. This can be done in very simple ways, for example, combining the concept *blue* and the concept *book*, we obtain a blue book. How can we formalize this? One approach to formalizing concept combination used fuzzy set theory. Under this view, a concept is a fuzzy set, that is, a set whose members have membership values in the interval [0, 1] rather than in the set {0, 1}. We write

$$\mu_A : X \to [0, 1] \tag{4}$$

and use the notation $\mu_A(x)$ to describe the membership of the item x in the concept A. In fuzzy set theory, we can form the conjunction or disjunction of two sets as follows:

$$\mu_{A \text{ and } B}(x) = min(\mu_A(x), \mu_B(x)) \tag{5}$$
$$\mu_{A \text{ or } B}(x) = max(\mu_A(x), \mu_B(x)) \tag{6}$$

Under these rules, combinations quickly become problematic. The classic example is that of a pet fish (Smith & Osherson, 1984). Consider a fish like a goldfish. A goldfish is not a particularly typical fish, nor a very typical pet, so we might assign low memberships to goldfish in each of these concepts:

$$\mu_{pet}(goldfish) = 0.3 \tag{7}$$
$$\mu_{fish}(goldfish) = 0.25 \tag{8}$$

However, it is a very typical pet fish, i.e. we would assign something like

$$\mu_{pet\,fish}(goldfish) = 0.9 \tag{9}$$

The question then is what is going on in the combination of these concepts to allow the typicality of the instance in the combined concept to be higher than it is in either of the individual concepts.

This kind of phenomenon was examined in Hampton (1988a, 1988b). In the 'pet fish' kind of example, where the typicality of an item in a conjunction of concepts is higher than the typicality of an item in one or more of its constituent concepts, the phenomenon is called 'overextension'. On the other hand, in the case of disjunction of concepts, a phenomenon termed underextension is also seen. In this case, the typicality of an item in the disjunction of two concepts is lower than typicality in one or more of the constituent concepts.

2.7 Negation

Another problematic phenomenon in human concept use is the use of negation. In Alxatib and Pelletier (2011), an experiment is performed where the membership of humans in the category 'tall' is measured. Imagine we have five suspects in a robbery, all men, with heights 5′ 4″, 5′ 7″, 5′ 11″, 6′ 2″, and 6′ 6″. Participants are asked to judge, of each suspect, whether they are (a) tall, (b) not tall, (c) tall and not tall, or (d) neither tall nor not tall. The percentages of people agreeing to these statements are given in the table below. Note that in the experiment, the suspects were shown in a randomized order; we present them here in height order for ease of presentation.

Suspect	Tall		Not tall		Tall and not tall		Neither tall nor not tall	
	True	False	True	False	True	False	True	False
5′ 4″	1.3	98.7	94.7	3.9	14.5	76.3	27.6	65.8
5′ 7″	5.3	93.4	78.9	17.1	21.1	65.8	31.6	57.9
5′ 11″	46.1	44.7	25	67.1	44.7	40.8	53.9	42.1
6′ 2″	80.3	10.5	9.2	82.9	28.9	56.6	36.9	55.3
6′ 6″	98.7	1.3	0	100	5.3	81.6	6.6	89.5

There is a strong correlation between the values for the borderline statements 'tall and not tall' and 'neither tall nor not tall'. Participants were willing to agree to both of these statements. In the first statement 'tall and not tall', a participant would seem to ascribe the value *true* both to 'tall' and to 'not tall', whereas in the second, the participant would also judge that he is neither.

The problem of negation is also considered by Hampton, in a setting combined with conjunction (Hampton, 1997). In this setting, Hampton considers the typicality of items to conjunctions of concepts. For example, we can consider the typicality of *parrot* to the conjunction *Pets which are also birds* and *Pets which are not birds*. Of course, we would consider the typicality of *parrot* in the first to be high and in the second to be low. However, if concepts are being combined in line with logical reasoning, then the sum of the typicality in *Pets which are also birds* and in *Pets which are not birds* should not exceed the typicality of *parrot* in *Pets*. This is another form of overextension, as we have seen in previous descriptions.

3   Quantum Models

The problems for modelling cognitive phenomena outlined in Sect. 2 above have all been tackled using methods from quantum theory. Below, in Sect. 3.1 we give an overview of the key ideas before explaining the models in detail. After this overview, we discuss approaches to modelling similarity, diagnosticity, and the conjunction fallacy. These approaches have been developed over a research programme laid out in Busemeyer et al. (2011), Franco (2009), Pothos et al. (2013) and other papers, and crucially depend on the notion of interference. Following this, we consider the sure thing principle and question order models. Again, these models depend on the notion of interference. We go on to discuss models of concept combination and negation. These approaches use the quantum phenomenon of entanglement as well as interference to describe how humans combine concepts. Linked to these is the notion of contextuality. Finally, we talk about quantum models that are applied at a neuronal level.

3.1   Overview of Quantum Models in Psychology

The following gives an overview of some of the key quantum concepts used in approaches such as those in Busemeyer et al. (2011), Pothos et al. (2013). Within applications of quantum theory to Psychology, questions are considered from the point of view of an agent. We consider an agent as having a certain state $|\psi\rangle$, and as questions are asked of the agent, or the agent is exposed to a stimulus, the state of the agent changes. Within quantum theory, a question about a system is termed an *observable*: a self-adjoint linear operator. The eigenvectors $\{|a_i\rangle\}_i$ of an observable A form an orthonormal basis of the Hilbert space, and the eigenvalues $\{a_i\}$ associated with the $\{|a_i\rangle\}_i$ form the values that the observable can take, i.e. the answers to the questions. To give a simple example: a yes/no question such as "is the sky blue?" might be encoded in a two-dimensional vector space. An observable B on the space will have two orthonormal eigenvectors $|blue\rangle$ and $|not\ blue\rangle$. We can set the eigenvalues associated with $|blue\rangle$ and $|not\ blue\rangle$ to be 1 and 0, respectively.

Now, we represent an agent's belief about the state of the sky as a vector $|s\rangle$. Given that the agent is ambivalent about what the weather is (possibly clear and blue, possibly overcast and grey), the probability that we receive the answer 'yes!' to the question "Is the sky blue" is given by the following calculation:

$$P(yes) = |\langle blue|s\rangle|^2 \tag{10}$$

This quantity is the square of the projection onto the *blue* axis, illustrated in Fig. 4.

This notion can be extended. Firstly, there may be more than one basis vector corresponding to a given answer, so that that answer corresponds to a subspace of

Fig. 4 A two-dimensional space with orthonormal basis $|blue\rangle$, $|not\ blue\rangle$. The state $|s_0\rangle$ represents belief in a mostly blue sky with a couple of clouds, and the state $|s_1\rangle$ represents belief in a mostly grey sky. The projection onto the *blue* axis is high for $|s_0\rangle$ and low for $|s_1\rangle$, whereas the projection onto the *not blue* axis is low for $|s_0\rangle$ and higher for $|s_1\rangle$.)

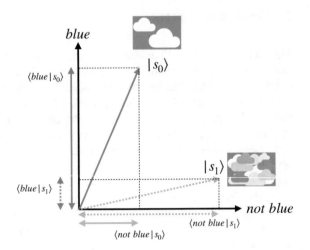

the vector space of dimension greater than 1. For example, we could have a three-dimensional vector space with one dimension $|blue\rangle$ corresponding to blue sky, one dimension $|cloudy\rangle$ corresponding to cloudy sky, and one dimension $|night\rangle$ corresponding to nighttime, and suppose we define an observable D to ask the question "is it daytime?" D has eigenvectors $|blue\rangle$ and $|cloudy\rangle$ with eigenvalue 1 (corresponding to yes) and eigenvector $|night\rangle$ with eigenvalue 0 (corresponding to no). Then,

$$D = 1 \cdot |blue\rangle\langle blue| + 1 \cdot |cloudy\rangle\langle cloudy| + 0 \cdot |night\rangle\langle night| = |blue\rangle\langle blue| + |cloudy\rangle\langle cloudy| \tag{11}$$

The probability of an answer 'yes' to the question "is it daytime?" is again the square of the probability amplitude, which is the magnitude of the projection onto the 'daytime' subspace spanned by the eigenvectors $|blue\rangle$ and $|cloudy\rangle$. If the current state of affairs is given by $|s\rangle$, then this is calculated by

$$P(yes_{day}) = \langle s|D|s\rangle \tag{12}$$

illustrated in Fig. 5.

More generally again, the current state of affairs may be uncertain. Quantum theory models this as a mixture of states

$$\rho = \sum_i s_i |s_i\rangle\langle s_i| \tag{13}$$

where $\sum_i s_i = 1$.

Then the probability of the answer 'yes!' to the question "is it daytime?" is given by

$$Tr(D\rho) \tag{14}$$

Fig. 5 This figure shows a three-dimensional space with orthonormal basis $|blue\rangle, |cloudy\rangle, and$ $|night\rangle$. The state $|s_0\rangle$ represents belief that the sky is slightly cloudy and just starting to get dark. The probability amplitude of the answer 'yes!' to the question "is it daytime?" is given by the projection of $|s_0\rangle$ onto the plane spanned by $|blue\rangle, |cloudy\rangle$

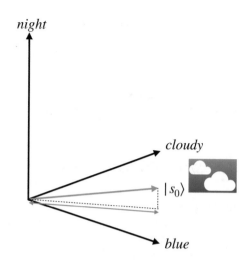

where Tr stands for matrix trace. If we are not at all uncertain about the state, then $\rho = |s\rangle\langle s|$ and this is equivalent to the formula (12).

A key difference between classical probability theory and quantum probability theory is the treatment of conjunctions of events. In classical probability theory, the probability $P(AB)$ conjunction of two events A and B is equal to the probability $P(BA)$ of the two events in the opposite order. In quantum probability, this does not hold, due to the fact that in general the matrices corresponding to the observables $\hat{A}$ and $\hat{B}$ will not commute.

This means that conjunction in quantum probability needs to be defined in sequential terms. Specifically, suppose we want to calculate the probability of an agent in state $|s\rangle$ judging A and then B. We have

$$P(A \text{ and then } B) = P(A) \cdot P(B|A) \tag{15}$$

The first term $P(A)$ is expressed as

$$P(A) = ||A|s\rangle||^2 \tag{16}$$

and the second term $P(B|A)$ is given by

$$P(B|A) = ||B|s_A\rangle||^2 \tag{17}$$

where $|s_A\rangle$ is the state of knowledge on the agent after judging that A holds, i.e.

$$|s_A\rangle = \frac{A|s\rangle}{||A|s\rangle||} \tag{18}$$

Putting this all together, we see that

$$P(A \text{ and then } B) = ||A|s\rangle||^2 ||B|s_A\rangle||^2 \tag{19}$$

$$= ||A|s\rangle||^2 \left(\frac{BA|s\rangle}{||A|s\rangle||} \right)^2 = ||BA|s\rangle||^2 \tag{20}$$

i.e. the agent projects first to subspace A, and then to subspace B. In general, this is not equal to $||AB|s\rangle||^2$, as it would be in classical probability.

3.2 Similarity Judgements

In Pothos et al. (2013), the model described above is used to account for the asymmetry of similarity judgements. Recall that Tversky (1977) showed that when asked to consider the similarity of one entity, such as a country, to another, the judgements of similarity are higher if the entity that is more salient is the one being compared to. So, for example, more participants agreed with the statement "Korea is similar to China" than "China is similar to Korea" (the participants were all Israeli college students).

Pothos et al. (2013) explicate this as follows. The entities being judged, for example, *China* and *Korea* are modelled as subspaces of a given vector space. Each of these subspaces has an associated projector P_{China}, P_{Korea}. The state of belief of the agent is modelled as a quantum state $|\psi\rangle$. The probability that the agent's belief state is consistent with the concept *China* or the concept *Korea* is given by the square of the projection onto the relevant subspace: $||P_{China}|\psi\rangle||^2 = \langle\psi|P_{China}\psi\rangle$. The initial belief state $|psi\rangle$ is set up so that $\langle\psi|P_{China}\psi\rangle = \langle\psi|P_{Korea}\psi\rangle$.

Then, the similarity of Korea to China is calculated but first projecting $|psi\rangle$ onto the Korea subspace, and then projecting the resulting vector onto the China subspace:

$$Sim(Korea, China) = ||P_{China}P_{Korea}|\psi\rangle||^2 \tag{21}$$

This quantity calculates the probability that the original state $|\psi\rangle$ is consistent with Korea, multiplied by the probability that this new state is consistent with China. The agent projects their belief state onto Korea, after which their new belief state is normalized, so that $|\psi_{Korea}\rangle = \frac{P_{Korea}|\psi\rangle}{||P_{Korea}|\psi\rangle||}$. This new state is projected onto China, and is consistent with the concept China with probability $||P_{China}|\psi_{Korea}\rangle$. So, we have that

$$||P_{China}|\psi_{Korea}\rangle||^2 \cdot ||P_{Korea}|\psi\rangle||^2 ||P_{China}P_{Korea}|\psi\rangle||^2 \tag{22}$$

Now, as long as the projectors P_{China} and P_{Korea} do not commute, we have that $Sim(Korea, China) \neq Sim(China, Korea)$. Moreover, Pothos et al. (2013) show that the similarity of Korea to China can be made to be higher than the similarity of China to Korea by modelling the more salient entity, in this case China, as a larger (higher dimensionality) subspace.

3.3 Diagnosticity

Pothos et al. (2013) also show that the phenomenon of diagnosticity (Sect. 2.2) can be modelled. In this model, each country is modelled as a single-dimensional subspace. The diagnosticity effect is produced by the order in which the projectors for the countries are applied to the agent's state vector $|\psi\rangle$.

Recall from Sect. 2.2 that the diagnosticity effect says that similarity judgements can be affected by the context they are made in. An example from Tversky (1977) is the following. Take the sets of countries {Austria, Sweden, Poland, Hungary} and {Austria, Sweden, Norway, Hungary}. Participants see only one quadruple and are asked to judge which country Austria is most similar to. In the quadruple {Austria, Sweden, Poland, Hungary}, Austria is judged most similar to Sweden by a majority of participants. However, in {Austria, Sweden, Norway, Hungary}, Austria is judged most similar to Hungary. Given that both are available in each set of countries, why does the similarity change?

Pothos et al. (2013) set each country to be represented as a ray. The similarity of Austria to any of the other countries in the set is modelled by saying that the agent's belief state $|\psi\rangle$ first projects onto the subspaces corresponding to the other countries. There are two orders in which this can happen, since there are two other countries. Therefore, these orders are averaged. Then, depending on the placing of the rays in the Hilbert space, the required effect can be produced. Pothos et al. (2013) show that with some reasonable assumptions about the similarities between the pairs of countries (i.e. we assume that Sweden is similar to Norway, and that Hungary is similar to Poland), the effect can be reliably reproduced over a number of simulations.

3.4 Conjunction Fallacy and Over/Underextension

Franco (2009) provide an account of the conjunction fallacy using the notion of interference effect. The model works as follows. Recall that in the conjunction fallacy (Sect. 2.3), the problem is that the conjunction of two events (that Linda is a bank teller and a feminist) is judged more probable than the single event that Linda is a feminist. According to classical probability, this cannot hold.

Busemeyer et al. (2011), Franco (2009) model this using the quantum formalism described in Sect. 3.1. They consider the two questions A: "is Linda a feminist?" and B: "is Linda a bank teller?". The question A is represented by an observable on a two-dimensional Hilbert space with eigenvectors $|a_0\rangle$ and $|a_1\rangle$ which have the eigenvalues $a_0 = 0$ (Linda is not a feminist) and $a_1 = 1$ (Linda is a feminist), respectively. Question B is similarly represented. Furthermore, Franco (2009) represent the probability $P(b_j|a_i)$ as the probability that the answer to question B is b_j, given that the answer to A has already been determined to be a_i. If this latter condition is the case, then within this two-dimensional setting this means that the current state we are in is $|a_i\rangle$ itself, and therefore $P(b_j|a_i) = |\langle b_j|a_i\rangle|^2$, following Eq. (10). Notice that

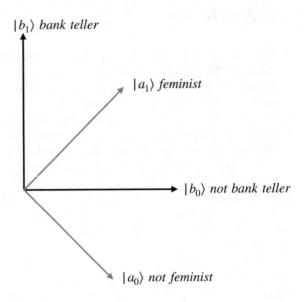

Fig. 6 Orthonormal bases in a two-dimensional space representing the questions "Is Linda a bank teller?" and "Is Linda a feminist?"

this means that $P(b_j|a_i) = |\langle b_j|a_i\rangle|^2 = |\langle a_i|b_j\rangle|^2 = P(a_i|b_j)$. Now as in Eq. (20), the probability of the answer that Linda is a feminist and a bank teller, $P(a_1\text{-}and\text{-}b_1)$ is modelled as $P(a_1)P(b_1|a_1)$.

Franco (2009) then lays out a specific relation between the eigenvectors of $\hat{A}$ and the eigenvectors of $\hat{B}$, as follows.

The two observables $\hat{A}$ and $\hat{B}$ both operate on the same two-dimensional Hilbert space, and the eigenvectors of $\hat{A}$ can be transformed into the eigenvectors of $\hat{B}$ via a unitary transformation

$$U = \begin{pmatrix} e^{-i\xi}\sqrt{P(b_1|a_1)} & e^{-i\phi}\sqrt{1-P(b_1|a_1)} \\ -e^{i\phi}\sqrt{1-P(b_1|a_1)} & e^{i\xi}\sqrt{P(b_1|a_1)} \end{pmatrix}$$

Visually, we have a pair of orthonormal bases as given in Fig. 6, where $|b_i\rangle = U|a_i\rangle$.

Now, given this representation, the state of knowledge of the agent is represented as a vector:

$$|s\rangle = \sqrt{P(a_0)}|a_0\rangle + e^{i\phi_a}\sqrt{P(a_1)}|a_1\rangle$$

which gives the correct probabilities for the agent to judge whether or not Linda is a feminist.

Expressing this vector in the basis of $\hat{B}$, we obtain

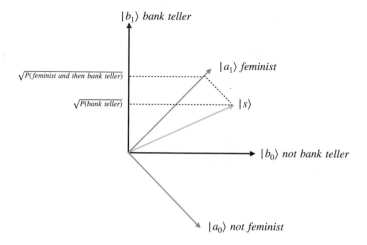

Fig. 7 Starting in the state $|s\rangle$, the probability that Linda is judged to be a bank teller is less than that the probability that Linda is judged first to be a feminist and then to be a bank teller

$$|s\rangle = \sqrt{P(a_0)}U^T|b_0\rangle + \sqrt{P(a_1)}U^T|b_1\rangle \tag{23}$$

$$= \left(\sqrt{P(a_0)}\sqrt{P(b_1|a_1)}e^{i\xi} - \sqrt{P(a_1)}\sqrt{P(b_1|a_0)}e^{i\phi_a-\phi}\right)|b_0\rangle \tag{24}$$

$$+ \left(\sqrt{P(a_0)}\sqrt{P(b_1|a_0)}e^{i\phi} + \sqrt{P(a_1)}\sqrt{P(b_1|a_1)}e^{i\phi_a-\xi}\right)|b_1\rangle \tag{25}$$

Now, calculating $P(b_1)$, i.e. the probability that the agent judges that Linda is a bank teller, we get

$$P(b_1) = |\sqrt{P(a_0)}\sqrt{P(b_1|a_0)}e^{i\phi} + \sqrt{P(a_1)}\sqrt{P(b_1|a_1)}e^{i\phi_a-\xi}|^2 \tag{26}$$

$$= P(a_0)P(b_1|a_0) + P(a_1)P(b_1|a_1) + 2\sqrt{P(a_0)P(a_1)P(b_1|a_0)P(b_1|a_1)}cos(\phi + \xi + \phi_a) \tag{27}$$

The last term in this sum is called the *interference term* $I(s, A)$. If $P(a_0)P(b_1|a_0) + I(s, A)$ is negative, then we obtain that $P(b_1) < P(a_1 \text{and then } b_1)$, i.e. that the probability that Linda is a bank teller is judged less likely than the probability that Linda is a feminist and a bank teller. This can be understood visually in Fig. 7.

This calculation is expressed succinctly in Busemeyer et al. (2011) as follows. Consider that the event of answering 'yes' to "is Linda a feminist?" is represented by the projector P_F, and that the event of answering 'yes' to "is Linda a bank teller?" is given by P_B. The projectors corresponding to the answer 'no' to each question are given by $P_{\neg F} = 1 - P_F$ and $P_{\neg B} = 1 - P_B$, respectively.

Now, as in Eq. (20), given a state $|s\rangle$ we have

$$P(F \text{ and then } B) = ||P_B P_F |s\rangle||^2 \tag{28}$$

We can compare this value with the value $P(B) = ||P_B|s\rangle||^2$ as follows:

$$||P_B|s\rangle||^2 = ||P_B \cdot I|s\rangle||^2 = ||P_B(P_F + P_{\neg F}|s\rangle||^2 \tag{29}$$

$$= ||P_B P_F|s\rangle + P_B P_{\neg F}|s\rangle||^2 \tag{30}$$

$$= ||P_B P_F|s\rangle||^2 + ||P_B P_{\neg F}|s\rangle||^2 + \langle s_{B\neg F}|s_{BF}\rangle + \langle s_{BF}|s_{B\neg F}\rangle \tag{31}$$

where $|s_{BF}\rangle = P_B P_F|s\rangle$ and $|s_{B\neg F}\rangle = P_B P_{\neg F}|s\rangle$.

This term $\langle s_{B\neg F}|s_{BF}\rangle + \langle s_{BF}|s_{B\neg F}\rangle$ is the interference term that we called $I(s, A)$ above, and we can construct this term to be negative. Again, if $\langle s_{B\neg F}|s_{BF}\rangle + \langle s_{BF}|s_{B\neg F}\rangle + ||P_B P_{\neg F}|s\rangle||^2$ is negative overall, then we have

$$||P_B|s\rangle||^2 < ||P_B P_F|s\rangle||^2 \tag{32}$$

and the conjunction fallacy is fulfilled.

A related phenomenon to the conjunction fallacy is that a similar phenomenon is observed when considering disjunctions. Morier and Bordiga (1984) show that in a setup where participants are asked to judge the probability that Linda is (a) a feminist, (b) a bank teller, (c) a feminist and a bank teller, and (d) a feminist or a bank teller, the order from least probable to most probable is given by

$P(bank\ teller) < P(feminist\ and\ bank\ teller) < P(feminist\ or\ bank\ teller) < P(feminist)$

Busemeyer et al. (2011) show that their probability model can also account for this phenomenon. They consider the probability of answering 'no' to the question "is Linda neither a bank teller nor a feminist?". This is modelled as the probability $1 - ||P_{\neg F}P_{\neg B}||^2$. Notice that here, the order of the projectors is such that the state is first projected onto $P_{\neg B}$—this is because this subspace is more probable. Consider

$$||P_{\neg F}|s\rangle||^2 = ||P_{\neg F}P_{\neg B}||^2 + ||P_{\neg F}P_B||^2 + \langle s_{\neg FB}|s_{\neg F\neg B}\rangle + \langle s_{\neg F\neg B}|s_{\neg FB}\rangle \tag{33}$$

Again, if the interference term $\langle s_{\neg FB}|s_{\neg F\neg B}\rangle + \langle s_{\neg F\neg B}|s_{\neg FB}\rangle$ is sufficiently negative, we have $||P_{\neg F}P_B||^2 + \langle s_{\neg FB}|s_{\neg F\neg B}\rangle + \langle s_{\neg F\neg B}|s_{\neg FB}\rangle < 0$ and then $||P_{\neg F}|s\rangle||^2 < ||P_{\neg F}P_{\neg B}||^2$, as needed for the disjunction fallacy.

3.4.1 Double Overextension

Whilst this model covers a number of kinds of probability judgement error, one phenomenon that is not covered is the phenomenon of double overextension. This occurs when the probability of an item belonging to a conjunction of concepts is judged to be higher than the probability of the item belonging to either of the constituent concepts. In the Linda story, this would mean that the probability of Linda being judged to be a feminist and a bank teller would not only be judged higher than the probability of her being a bank teller, but also higher than her being judged to be a feminist.

A quick look at the formalism shows that this cannot hold, at least as long as we assume the order of conjunction is fixed.

$$P(F \text{ and then } B) = P(F) \cdot P(B|F) = ||P_B P_F |s\rangle||^2 \tag{34}$$

$$= ||P_B|s_F\rangle||^2 \cdot ||P_F|s\rangle||^2 < ||P_F|s\rangle||^2 = P(F) \tag{35}$$

As long as we hold the order fixed, we cannot have that both $P(F)$ and $P(B)$ are lower than $P(F \text{ and then } B)$.

However, this kind of double overextension has been empirically observed, as described in Sect. 2.6. In Sect. 3.8, we will describe an approach in which the problem of double overextension can be modelled.

3.5 Violation of the 'Sure Thing' Principle

The 'sure thing' principle states that if an agent would take a specific course of action C in all possible states of the world, then they should take action C when they do not know what the state of the world is. However, as described in Sect. 2.5, humans usually do not behave like this. This is seen in a prisoner's dilemma situation, the situation where we are choosing whether to book a holiday, or a situation where we are choosing whether to make a bet, as well as many more. Pothos and Busemeyer (2009) show that this phenomenon can be modelled using the same kind of quantum similarity model described in Sect. 3.2. Other approaches include that in Khrennikov and Haven (2009).

The model is applied to a prisoner's dilemma situation. The situation can be modelled as the tensor product of one space modelling our opponents' actions, and one space modelling our actions. Each space is two-dimensional, with one dimension corresponding to the 'cooperate' (C) action, and the other corresponding to the 'defect' (D) action. The agent's state at the start of the game is therefore represented as the state

$$|\psi_0\rangle = \frac{1}{2}|DD\rangle + |DC\rangle + |CD\rangle + |CC\rangle \tag{36}$$

If the opponent is known to defect, then the state changes to

$$|\psi_D\rangle = \alpha_{DD}|DD\rangle + \alpha_{DC}|DC\rangle \tag{37}$$

where $\alpha_{DD}^2 + \alpha_{DC}^2 = 1$. On the other hand, if the opponent is known to cooperate, the state changes to

$$|\psi_C\rangle = \alpha_{CD}|CD\rangle + \alpha_{CC}|CC\rangle \tag{38}$$

where $\alpha_{CD}^2 + \alpha_{CC}^2 = 1$. Our decision whether or not to defect is modelled as a projection M onto our defect axis, i.e.

$$M = |DD\rangle\langle DD| + |CD\rangle\langle CD| \tag{39}$$

Now, if we know that our opponent defected, then our probability of defection is

$$\langle\psi_D|M|\psi_D\rangle = (\alpha_{DD}^*\langle DD| + \alpha_{DC}^*\langle DC|)(|DD\rangle\langle DD| + |CD\rangle\langle CD|)(\alpha_{DD}|DD\rangle + \alpha_{DC}|DC\rangle) \tag{40}$$

$$= |\alpha_{DD}|^2 \tag{41}$$

where $|\alpha|$ indicates the absolute value of α. However, if we know that our opponent cooperated, then our probability of defection is

$$\langle\psi_C|M|\psi_C\rangle = (\alpha_{CD}^*\langle CD| + \alpha_{CC}^*\langle CC|)(|DD\rangle\langle DD| + |CD\rangle\langle CD|)(\alpha_{CD}|CD\rangle + \alpha_{CC}|CC\rangle) \tag{42}$$

$$= |\alpha_{CD}|^2 \tag{43}$$

We model not knowing whether the opponent has defected or not as the superposition $|\psi\rangle = \frac{1}{\sqrt{2}}(|\psi_D\rangle + |\psi_C\rangle))$ When we are in this state of ignorance, we have

$$\langle\psi|M|\psi\rangle = \frac{1}{2}(\alpha_{DD}^*\langle DD| + \alpha_{DC}^*\langle DC| + \alpha_{CD}^*\langle CD| + \alpha_{CC}^*\langle CC|)(|DD\rangle\langle DD| + |CD\rangle\langle CD|) \tag{44}$$

$$(\alpha_{DD}|DD\rangle + \alpha_{DC}|DC\rangle + \alpha_{CD}|CD\rangle + \alpha_{CC}|CC\rangle) \tag{45}$$

$$= \frac{1}{2}(\alpha_{DD}^2 + \alpha_{DC}^2) + Re(\langle\alpha_{DD}DD|\alpha_{DC}DC\rangle) \tag{46}$$

Again, this last term is an interference term, and may be negative. This means that we can obtain the situation that the probability of defecting is smaller in the state of ignorance than it is in the case that we know what the action of our opponent is.

3.6 Question Order Models and Contextuality

The quantum probability theory of judgement proposed in Busemeyer et al. (2011) and described in Sect. 3.1 is applied to question order effects in Wang et al. (2013). The question order effects are those taken from Moore (2002), outlined in Sect. 2.4. Consider the consistency effect exemplified by the questions "Do you think Clinton is trustworthy?" and "Do you think Gore is trustworthy?". The consistency effect means that when the question is asked second (the comparative context), people change their responses to make them more similar to the response to the first question, in comparison to when the question is asked first. Wang et al. (2013) model this as follows.

A person's belief is represented by a state vector $|s\rangle$ in an n-dimensional Hilbert space. Each potential response x to a question Q corresponds to a subspace Qx that has a unique orthogonal projector P_{Qx}. The projectors to all responses to question Q sum to the identity: $\sum_x P_{Qx} = I$. The probability of responding x to the question Q is given by $||P_{Qx}||^2$, and the updated belief state after deciding on an answer to the question is $|s_{Qx}\rangle = \frac{P_{Qx}|s\rangle}{||P_{Qx}|s\rangle||}$. As in Sect. 3.1, the probability of responding x to question A and then y to question B is

$$p(Ax \text{ and then } By) = ||P_{Ax}|s\rangle||^2 ||P_{By}|s_{Ax}\rangle||^2 = ||P_{By} P_{Ax}|s\rangle||^2 \quad (47)$$

An *order effect* for a given question occurs when the probability of giving an answer to a question in the comparative context differs from the probability of giving that answer in the non-comparative context. In the Clinton/Gore setting, we have that the total probability TP_{Cy} of answering 'yes' to the question "is Clinton trustworthy?" in the comparative context is given by

$$TP_{Cy} = p(Gy \text{ and then } Cy) + p(Gn \text{ and then } Cy)$$
$$= ||P_C P_G|s\rangle||^2 + ||P_C P_{\neg G}|s\rangle||^2 \quad (48)$$

However, as we saw in Sect. 3.4, the probability of answering 'yes' in the non-comparative context is given by

$$p(Cy) = ||P_C|s\rangle||^2 = ||P_C P_G|s\rangle + P_C P_{\neg G})|s\rangle||^2 \quad (49)$$
$$= ||P_C P_G|s\rangle||^2 + ||P_{\neg G})|s\rangle||^2 + 2Re\langle s|P_G P_c(1 - P_G)|s\rangle$$
$$= TP_{Cy} + 2Re\langle s|P_G P_c(1 - P_G)|s\rangle \quad (50)$$

The order effect C_C is therefore given by

$$TP_{Cy} - p(Cy) = -2Re\langle s|P_G P_c(1 - P_G)|s\rangle = 2||P_C P_G|s\rangle||^2 - 2Re\langle s|P_G P_c|s\rangle \quad (51)$$
$$= 2||P_C P_G|s\rangle||^2 - 2|\langle s|P_G P_c|s\rangle|\cos\phi \quad (52)$$
$$= 2||P_C P_G|s\rangle||^2 - 2\theta||P_c|s\rangle|| \cdot ||P_G|s\rangle|| \quad (53)$$

where $\theta = \cos\phi|\langle s|P_G P_c|s\rangle|/(||P_c|s\rangle|| \cdot ||P_G|s\rangle||)$.

A similar calculation gives the order effect for Gore, C_G to be

$$C_G = 2||P_G P_c|s\rangle||^2 - 2\theta||P_c|s\rangle|| \cdot ||P_G|s\rangle|| \quad (54)$$

Now, since θ is the same in both cases, we have

$$0 = 2p(Gy \text{ and then } Cy) - C_C - (2p(Cy \text{ and then } Gy) - C_G) \tag{55}$$

$$= 2p(Gy \text{ and then } Cy) - p(Gy \text{ and then } Cy) - p(Gn \text{ and then } Cy) + p(Cy) \tag{56}$$

$$- (2p(Cy \text{ and then } Gy) - p(Cy \text{ and then } Gy) - p(Cn \text{ and then } Gy) + p(Gy)) \tag{57}$$

$$= p(Gy \text{ and then } Cy) - p(Gn \text{ and then } Cy) + p(Cy \text{ and then } Gy) + p(Cy \text{ and then } Gn) \tag{58}$$

$$- p(Cy \text{ and then } Gy) + p(Cn \text{ and then } Gy) - p(Gy \text{ and then } Cy) - p(Gy \text{ and then } Cn) \tag{59}$$

$$= p(Cy \text{ and then } Gn) + p(Cn \text{ and then } Gy) - p(Gy \text{ and then } Cn) - p(Gn \text{ and then } Cy) \tag{60}$$

This last quantity $p(Cy \text{ and then } Gn) + p(Cn \text{ and then } Gy) - p(Gy \text{ and then } Cn) - p(Gn \text{ and then } Cy)$ is termed q.

More generally, take the quantity $p_{AB} = p(Ay \text{ and then } Bn) + p(An \text{ and then } By)$ to be the probability that the answers to the two questions are different in the AB order, and take $p_{BA} = p(By \text{ and then } An) + p(Bn \text{ and then } Ay)$ in the BA order. Then the quantum model predicts that

$$q = p_{AB} - p_{BA} = 0 \tag{61}$$

Wang et al. (2013) test this theory on the data from Moore (2002) (described in Sect. 2.4) using the sample frequencies to compute the values p_{AB} and p_{BA}, and then using a z-test to compute whether q is significantly different from 0.

The conclusions are that the quantum question order model fits the data extremely well. The three datasets exemplifying the consistency, contrast, and additive effects have q-values that are not at all significantly different from 0. The fourth dataset (Rose-Jackson) was not well modelled by the QQ model. However, the authors argue that there is a violation of the assumptions in this model. Specifically, in this case, additional information is supplied about each of the characters, and this entails that the value θ is not the same in both cases. Hence, the cancellations worked out in Eq. (56) do not hold in this case.

A crucial part of the argument in this paper is also that other probabilistic models cannot account for both the question order effects seen and the QQ equality. In the case of a Bayesian probability model, the assumption is that the event Ay is a set contained in a sample space, and that By is another set contained in the same sample space. The probability of the conjunction is then defined by the intersection of the two events, which is commutative. In order to introduce order effects, two new events must be introduced: an event $O1$: "A was asked before B", and an event $O2$: "B was asked before A". However, in general, introducing these two events does not predict the QQ equality to obtain.

Markov models can incorporate question order effects, since they can encode the memory of previous events into the state representation. However, the authors argue that in general, QQ equality is not upheld.

3.7 Contextuality

A fundamental aspect of Quantum Mechanics is contextuality. The phenomenon of contextuality is as follows. Suppose we have a system with a number of agents who can perform measurements on the system, and suppose that we have determined joint distributions on the outcomes of pairs of measurements, by means of redoing the experiment a number of times. Contextuality occurs when it is not possible to form a joint distribution across all measurement outcomes such that this joint distribution agrees with the pairwise distributions. This means that before the measurement happens, the outcome of the measurement is undetermined. Contrast this with the outcomes of measurements in normal life. Suppose you wake up, and you wonder about the weather outside. You can take a measurement (open your curtains) and see that it is sunny, or cloudy. However, whether or not it was sunny or cloudy was determined before you opened the window.

We explain this below following the explanation in Nielsen and Chuang (2010). To show that the joint distribution does not exist, an analysis of the situation outside of quantum theory is made, and a *Bell inequality* is formed. Suppose that we have a system and two observers Alice and Bob. Alice can make two measurements, and each has the possibility of returning 1 or -1. Bob can also make two measurements, and each of these has the possibility of returning 1 or -1. The measurements are each modelled as random variables A, A', B, and B'. We now consider the quantity

$$AB + A'B + A'B' - AB' = (A + A')B + (A' - A)B' \tag{62}$$

Since $A, A' = \pm 1$, either $(A + A')B = 0$ or $(A' - A)B' = 0$. And then we have that $AB + A'B + A'B' - AB' = \pm 2$. We now consider what is termed a hidden variable. We suppose that it is possible to determine the outcome of each of the random variables individually so that $A = a$, $A' = a'$, $B = b$, and $B' = b'$, giving us a joint probability $p(a, a', b, b')$. Now, we can calculate the expected value of $AB + A'B + A'B' - AB'$ as

$$\mathbb{E}(AB + A'B + A'B' - AB') = \sum_{a,a',b,b'} p(a, a', b, b')(ab + a'b + a'b' - ab') \tag{63}$$

$$\leq 2 \sum_{a,a',b,b'} p(a, a', b, b') = 2 \tag{64}$$

However, within quantum theory this inequality is violated. Consider the quantum state $|\psi\rangle = \frac{|01\rangle + |10\rangle}{\sqrt{2}}$, and suppose Alice can perform measurements $A = Z_1$ or $A' = X_1$ on the first qubit and Bob can perform measurements $B = \frac{-Z_2 - X_2}{\sqrt{2}}$ or $B' = \frac{Z_2 - X_2}{\sqrt{2}}$ on the second qubit.[10] Calculating the average value $\langle M \rangle$ of an observable M as $\langle \psi | M | \psi \rangle$, we find that

---

[10] Z and X are the Pauli matrices.

$$\langle AB \rangle = \frac{1}{\sqrt{2}}, \quad \langle A'B \rangle = \frac{1}{\sqrt{2}}, \quad \langle A'B' \rangle = \frac{1}{\sqrt{2}}, \quad \langle AB' \rangle = -\frac{1}{\sqrt{2}} \qquad (65)$$

meaning that $\mathbb{E}(AB + A'B + A'B' - AB') = 2\sqrt{2}$, and violating the inequality in (64).

The violation of this inequality has been demonstrated experimentally, and shows that either there is no set of definite values of the random variables A, A', B, and B' that exist independent of our observation, or that the act of Alice's measurement affects the act of Bob's measurement.

The notion of contextuality also arises within cognitive science. Contextuality is crucially different from context-dependence. The phenomenon of context-dependence arises when the answer to a question depends on the context that the question is asked. This kind of context-dependence is called *direct influences* (Dzhafarov & Kujala, 2016). However, another kind of context-dependence that is analogous to quantum contextuality may be present. In Basieva et al. (2019), the following example is given. Suppose we have the three questions

1. Do you like chocolate?
2. Are you afraid of pain?
3. Do you see your dentist regularly?

Imagine that you are asked a pair of these questions. It may well be that the context in which you are asked a given question directly influences your answer to that question. So if you are asked (1) and (2) together, you may respond positively to (1), but if you are asked (1) and (3) together, the thought of going to the dentist may incline you to answer negatively to (1). This is the notion of direct influence. Basieva et al. (2019), following Dzhafarov and Kujala (2016), use the following notation: a response to question q_i asked in context c_j is a random variable denoted R_i^j, and described by

$$R_i^j : \frac{Yes \quad No}{p_i^j \quad 1 - p_i^j} \frac{\text{response}}{\text{probability}} \qquad (66)$$

These random variables can be laid out as follows (Dzhafarov & Kujala, 2016):

$$\begin{array}{cc|c} R_1^1 \ R_2^1 & & c_1 \\ \ R_2^2 \ R_3^2 & & c_2 \\ R_1^3 & R_3^3 & c_3 \\ \hline q_1 \ q_2 \ q_3 & \text{system } \mathcal{R}_3 \end{array} \qquad (67)$$

This says that in context 1 (c_1), questions 1 and 2 (q_1 and q_2) are asked, and so on.

Now, we might assume that all direct influences have somehow been eliminated, so that you have the same probability of answering yes or no to a given question no matter what the context. We then have that

$$R_1^1 = R_1^3 \tag{68}$$

$$R_2^1 = R_2^2 \tag{69}$$

$$R_3^2 = R_3^3 \tag{70}$$

However, we can also imagine a setup in which, if you answer yes to one question in the context, you always answer no to the other question. Representing yes and no by 1 and -1, we then have

$$R_1^1 = -R_2^1 \tag{71}$$

$$R_2^2 = -R_3^2 \tag{72}$$

$$R_1^3 = -R_3^3 \tag{73}$$

We can then easily derive a contradiction by following a chain of equalities:

$$R_1^1 = R_1^3 = -R_3^3 = -R_3^2 = R_2^2 = R_2^1 = -R_1^1 \tag{74}$$

This is saying that the pairs of random variables R_1^1 and R_1^3, R_2^1 and R_2^2, R_3^2 and R_3^3 cannot be identical. This kind of context-dependence is not reducible to direct influences.

Within quantum theory, contextuality applies only to systems in which there is no notion of direct influences. Measurements of a property in different conditions are described by the same random variable. The theory developed in Dzhafarov and Kujala (2016), called contextuality-by-default or CbD, can also take account of the presence of direct influences, and is therefore an extension of the quantum-theoretic notion. CbD is able to quantify how much the distributions of responses to different questions differ in different contexts, and how much the identities of the random variables differ in different contexts. By comparing these two quantities, it is possible to show how much contextuality is present over and above the effect of direct influences.

In previous work examining the contextuality of human judgements, the impact of direct influences was not always considered (Aerts et al., 2013; Dzhafarov & Kujala, 2014; Khrennikov et al., 2014), and so it was not clear the extent to which true contextuality was actually present in human decision-making. For example, in Bruza et al. (2012, 2015), whilst the effect of direct influences was taken into account, the presence of true contextuality was only minimally detected. However, in Basieva et al. (2019), Cervantes and Dzhafarov (2019), Dzhafarov (2019), experiments have been carried out to show that this phenomenon is indeed present in human judgement.

3.8   Concept Combination and Overextension

As described in Sect. 2.6, the phenomenon of overextension and underextension has been observed in experiments in Hampton (1987, 1988a, 1988b), Smith and Osherson (1984).

In Aerts and Gabora (2005a, 2005b), the authors set out a quantum model of concepts and their combination. The model of a concept is called the State-Context-Property or SCOP formalism. The idea behind this is that typicality ratings of exemplars in a concept can depend on the context in which they are considered. For example, in the context "The pet is chewing a bone", a set of examples might be ordered as *dog, cat, rabbit, hamster, guinea pig, mouse, hedgehog, bird, parrot, snake, canary, goldfish, spider, guppy*, whereas under the context "The pet is being taught", the same set of exemplars would be ordered *dog, parrot, cat, bird, hamster, canary, guinea pig, rabbit, mouse, hedgehog, snake, goldfish, guppy, spider*. The idea is that each context acts on the concept to bring it into another state.

A SCOP consists of a set Σ of states p_i, a set $\mathcal{M}$ of contexts e_i, such as "the pet is chewing a bone", and a set $\mathcal{L}$ of properties a_i, such as 'is feathered'. Together with these sets, there is a function

$$\mu : \Sigma \times \mathcal{M} \times \Sigma \to [0, 1] \tag{75}$$

which is a probability function that describes how a state p under the influence of context e changes into state q.

$$\mu(q, e, p) = \langle p | P_e | p \rangle \tag{76}$$

There is also a function

$$\nu : \Sigma \times \mathcal{L} \to [0, 1] \tag{77}$$

which gives the weight of a property a for the concept in the state p. So, for example, after considering the concept 'pet' in the context "the pet is chewing a bone", the weight of the property 'is furry' might increase.

$$\nu(p, a) = \langle p | P_a | p \rangle \tag{78}$$

In Aerts and Gabora (2005a), a Hilbert space representation of the theory is given. Concepts are represented in a Hilbert space in which the basis dimensions of the space represent basic states of the concept. The basic states of a concept are defined via basic contexts of a concept, which roughly means that the basic concept contexts are the most specific. For example, we might have the contexts $a_1 = $ "the pet is chewing a bone", and $a_2 = $ "the pet is chewing a bone in the garden", and we would say that a_2 is more specific than a_1. The set of atomic contexts is denoted by X, and these are considered to be rank-1 projectors. The basic states are then the states corresponding to these rank-1 projectors. The ground state $\hat{p}$ of a concept is given by the following superposition of the basis states of the Hilbert space:

$$|\hat{p}\rangle = \sum_{u \in X} \frac{1}{|X|} |u\rangle \tag{79}$$

so that the ground state has an equal probability of changing to any basic state under one of the basic contexts.

General contexts are represented by projectors that are not merely of rank 1. So suppose that we have 100 basic contexts, and 25 of these basic contexts are more specific than $e_{bone} = $ *the pet is chewing a bone*, i.e. they are more detailed contexts in which the pet is chewing a bone.[11] Then we would build a projector corresponding to the context e_{bone} by summing together the projectors for the basic contexts in the set $E_{bone} = \{u \in X | u$ is more specific than $e_{bone}\}$:

$$P_{bone} = \sum_{u \in E_{bone}} |u\rangle\langle u| \tag{80}$$

The new state under the influence of this context is calculated in the standard way, i.e. in the same way that the knowledge state of the agent is evolved in Sect. 3.1, Eq. (18):

$$|p_{bone}\rangle = \frac{P_{bone}|\hat{p}\rangle}{||P_{bone}|\hat{p}\rangle||} \tag{81}$$

In our example with 100 basic contexts and 25 of these basic contexts are more specific than e_1, we have that

$$P_{bone}|\hat{p}\rangle = \sum_{u \in E_{bone}} |u\rangle\langle u||\hat{p}\rangle = \sum_{u \in E_{bone}} \frac{1}{\sqrt{100}} |u\rangle \tag{82}$$

$$||P_{bone}|\hat{p}\rangle||^2 = \langle \hat{p}|P_{bone}|\hat{p}\rangle = \sum_{u \in E_{bone}} |\langle \hat{p}|u\rangle|^2 = \sum_{u \in E_{bone}} \frac{1}{100} = \frac{25}{100} \tag{83}$$

so that

$$|p_{bone}\rangle = \sum_{u \in E_{bone}} \frac{1}{\sqrt{25}} |u\rangle = \sum_{u \in E_{bone}} \frac{1}{5} |u\rangle \tag{84}$$

Recall that in the pet fish problem, the problem is that the exemplar 'goldfish' is viewed as more typical to the 'pet fish', conceived as the conjunction of two concepts 'pet' and 'fish' than it is to either of the constituent concepts. To describe the notion of the typicality of an exemplar to a concept, Aerts and Gabora (2005a) use the function μ described in Eq. (75). Aerts and Gabora (2005a) explicate the notion

---

[11] Note that there might be basic contexts that are unrelated to "the pet is chewing a bone", for example, "the pet is chewing a cracker in a cage and scratching its wing"—although this might be a basic context, it is not related to the context "the pet is chewing a bone".

Table 6 Numbers of basic states assigned to each exemplar in the concepts *pet* and *fish*

	Pet	Fish
Dog	50	0
Cat	35	0
Goldfish	10	10
Shark	5	20
Tuna	0	20

of typicality as follows. Each exemplar is assigned to a context, so if we want to consider the exemplar $p_{gf} = goldfish$, we form the context $e_{gf} = the\ pet\ is\ a\ goldfish$. Now, consider the concept *pet* in the context $e_{bone} = the\ pet\ is\ chewing\ a\ bone$. As described by Eqs. (18) and (84) in the example above, we apply the context to the state *pet* to obtain the state p_{bone}.

To calculate the weight of the exemplar *goldfish* in the context *the pet is chewing a bone*, we calculate the probability that p_{bone} changes to p_{gf} in the context e_{gf}, i.e. we calculate

$$\mu(p_{gf}, e_{gf}, p_{bone}) = \langle p_{bone} | P_{gf} | p_{bone} \rangle \tag{85}$$

Earlier, we said that there are 25 out of 100 basic contexts in which the pet is chewing a bone. Suppose there are 10 contexts where the pet is a goldfish, but only 1 context where the pet is chewing a bone and is a goldfish. Then we will have

$$\mu(p_{gf}, e_{gf}, p_{bone}) = \langle p_{bone} | P_{gf} | p_{bone} \rangle = \sum_{u \in E_{gf}} \sum_{v \in E_{bone}} \sum_{w \in E_{bone}} \frac{1}{25} \langle v | u \rangle \langle u | w \rangle \tag{86}$$

$$= \sum_{u \in E_{bone} \cap E_{gf}} \frac{1}{25} = \frac{1}{25} \tag{87}$$

So the weight of the exemplar *goldfish* in the context *the pet is chewing a bone* is $\frac{1}{25}$. This quantity is what Aerts et al. (2015) use to formalize the notion of typicality of an exemplar to a concept.

To examine the *pet fish* situation, we will first set up an example situation. In Aerts and Gabora (2005a), these figures were generated from a psychological experiment involving asking humans for the typicality figures. In this chapter, we simplify the figures for explanatory purposes.

Suppose that we have a pool of 100 basic states for *pet* and a pool of 50 basic states for *fish*. The basic states are distributed amongst the following exemplars as in Table 6.

For each concept, we outline how many basic contexts are included in each of the contexts $e^x_{pet} = the\ x\ is\ a\ pet$, $e^x_{fish} = the\ x\ is\ a\ fish$, $e^x_{furry} = the\ x\ is\ furry$, and $e^x_{swim} = the\ x\ is\ swimming\ around\ in\ a\ glass\ tank\ in\ the\ house$.

Table 7 Number of basic contexts assigned to each context for each exemplar in the concept *pet*

	e_{pet}^{pet}	e_{fish}^{pet}	e_{furry}^{pet}	e_{swim}^{pet}
Dog	50	0	45	0
Cat	35	0	30	0
Goldfish	10	10	0	9
Shark	5	5	0	5
Tuna	0	0	0	0

Table 8 Numbers of basic contexts assigned to each context for each exemplar in the concept *fish*

	e_{pet}^{fish}	e_{fish}^{fish}	e_{furry}^{fish}	e_{swim}^{fish}
Dog	0	0	0	0
Cat	0	0	0	0
Goldfish	10	10	0	9
Shark	5	20	0	5
Tuna	0	20	0	0

The typicality of *goldfish* in the concept *pet* is then

$$\mu(goldfish, goldfish, pet) = \langle pet|P_{goldfish}|pet\rangle = \langle pet| \sum_{u \in E_{goldfish}} |u\rangle\langle u||pet\rangle \quad (88)$$

$$= \frac{1}{100} \sum_{u \in E_{goldfish}} \sum_{v \in E_{pet}} \sum_{w \in E_{pet}} \langle v|u\rangle\langle u|w\rangle = \frac{10}{100} \quad (89)$$

A similar calculation gives us that the typicality of *goldfish* in the concept *fish* is $\frac{20}{100}$.

The shared contexts of *pet* and *fish* are $e_{fish}^{pet}, e_{pet}^{fish}, e_{swim}^{pet}$, and e_{swim}^{fish}. Now, to form combinations of concepts, such as *pet fish*, the concept is represented in the tensor product space of the two concepts *pet* and *fish*. The basic contexts of the concept *pet fish* are denoted as $E^{pet\,fish}$ which is the intersection of the set E^{pet} of the basic contexts of *pet*, and the set E^{fish} of the basic contexts of *fish*. The assumption is made for the example that $E^{pet\,fish} = E_{fish}^{pet} E_{pet}^{fish}$, where $E_{fish}^{pet} = \{u \in E^{pet}|u$ is more specific than $e_{fish}^{pet}\}$, i.e. it is the set of basic contexts of the concept *pet* that are more specific than *the pet is a fish*. Note that in Tables 7 and 8, the cardinality of that set is 15.

The concept *pet fish* is modelled as

$$|pet\,fish\rangle = \sum_{u \in E^{pet\,fish}} \frac{1}{\sqrt{15}}|u\rangle \otimes |u\rangle \quad (90)$$

To look at the membership of *goldfish* in *pet fish*, we consider the context *the pet is a goldfish*, modelled as

$$G = P^{pet}_{goldfish} \otimes 1^{fish} = \sum_{u \in E^{pet}_{goldfish}} |u\rangle\langle u| \otimes \mathbb{I} \qquad (91)$$

Then the typicality of *goldfish* in *pet fish* is given by

$$\mu(goldfish, goldfish, pet\,fish)$$
$$= \langle pet\,fish|G|pet\,fish\rangle \qquad (92)$$
$$= \frac{1}{15} \sum_{v \in E^{pet\,fish}} \langle v| \otimes \langle v| \sum_{u \in E^{pet}_{goldfish}} |u\rangle\langle u| \otimes \mathbb{I} \sum_{w \in E^{pet\,fish}} |w\rangle \otimes |w\rangle \qquad (93)$$
$$= \frac{1}{15} \sum_{v \in E^{pet\,fish}} \sum_{u \in E^{pet}_{goldfish}} \sum_{w \in E^{pet\,fish}} \langle v|u\rangle\langle u|v\rangle\langle v|w\rangle = \frac{10}{15} \qquad (94)$$

We can see that the typicality of *goldfish* in *pet fish* at $\frac{2}{3}$ is much higher than the typicality of *goldfish* in either *pet* ($\frac{1}{10}$) or *fish* ($\frac{2}{10}$).

Moreover, although we have only applied the context *goldfish* to the *pet* subsystem, the *fish* subsystem has changed as well:

$$\nu(goldfish, pet\,fish)$$
$$= \frac{G|pet\,fish\rangle}{\sqrt{\langle pet\,fish|G|pet\,fish\rangle}} = \frac{\sqrt{15}}{\sqrt{10}} \sum_{u \in E^{pet}_{goldfish}} |u\rangle\langle u| \otimes \mathbb{I} \sum_{w \in E^{pet\,fish}} \frac{1}{\sqrt{15}}|w\rangle \otimes |w\rangle \qquad (95)$$
$$= \frac{1}{\sqrt{10}} \sum_{u \in E^{pet}_{goldfish}} \sum_{w \in E^{pet\,fish}} |u\rangle\langle u|w\rangle \otimes |w\rangle = \frac{1}{\sqrt{10}} \sum_{u \in E^{pet}_{goldfish}} |u\rangle \otimes |u\rangle \qquad (96)$$

If we constructed a similar context saying that the fish is a goldfish, the concept would collapse to the same state. In the *pet fish* concept, saying that the pet is a goldfish has the same effect as saying that the fish is a goldfish—and this is the desired outcome. In contrast, if the concept pet fish was modelled without entanglement, for example, as

$$|pet\,fish'\rangle = |pet_{fish}\rangle \otimes |fish_{pet}\rangle \qquad (97)$$
$$= \sum_{u \in E^{pet}_{fish}} \sum_{v \in E^{fish}_{pet}} |u\rangle \otimes |v\rangle \qquad (98)$$

Then

$$G|pet\,fish'\rangle = \sum_{u\in E^{pet}_{goldfish}} |u\rangle\langle u| \otimes \mathbb{I} \sum_{v\in E^{pet}_{fish}} \sum_{w\in E^{fish}_{pet}} |v\rangle \otimes |w\rangle \qquad (99)$$

$$= \sum_{u\in E^{pet}_{goldfish}} |u\rangle \otimes \sum_{w\in E^{fish}_{pet}} |w\rangle \qquad (100)$$

where although the pet subsystem has changed to a goldfish, the fish subsystem has not. If we constructed a similar context saying that the fish was a goldfish, we would not obtain the same state. We would instead have a state consisting of a generic pet, and a goldfish. As such, the entanglement is essential to correctly model this problem.

Aerts and Gabora (2005b) go on to describe that this approach could be extended to model the composition of concepts in more complicated structures, up to the sentence level.

Whilst the approach given in Aerts and Gabora (2005b) is able to model the phenomenon of overextension, it suffers from the problem that the space in which concepts must be modelled gets longer dependent on the phrase or sentence length. In Coecke et al. (2010), an approach to modelling language was proposed that combines the very successful distributional models of meaning with the compositional approaches from formal semantics. In brief, this approach models words in differing vector spaces based on their grammatical type. Nouns are modelled in a vector space N for nouns, and sentences are modelled in a vector space S for sentences. Adjectives are modelled as linear maps on the noun space, i.e. matrices in $N \otimes N$, and transitive verbs are bilinear maps from two copies of the noun space to the sentence space, or tensors in $N \otimes S \otimes N$. The benefit of this approach is that arbitrarily long sentences can all be mapped into the same vector space S. Whilst this approach was originally developed with an eye to natural language processing, it has also been used to tackle problems in cognitive science.

In Coecke and Lewis (2015), the compositional distributional model is used to examine the pet fish problem, and more generally the problem of overextension. The first approach to tackling the pet fish problem is to view the word *pet* as an adjective, rather than a noun. Then the parameters of the matrix can be tuned so that the similarity of the vector *goldfish* to *pet fish* is greater than the similarity to either *pet* or *fish*. The sentence can further be expanded out to look at the conjunctions *pet which is a fish* and *fish which is a pet*, and this format is used to model the importance of attributes in the conjunction of concepts. In Hampton (1987), data is collected on the importance of attributes in concepts and their combination. Concepts are considered in pairs that are related to some degree, for example, 'Pets' and 'Birds'. Six pairs are considered in total. Participants are asked to generate attributes for each concept and for their conjunctions, where conjunction in this case is rendered as "Pets which are also Birds", or "Birds which are also Pets". For example, attributes such as 'lives in the house', 'is an animal', and 'has two legs' are generated for 'Pets', 'Birds'. For each pair of concepts and their conjunction, attributes that had been generated by at least 3 out of the 10 participants were collated. Participants were then asked to rate the importance of each attribute to each concept and to each conjunction. Importance ratings were made on a 7-point verbal scale ranging from "Necessarily true of all examples of the concept" to "Necessarily false of all examples of the concept". Numerical ratings were subsequently imposed ranging from 4 to −2, respectively.

The question then arises of how the importance of attributes in the conjunction of the concepts is related to the importance of attributes in the constituent concepts. Key phenomena are that conjunction is not commutative, that inheritance failure can occur (i.e. an attribute that is important in one of the concepts is not transferred to the conjunction), that attribute emergence can occur, where an attribute that is important in neither of the conjuncts becomes important in the conjunction, and further, that necessity and impossibility are preserved. In order to model this data, Hampton uses a multilinear regression.

In Coecke and Lewis (2015), the phrase *A which is a B* is modelled as

$$|A \text{ which is a } B\rangle = |A\rangle \odot \underline{is}|B\rangle$$

where $\underline{is}$ is a matrix modelling the word *is* as a transitive verb.

To model the data from Hampton (1987), parameters for the verb *is* are found using a numerical solver. Results are that it is essentially possible to model the data exactly. We should note that since a large matrix is used, the numbers of parameters are very large, however, as a demonstration of how the compositional model can be applied to psychological data it is still useful.

3.9   Modelling of Over- and Underextension in Psychological Data

The phenomenon of over- and underextension in human concept combination has been extensively investigated in Hampton (1988a, 1988b). In Hampton (1988a), Hampton looks at how the membership of exemplars in conjunctions of concepts compares that the membership of exemplars in the individual concepts. In Hampton (1988b), he examines how the membership of exemplars in disjunctions of concepts compares to that in the individual concepts. The overall findings are that in the conjunctive case, overextension is often seen, and in the disjunctive case, underextension is often seen. Moreover, membership in the conjunction is dependent on the order in which the conjunction is made. As we saw in Sect. 3.2, if concepts are modelled as regions of a feature space, the order of conjunction should not matter.

These phenomena are given a quantum modelling in Aerts (2009). The model is based on a Fock space. The combination of two concepts is considered in two differing ways. Concepts A and B are modelled as states $|A\rangle$ and $|B\rangle$ of a Hilbert space $\mathcal{H}$. Firstly, the disjunction of two concepts A and B is considered as the superposition of their two states: $A \text{ or } B = \frac{1}{\sqrt{2}}(|A\rangle + |B\rangle)$. The membership $\mu_{A \text{ or } B}(X)$ of an item X in the concept $A \text{ or } B$ is given by an orthogonal projection M_X, i.e.

$$\mu_{A\ or\ B}(X) = \frac{1}{2}((\langle A| + \langle B|)M_X(|A\rangle + |B\rangle))$$
$$= \frac{1}{2}(\langle A|M_X|A\rangle + \langle B|M_X|B\rangle + Re(\langle A|M_X|B\rangle))$$
$$= \frac{1}{2}(\mu_A(X) + \mu_B(X)) + Re(\langle A|M_X|B\rangle)$$

where this last term is the interference term, as in Sect. 3.4.

On the other hand, the disjunction of the two concepts can be modelled in two copies of $\mathcal{H}$, as $|A\rangle \otimes |B\rangle$. The operator $M_X \otimes M_X$ is a projection operator over the space that asks whether X is a member of A and whether X is a member of B. To determine the membership of X in $A\ or\ B$, we therefore calculate

$$\mu_{A\ or\ B}(X) = 1 - ((\langle A| \otimes \langle B|)(\mathbb{I} - M_X) \otimes (\mathbb{I} - M_X)(|A\rangle \otimes |B\rangle))$$
$$= 1 - (1 - \mu_A(X))(1 - \mu_B(X)) = \mu_A(X) + \mu_B(X) - \mu_A(X)\mu_B(X)$$

which gives us a classical-style disjunction.

For the case of conjunction, a classical-style conjunction can be modelled in $\mathcal{H} \otimes \mathcal{H}$ simply by taking

$$\mu_{A\ and\ B}(X) = \langle A| \otimes \langle B|)(M_X \otimes M_X)(|A\rangle \otimes |B\rangle)$$
$$= \mu_A(X)\mu_B(X)$$

and the conjunction in one copy of $\mathcal{H}$ is modelled as $\frac{1}{\sqrt{2}}(|A\rangle + |B\rangle)$ again. This means that in the single space, conjunction and disjunction are the same.

The single space and the double space are combined in a Fock space $\mathcal{H} \otimes \mathcal{H} \oplus \mathcal{H}$. The concept overall is represented as

$$\psi(A, B) = me^{i\theta}|A\rangle \otimes |B\rangle + \frac{ne^{i\phi}}{\sqrt{2}}(|A\rangle + |B\rangle)$$

where $m^2 + n^2 = 1$. Then,

$$\mu_{A\ and\ B}(X) =$$
$$\left(me^{i\theta}\langle A| \otimes \langle B| + \frac{ne^{i\phi}}{\sqrt{2}}((\langle A| + \langle B|)\right) M_X \otimes M_X \oplus M_X \left(me^{i\theta}|A\rangle \otimes |B\rangle + \frac{ne^{i\phi}}{\sqrt{2}}(|A\rangle + |B\rangle)\right)$$
$$= m^2\mu_A(X)\mu_B(X) + n^2(\frac{1}{2}(\mu_A(X) + \mu_B(X)) + Re(\langle A|M_X|B\rangle))$$

A similar calculation gives us that

$$\mu_{A\ or\ B}(X) = m^2(\mu_A(X) + \mu_B(X) - \mu_A(X)\mu_B(X)) + n^2(\frac{1}{2}(\mu_A(X) + \mu_B(X)) + Re(\langle A|M_X|B\rangle))$$

Essentially, the part of the concept that is modelled as a superposition contributes to the combinations *A and B* or *A or B* to move the combination away from a classical combination of concepts.

Aerts (2009) then shows that the data from Hampton (1988a, 1988b) can be modelled in a space $\mathbb{R}^2 \otimes \mathbb{R}^2 \oplus \mathbb{R}^4$, in which the angles θ and ϕ are tuned.

In Sozzo (2014), the same sort of Fock space modelling is used to describe problematic borderline cases in negation. Sozzo (2014) models the data from Alxatib and Pelletier (2011) by modelling the concepts *tall* and *not tall* as separate states in the one-system part of the space, $\mathcal{H}$. The combinations *tall and not tall* and *neither tall nor not tall* are then modelled in the two-system part of the space $\mathcal{H} \otimes \mathcal{H}$, and combined with the superposition of $|A\rangle$ and $|A'\rangle$. As such, the superposition of the two states combines with their logical combination to produce the effects seen in Alxatib and Pelletier (2011).

In Aerts et al. (2015), Sozzo (2015), this Fock space modelling of conjunction and negation are combined. If a concept A is modelled by $|A\rangle$, then the negation of that concept is modelled by another state $|A'\rangle$. Within the first part of the Fock space, in which the concepts A and B are modelled within a single Hilbert space $\mathcal{H}$, negations are modelled using $|A'\rangle$ and $|AB\rangle$. However, within the two-system part of the Fock space, the negations of concepts can be represented classically, as *A and B* or *A or B* are. Specifically, the membership of X in the negation of A is modelled as

$$\langle A | \mathbb{I} - M_X | A \rangle = 1 - \mu_A(X)$$

This system is able to successfully model data from Hampton (1997) and from Sozzo (2015).

3.9.1 Ambiguity, Entanglement, and Concept Combination

A key part of human concept use is how we express these concepts in language. However, when concepts are expressed in language, the language can have one word which refers to the same concept. A well-used example is the word *bank*. If I say "I'm going to the bank to deposit some money", then *bank* has a different meaning than if I say "I'm going to the bank to fish". In Bruza et al. (2012), an experiment is performed that looks into pairs of such ambiguous words and their combinations. An example is *apple chip*. This could mean a part of a Macintosh computer, or a slice of dried fruit. If I interpret *apple* in the sense *computer*, then it is very unlikely that I should interpret *chip* as a slice, or part, or the computer. Similarly, if I interpret *apple* in the sense *fruit*, then it is very unlikely that I will interpret *chip* in the sense of a computer chip.

A quantum modelling of this phenomenon would seem very appropriate: if we view each individual concept as modelled in a Hilbert space $\mathcal{H}$, and I disambiguate the meanings so that I have 4 vectors $|apple_{computer}\rangle$, $|apple_{fruit}\rangle$, $|chip_{computer}\rangle$, and $|chip_{slice}\rangle$, then we can model the combined concept in $\mathcal{H} \otimes \mathcal{H}$ as

$$apple\ chip = \frac{|apple_{computer}\rangle|chip_{computer}\rangle + |apple_{fruit}\rangle|chip_{fruit}\rangle}{\sqrt{2}}$$

Given a state $|\psi\rangle$ that the agent is in, the interpretation of the combination will collapse to either the computer sense or the fruit sense.

However, another form of combination is as follows. We could model *apple* as $|apple\rangle = \frac{1}{\sqrt{2}}|apple_{computer}\rangle + |apple_{fruit}\rangle$, and $|chip\rangle = \frac{1}{\sqrt{2}}|chip_{computer}\rangle + |chip_{slice}\rangle$

$$apple\ chip = |apple\rangle \otimes |chip\rangle \tag{101}$$

$$= \frac{1}{2}(|apple_{computer}\rangle|chip_{computer}\rangle + |apple_{computer}\rangle|chip_{fruit}\rangle \tag{102}$$

$$+ |apple_{fruit}\rangle|chip_{computer}\rangle + |apple_{fruit}\rangle|chip_{fruit}\rangle) \tag{103}$$

In this case, a participant could interpret the combination *apple chip* in any of the four possible combinations.

This kind of phenomenon is investigated in Bruza et al. (2015) In these experiments, the participant is first primed for one of the senses of the combination—for example, the word *banana* is shown to them before they are asked to make a decision on the interpretation of the compound. The participants are then asked to describe the interpretation they have made. These interpretations are used to build a set of probabilities for the interpretations of the compounds, and these probabilities are analysed to determine whether the compound is separable or entangled. There are a number of factors which need to be accounted for in this, which are discussed in Dzhafarov and Kujala (2016) and which we discussed in Sect. 3.7.

The results are that in general, there is no clear evidence for the entanglement of these concepts in the human mental lexicon. However, in a number of cases, there is some evidence of phenomena occurring which are not completely explained by a classical probabilistic model.

Another approach to modelling the ambiguity of words is to model words as density matrices (Meyer & Lewis, 2020; Piedeleu et al., 2015). This was proposed as an extension of the compositional distributional model of Coecke et al. (2010). As mentioned in Sect. 3.1, a density matrix is a probability distribution over a number of quantum states:

$$\rho = \sum_i \alpha_i |v_i\rangle\langle v_i| \tag{104}$$

where $\forall i,\ \alpha_i > 0$ and $\sum_i \alpha_i = 1$. The idea of using density matrices to model ambiguity runs as follows. Each sense of a word is modelled as a vector, and then these senses are mixed together following Eq. (104) above. Crucially, the same compositional framework as in the standard vector-based model is available. This means that as words are used in context, their meaning can be disambiguated, in the same way that humans are able to. In Meyer and Lewis (2020), a neural approach to building density matrices for words is proposed and tested on a set of sentence similarity data that requires words to be disambiguated correctly. The representations learnt by Meyer and Lewis (2020) perform very well, outperforming some of the strongest neural network systems for language processing currently available.

3.10 Cognitive and Neuronal Structures

In the foregoing sections, the level of modelling was at a high, cognitive level. There is no attempt to form any match between these models and actual processes in the brain. However, there has been a range of research into using quantum theory to model cognitive processes at a level that is closer to a neuronal level.

Fuss and Navarro (2013) describe a quantum model of two-alternative forced choice (2AFC) experiments. In this sort of experiment, the participant is asked to make a choice about a stimulus. For example, there may be two shapes on a computer screen and the participant is asked which one is a square. The participant must choose one or the other, and does not have to choose to say neither. Information about the nature of processing is inferred from reaction times. For example, the average reaction time to the image in Fig. 8a may be faster than the average reaction time to the image in Fig. 8b.

A standard way of modelling reaction times in this sort of experiment is to say that a pair of neurons, or groups of neurons, is responsible for accumulating evidence for one choice or the other, and that when the evidence hits a threshold, the participant reacts. The two accumulators are in a race to get to the threshold. So in the examples in Fig. 8, evidence is being accumulated by one accumulator for *left* and one for *right*. In Fig. 8a, evidence for *right* would only accumulate slowly, and so the reaction *left* would happen more quickly, whereas in Fig. 8b, more of the evidence would be added to the *right* accumulator, and so overall the threshold would take longer to reach. This is usually modelled as a random walk with drift. An extension to a simple random walk model is to consider that multiple random walks could be happening in parallel. However, having multiple random walks happening in parallel can still only produce a model where evidence is accumulated. On the other hand, a model that explicitly models neurons and the connections between them (connectionist models) can account for inhibitory connections, i.e. that evidence for one choice can inhibit the accumulation of evidence for the other choice.

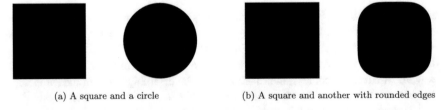

(a) A square and a circle　　　　　　　(b) A square and another with rounded edges

Fig. 8 Differing stimuli in a 2AFC experiment

In Fuss and Navarro (2013), a parallel random walk model using quantum theory is proposed. This allows for the kinds of inhibition we just mentioned to be modelled, by means of interference. This kind of model can produce qualitatively different probability distributions over the search space, allowing for more efficient decision-making to be carried out. The authors show that this kind of model can fit 2AFC experimental data more closely than a simple standard random walk with drift (although there are more complex models which may do better).

More recently, Khrennikov et al. (2018) have begun to formulate a description of neuronal substrates within quantum terms. This does not claim that neuronal and cognitive activities are being modelled at the quantum level (although models to explain cognitive activity through actual quantum modelling has been proposed in Hameroff (2014), Penrose (1990)). Rather, the idea is that quantum theory can be used to represent the information and activity present in neuronal assemblies.

3.11   Tensor Product Representations

Whilst not explicitly quantum-inspired, the tensor product representations of Smolensky (1990) use structures which are native to quantum systems. The aim in Smolensky (1990) is to describe a means of representing symbolic structures and their composition within a connectionist substrate. Connectionist networks (nowadays usually called neural networks) are based on the idea that cognitive systems can be modelled by the interconnection of small artificial neurons, and indeed the motivation for this is clear—as our cognitive system is based on the interconnection of many interacting neurons. There are other aspects to the basis of our cognitive system, such as the physical bodies that each of us inhabits, but undeniably our (actual) neural network plays a part in cognition. Also, there is no claim here that artificial neural networks bear any resemblance to real neural networks. However, artificial neural networks are extremely successful at various tasks.

An aspect of cognition that humans excel at is symbolic manipulation. This doesn't mean complex mathematics, but the ability that if we understand the sentence *Jen is Junpa's mother*, we can also understand the sentence *Junpa is Jen's mother*, even if we judge it to be false.

Artificial neural networks don't have an obvious means of representing symbols so that we can manipulate them in this way, and in Smolensky (1990), Smolensky and Legendre (2006) a method for doing so is proposed.

Vectors and tensor products are used to represent objects, roles, and structures. The underlying idea is that roles, such as subject, or object, are represented as vectors, and fillers, also represented as vectors, can be bound to these roles using the tensor product, and collections of roles and fillers are combined using vector addition. Recursive structures such as trees can be represented. The contents of the leaves are encoded in the fillers, and role vectors encode the tree structure (Table 9).

To be able to represent trees, we view the binding of a role and a filler as itself a filler, i.e. the filler is itself a complex structure, and not an atomic symbol. The binding

Table 9 Space of descriptions

Structure	Symbolic	Connectionist
Set	$\mathbf{f}$	$\mathbf{f} \in V_F$
String	$\mathbf{f}_i / r_i$	$\mathbf{f}_i \otimes \mathbf{r}_i$
Tree	$\mathbf{s} = \{\mathbf{f}_i / r_i\}$	$\mathbf{s} = \sum_i \mathbf{f}_i \otimes \mathbf{r}_i \in \mathcal{S}^*$

$\mathbf{f}/r$ of a filler $\mathbf{f}$ to a role r is represented as a vector $\mathbf{f} \otimes \mathbf{r}$. A sentence $\mathbf{s}$ is represented as a sum of filler/role bindings $\sum_i \mathbf{f}_i \otimes \mathbf{r}_i$, and these can be applied recursively. For example, let $\mathbf{s}$ = [NP VP] be a binary tree with left and right subtrees NP and VP. Let $\mathbf{s}, \mathbf{v}_1, \mathbf{v}_2$ be the vectors representing $\mathbf{s}$, NP, VP. The connectionist representation of $\mathbf{s}$ is

$$\mathbf{s} = \mathbf{v}_1 \otimes \mathbf{r}_0 + \mathbf{v}_2 \otimes \mathbf{r}_1$$

If VP is a tree rather than an atomic symbol, it can be expressed in terms of its left and right subtrees VP = [Vt NP]. If $\mathbf{v}_3, \mathbf{v}_4$ represent the trees Vt, NP, then the structure $\mathbf{s}$ = [NP [Vt NP]] has the following representation:

$$\mathbf{s} = \mathbf{v}_1 \otimes \mathbf{r}_0 + (\mathbf{v}_3 \otimes \mathbf{r}_0 + \mathbf{v}_4 \otimes \mathbf{r}_1) \otimes \mathbf{r}_1$$
$$= \mathbf{v}_1 \otimes \mathbf{r}_0 + \mathbf{v}_3 \otimes (\mathbf{r}_0 \otimes \mathbf{r}_1) + \mathbf{v}_4 \otimes (\mathbf{r}_1 \otimes \mathbf{r}_1)$$
$$\equiv \mathbf{v}_1 \otimes \mathbf{r}_0 + \mathbf{v}_3 \otimes \mathbf{r}_{01} + \mathbf{v}_4 \otimes \mathbf{r}_{11}$$

A notable feature of this representation is that the vector space in which concepts live must be arbitrarily large, depending on the size of the structure to be represented. Symbols at depth d in a binary tree are realized by $\mathcal{S}_{(d)}$, the FR^d-dimensional vector space formed from vectors of the form $\mathbf{f} \otimes \mathbf{r}_i \otimes \mathbf{r}_j \otimes \cdots \otimes \mathbf{r}_k$ with d role vectors, where F is the dimension of the filler vectors $\mathbf{f}$ and R is the dimension of the individual role vectors $\mathbf{r}_i$. A vector space containing all vectors in $\mathcal{S}_{(d)}$ for all d is

$$\mathcal{S}^* \equiv \mathcal{S}_{(0)} \oplus \mathcal{S}_{(1)} \oplus \mathcal{S}_{(2)} \oplus \cdots$$

Vectors $\mathbf{s}_{(i)}$ are embedded into this space, meaning that the normal operation of vector addition can be used to combine sentence components.

Once a set of concepts (e.g. *Jen*, *loves*, and *Junpa*) have been combined into one vector representation, it is also possible to pull the representation apart, and recombine it. This is done by means of an unbinding mechanism. Suppose we form the sentence representation

$$|S\rangle = |Jen\rangle \otimes |subj\rangle + |runs\rangle \otimes |verb\rangle$$

we can ask "who is the subject of the sentence" by unbinding $|subj\rangle$ from the sentence. If role vectors are all orthogonal, then the unbinding is simply done by taking the inner product of $|S\rangle$ with the role vector: $\langle S|subj\rangle = |Jen\rangle$.

Tensor product networks have recently been used extensively to examine the structure of recurrent neural networks (McCoy et al., 2019), and also as a potential substrate for quantum language processing (Wiebe et al., 2019). Moreover, very similar structures have been used in the neural-based computing framework Nengo (Bekolay et al., 2014). These latter structures are different in that they use circular convolution (Plate, 1995) rather than tensor product to bind roles and fillers. The benefit of using circular convolution rather than the tensor product is that all vectors are mapped into one shared space. An approximate unbinding mechanism can be employed by using circular correlation with a query vector.

Tensor product representations have also been used in Pothos and Truebloodm (2015) to model how the similarity of objects can depend on where the similarity is to be found. For example, given two birds with a spotted pattern, if the spotted pattern is found on the wings of one bird and the back of the other, they will not be judged as similar as if they both have the spotted pattern on their wings. The use of role vectors in the tensor product representations of Smolensky (1990), Smolensky and Legendre (2006) is leveraged to account for this phenomenon.

The compositional distributional model Coecke et al. (2010), mentioned in Sect. 3.8, has the property that all representations are mapped into one shared space, and shares some structures similar to the ICS model. In Al-Mehairi et al. (2017), it is shown that the ICS model can be mapped onto the compositional distributional model, to form a categorical compositional model of cognition.

A further mapping from ICS to the compositional distributional model can be considered. One of the most useful aspects of the compositional distributional model is the way that functional words such as adjectives and verbs modify the nouns they are applied to, either by mapping them to another vector in the noun space, or by mapping them into a sentence space. Within the ICS model, this does not happen in the same way. Consider the following problem, as investigated in Martin and Doumas (2020). Under the assumption that the concepts we use are expressed by the language we use, the way that we compose concepts can be examined by looking at the properties of words and phrases. Within the field of neural modelling, a key question is how neural representations of two concepts can be composed into a single combined representation. Neural representations of concepts are commonly modelled as (normalized) vectors, and similarity between concept representations is calculated by the inner product. Two common proposals for how to combine neural representations are via firstly tensor product binding or holographic reduced representations, as described above, or secondly vector addition (Hummel, 2011).

The problem with tensor product binding is that it does not transform the argument of a relation in the necessary way. Consider vectors $|fluffy\rangle$, $|cat\rangle$, and $|dog\rangle$. If the combinations *fluffy dog* and *fluffy cat* are represented by the tensor product of the two component vectors:

Fig. 9 Fluffy dogs and
fluffy cats are more similar
than dogs and cats

(a) Dog[12]

(b) Cat[13]

(c) Fluffy Dog[14]

(d) Fluffy Cat[15]

$$fluffy\ dog = |fluffy\rangle \otimes |dog\rangle$$
$$fluffy\ cat = |fluffy\rangle \otimes |cat\rangle$$

then the similarity between *fluffy cat* and *fluffy dog* is just the similarity between *cat* and *dog*, since

$$((\langle fluffy| \otimes \langle dog|)(|fluffy\rangle \otimes |cat\rangle)) = \langle fluffy|fluffy\rangle \otimes \langle dog|cat\rangle$$
$$= \langle dog|cat\rangle \tag{105}$$

This is undesirable, as can be seen in Fig. 9.

On the other hand, vector addition has been shown to be successful in modelling concept combination Hummel (2011), Martin and Doumas (2020), and has a clear neural correlate, termed *synchronous binding*.[12,13,14,15]

However, it is arguably not able to express gradations in meaning that are needed in terms like *red car* versus *red wine*. For example, given $|red\rangle$, $|wine\rangle$, and $|car\rangle$, suppose that *wine* and *car* are not very similar, say $\langle wine|car\rangle = 0.1$, and they are each fairly similar to *red*, say $\langle red|car\rangle = \langle red|wine\rangle = 0.5$. Then, adding vectors and normalizing:

$$red\ car = \frac{|red\rangle + |car\rangle}{\sqrt{3}}$$

$$red\ wine = \frac{|red\rangle + |wine\rangle}{\sqrt{3}}$$

and

$$\langle red\ car|red\ wine\rangle$$
$$= \frac{1}{3}(\langle red|red\rangle + \langle red|car\rangle + \langle red|wine\rangle$$
$$+ \langle wine|car\rangle)$$
$$= \frac{2}{3} + \frac{\langle wine|car\rangle}{3}$$

i.e. the similarity has increased from 0.1 to just over two-thirds. However, this again does not accord with our intuitions: the similarity between *wine* and *car* should not increase dramatically when the adjective *red* is applied, at least in part because applying this adjective does not even make the colours very similar. It might be argued that this can be overcome by downweighting the importance of the adjective in an adjective-noun combination. However, there are situations when we *would* like similarity to increase when the adjective is applied: the adjective *red* in *red car* and *red van* should increase similarity.

A approach than can address both these aspects is the quantum-inspired approach of Coecke et al. (2010). Since adjectives and verbs are represented as matrices rather than vectors, the combination of an adjective and a verb transforms the argument so

---

[12] https://commons.wikimedia.org/wiki/File:Wayfield%27s_Young_Argos, _the_Labrador_Retriever.jpg. Attribution: Andrew Skolnick, en:User:Askolnick, CC BY-SA 3.0. http://creativecommons.org/licenses/by-sa/3.0/, via Wikimedia Commons.

[13] https://commons.wikimedia.org/wiki/File:Cat_November_2010-1a.jpg. Attribution: Alvesgaspar, CC BY-SA 3.0. https://creativecommons.org/licenses/by-sa/3.0, via Wikimedia Commons.

[14] https://commons.wikimedia.org/wiki/File:Cute_dog_on_beach,_Los_Angeles, _California_LCCN2013634674.tif. Attribution: Carol M. Highsmith, Public domain, via Wikimedia Commons.

[15] https://commons.wikimedia.org/wiki/File:White_Persian_Cat.jpg. Attribution: Optional at the Persian language Wikipedia, GFDL. http://www.gnu.org/copyleft/fdl.html, via Wikimedia Commons.

that fluffy cats can be moved closer to fluffy dogs, i.e. the problem in Eq. (105) is not encountered. On the other hand, the application of an adjective like *red* does not force *red car* to become similar to *red wine*. How? Essentially, the use of a matrix gives enough 'space' for different uses of words to be represented.

3.11.1   Reasoning in Vector Spaces

Another line of work looks at using vector space methods to model structure in concepts. A key characteristic of the concepts we use is that they are hierarchical. Besides categorizing individual entities, we want to talk about types of entities—for example, we want to simply say that tigers are dangerous, rather than "that tiger is dangerous. And that tiger is dangerous. And that one. And...". Within a quantum model, we can start doing this by considering concepts as subspaces of a vector space, and using the lattice structure of the subspaces, or equivalently their projectors, to reason with. This was originally proposed by Birkhoff and Von Neumann (1936), and was applied in an information retrieval setting by Widdows and Peters (2003). In that work, the authors describe a means of applying negation and conjunction in document queries, based on quantum logic.

The density matrix variant of the compositional distributional model of Coecke et al. (2010) forms an extension of the quantum logic formed by the lattice of projectors. Words are modelled as positive operators, and hierarchy in concepts is modelled via the Löwner order. In Bankova et al. (2019), Sadrzadeh et al. (2018), it is shown that the Löwner order is compatible with the compositional model used. In Lewis (2019), a concrete modelling of words as positive operators is given which is shown to model human judgements of entailment very well. Further work into modelling negation is underway (Lewis, 2020), and work on modelling other forms of logical structure such as generalized quantifiers has also been undertaken by Hedges and Sadrzadeh (2019), Dostál et al. (2021).

4   Concluding Remarks

There are clearly a number of applications of quantum theory to problems in cognitive science. We have seen that human use of concepts has some unusual features, which seem to be well modelled using a quantum framework.

In the approach proposed by Busemeyer et al. (2011), Pothos et al. (2013), Wang et al. (2013) and used in numerous other papers, a concept is modelled as a subspace of a Hilbert space, and the knowledge state of an agent is modelled as a state in the Hilbert space. The theory therefore crucially incorporates the way in which decisions are dependent on the state of the agent, and how the presence or absence of information and the order in which stimuli occur can affect judgements. Within this model, the phenomenon of interference does a lot of work. In the modelling of the asymmetry of similarity judgements, $||P_{China}|\psi_{Korea}\rangle||^2 \cdot ||P_{Korea}|\psi\rangle||^2$ is interpreted as the similarity of Korea to China. However, in the model of the conjunction fallacy, this is interpreted as the conjunction of A and then B, as it also is in the question order model.

The conjunction and negation model proposed by Aerts and Gabora (2005b) also models concepts as subspaces of a vector space. Here, the subspaces have very high dimensionality and are based on very specific contexts that objects are seen in. The notion of entanglement plays a key part in modelling the combination of concepts such as *pet fish*. Aerts (2009), Aerts et al. (2015), Sozzo (2015) model concepts as states within a vector space, and again use the notion of entanglement. In this latter work, the vector spaces used are smaller—for example, around eight-dimensional. A drawback of these compositional methods is that in all cases, the size of the underlying space grows as more concepts are added to the combination. This has at least two drawbacks: for more complex concepts, spaces may grow unfeasible large, and furthermore it means that combinations of unequal numbers of concepts are not easy to compare.

Various papers have examined whether a form of contextuality is present in human concept use. In some cases, a claim of contextuality is erroneous, as argued by Dzhafarov and Kujala (2016). On the other hand, Basieva et al. (2019), Dzhafarov (2019) have shown that experiments can be designed in which human decision-making does show contextuality. If human decision-making does indeed display true contextuality, then the use of quantum computing to model these forms of decisions may provide a very useful substrate. However, in many cases there is no evidence of contextuality over and above the effect of direct influences, a more straightforward form of context-dependence.

Vector-based models of meaning and compositionality have been extremely successful in modelling language (Coeckeetal., 2010; Kartsaklis & Sadrzadeh, 2016; McCoy et al., 2019; Sadrzadeh et al., 2018; Smolensky, 1990). These models use the same fundamental structures as quantum theory, and extensions of the compositional distributional model have incorporated crucially quantum phenomena such as density matrices Bankova et al. (2019), Meyer and Lewis (2020), Piedeleu et al. (2015), Sadrzadeh et al. (2018). The models proposed in Smolensky (1990) can be argued to have relevance for realistic neuronal modelling (Bekolay et al., 2014). Moreover, theory from Smolensky (1990) has been developed for potential implementation on quantum computers in Wiebe et al. (2019), and the model from Coecke et al. (2010) has actually been implemented on a modern quantum computer (Lorenz et al., 2021). This implementation takes advantage of the fact that the fundamental structures in this model are the same as those used in quantum theory. In Lorenz et al. (2021), experiments on sentence classification are run, one to classify sentences as belonging to one of two topics, and a second to classify noun phrases as containing either a subject relative clause or an object relative clause. The general methodology of the experiments is to first generate a linguistic parse of a given sentence, and to express the sentence in terms of a DisCoCat diagram (as in Coecke et al. (2010)). The diagram is rewritten to eliminate complex aspects of the diagram—namely by straightening wires in the diagram in a certain way. Following this step, it is then possible to build a quantum circuit whose structure matches the grammatical structure of the sentence.

A quantum compiler translates the circuit into instructions for a physical quantum computer, which runs the circuit a certain number of times, and the relative frequency of given answers can be recorded.

4.1 Potential for Quantum Computing to Contribute to Cognitive Science

So what is the potential for cognitive science within the realm of quantum computing? On the one hand, we can simply say that human cognitive faculties are not computers, whether classical or quantum. However, this is unnecessarily negative. There are a wide range of cognitive processes that are well modelled using quantum theory. Models of similarity, judgement, and decision-making moreover seem to be well modelled even in fairly low-dimensional spaces.

At the same time, models that incorporate compositionality at a more complex level, using the grammar of natural language, have been implemented on quantum computers, and the potential for the benefits of this are large. The successes in Lorenz et al. (2021) have the potential to be extended to the cognitive domain. Drawing together some of the theory from the similarity and decision-making models to form a unified whole with models of language is an important way forward for this endeavour. One way to progress this would be to consider a form of compositionality that can be applied at the cognitive, rather than the language level, like a language of thought (Fodor, 1975).

Fodor's language of thought hypothesis is summarized (by himself (Fodor & Pylyshyn, 1988)) as the idea that cognition is reliant on mental representations, and that those mental representations have a combinatorial syntax and semantics. This argument is based on the fact that human cognitive abilities have three key properties:

- **Productivity**: We are able to entertain a large number of thoughts and to generate these on the fly, as combinations of representations that we already have.
- **Systematicity**: This is the property that if, for example, "Junpa loves Jen" can be understood, then "Jen loves Junpa" should also be understood, whether or not it is true.
- **Compositionality**: This is the property that the meaning of complexes of representations should be understandable from the meanings of the individual representations and how they are combined.

Whilst natural language arguably has these properties, and furthermore is intrinsically related to our thoughts, we do not have a 'grammar' that we can apply to thoughts to combine them in the same way that grammar exists for natural languages. Developing such a model together with the right sort of representations and notions of similarity and decision-making has the potential to form a very powerful model of cognition. Utilizing the compositional approach proposed in Coecke et al. (2010) may provide a way forward for a form of cognitive architecture that can be implemented on a quantum computer.

In this chapter, we have covered a range of quantum approaches to cognitive tasks. These all have the capacity to model certain aspects of human cognitive processes more faithfully than other approaches. The compositional models of Coecke et al. (2010) and Smolensky (1990) may be able to extend the purely cognitive approaches to model more complex phenomena in an intrinsically quantum way. Moreover, these compositional models are already being implemented on quantum computers. Hence, using quantum computing for cognitive simulation, and AI more generally, has a bright future ahead of it.

References

Aerts, D. (2009, October). Quantum structure in cognition. *Journal of Mathematical Psychology, 53*(5), 314–348. ISSN 00222496. https://doi.org/10.1016/j.jmp.2009.04.005. https://linkinghub. elsevier.com/retrieve/pii/S0022249609000558

Aerts, D., & Gabora, L. (2005a, January). A theory of concepts and their combinations I: The structure of the sets of contexts and properties. *Kybernetes, 34*(1/2), 167–191. ISSN 0368-492X. https://doi.org/10.1108/03684920510575799

Aerts, D., & Gabora, L. (2005b, January). A theory of concepts and their combinations II: A Hilbert space representation. *Kybernetes, 34*(1/2), 192–221. ISSN 0368-492X. https:// doi.org/10.1108/03684920510575807. https://www.emerald.com/insight/content/doi/10.1108/ 03684920510575807/full/html

Aerts, D., Gabora, L., & Sozzo, S. (2013). Concepts and their dynamics: A quantum-theoretic modeling of human thought. *Topics in Cognitive Science, 5*(4), 737–772. ISSN 1756-8765. https:// doi.org/10.1111/tops.12042. https://onlinelibrary.wiley.com/doi/abs/10.1111/tops.12042

Aerts, D., Sozzo, S., & Veloz, T. (2015). Quantum structure of negation and conjunction in human thought. *Frontiers in Psychology, 6*. ISSN 1664-1078. https://doi.org/10.3389/fpsyg.2015.01447. https://www.frontiersin.org/articles/10.3389/fpsyg.2015.01447/full

Al-Mehairi, Y., Coecke, B., & Lewis, M. (2017). Categorical compositional cognition. In J. A. de Barros, B. Coecke, & E. Pothos (Eds.), *Quantum interaction*. Lecture notes in computer science (pp. 122–134). Springer International Publishing. ISBN 978-3-319-52289-0. https://doi.org/10. 1007/978-3-319-52289-0_10

Alxatib, S., & Pelletier, F. J. (2011). The psychology of Vagueness: Borderline cases and contradic- tions. *Mind & Language, 26*(3), 287–326, 2011. ISSN 1468-0017. https://doi.org/10.1111/j.1468- 0017.2011.01419.x. https://onlinelibrary.wiley.com/doi/abs/10.1111/j.1468-0017.2011.01419.x

Ashby, F. G., & Gott, R. E. (1988). Decision rules in the perception and categorization of multi- dimensional stimuli. *Journal of Experimental Psychology: Learning, Memory, and Cognition, 14*(1), 33–53. ISSN 1939-1285 (Electronic), 0278-7393 (Print). https://doi.org/10.1037/0278- 7393.14.1.33

Bankova, D., Coecke, B., Lewis, M., & Marsden, D. (2019, March). Graded hyponymy for com- positional distributional semantics. *Journal of Language Modelling, 6*(2), 225. ISSN 2299- 8470, 2299-856X. https://doi.org/10.15398/jlm.v6i2.230. http://jlm.ipipan.waw.pl/index.php/ JLM/article/view/230

Basieva, I., Cervantes, V. H., Dzhafarov, E. N., & Khrennikov, A. (2019, November). True contextu- ality beats direct influences in human decision making. *Journal of Experimental Psychology: Gen- eral, 148*(11), 1925–1937. ISSN 1939-2222, 0096-3445. https://doi.org/10.1037/xge0000585. http://doi.apa.org/getdoi.cfm?doi=10.1037/xge0000585

Bekolay, T., Bergstra, J., Hunsberger, E., DeWolf, T., Stewart, T. C., Rasmussen, D., Choo, X., Voelker, A. R., & Eliasmith, C. (2014). Nengo: A Python tool for building large-scale functional brain models. *Frontiers in Neuroinformatics, 7.* ISSN 1662-5196. https://doi.org/10.3389/fninf. 2013.00048. http://journal.frontiersin.org/article/10.3389/fninf.2013.00048/abstract

Birkhoff, G., & Von Neumann, J. (1936). The logic of quantum mechanics. *Annals of Mathematics, 37*(4), 823–843, 1936. ISSN 0003-486X. https://doi.org/10.2307/1968621. https://www.jstor. org/stable/1968621

Bruza, P. D., Kitto, K., Ramm, B., Sitbon, L., Song, D., & Blomberg, S. (2012, April). Quantum-like non-separability of concept combinations, emergent associates and abduction. *Logic Journal of the IGPL, 20*(2), 445–457. ISSN 1367-0751. https://doi.org/10.1093/jigpal/jzq049

Bruza, P., Busemeyer, J. R., & Gabora, L. (2009, October). Introduction to the special issue on quantum cognition. *Journal of Mathematical Psychology, 53*(5), 303–305. ISSN 0022-2496. https://doi.org/10.1016/j.jmp.2009.06.002. http://www.sciencedirect.com/science/article/pii/S0022249609000637

Bruza, P. D., Kitto, K., Ramm, B. J., & Sitbon, L. (2015, August). A probabilistic framework for analysing the compositionality of conceptual combinations. *Journal of Mathematical Psychology, 67*, 26–38. ISSN 00222496. https://doi.org/10.1016/j.jmp.2015.06.002. https://linkinghub. elsevier.com/retrieve/pii/S002224961500036X

Busemeyer, J. R., Pothos, E. M., Franco, R., & Trueblood, J. S. (2011). A quantum theoretical explanation for probability judgment errors. *Psychological Review, 118*(2), 193–218. ISSN 1939-1471, 0033-295X. https://doi.org/10.1037/a0022542. http://doi.apa.org/getdoi.cfm? doi=10.1037/a0022542

Cervantes, V. H., & Dzhafarov, E. N. (2019, August). True contextuality in a psychophysical experiment. *Journal of Mathematical Psychology, 91*, 119–127. ISSN 00222496. https://doi.org/10. 1016/j.jmp.2019.04.006. https://linkinghub.elsevier.com/retrieve/pii/S002224961830186X

Coecke, B., & Lewis, M. (2015, September). A compositional explanation of the pet fish phenomenon. arXiv:1509.06594 [cs, math]. http://arxiv.org/abs/1509.06594

Coecke, B., Sadrzadeh, M., & Clark, S. (2010, March). Mathematical foundations for a compositional distributional model of meaning. arXiv:1003.4394 [cs, math]. http://arxiv.org/abs/1003. 4394

Dostál, M., Sadrzadeh, M., & Wijnholds, G. (2021). Fuzzy generalised quantifiers for natural language in categorical compositional distributional semantics. In M. Mojtahedi, S. Rahman, & M. S. Zarepour (Eds.), *Mathematics, logic, and their philosophies: Essays in honour of Mohammad Ardeshir.* Logic, epistemology, and the unity of science (pp. 135–160). Springer International Publishing. ISBN 978-3-030-53654-1. https://doi.org/10.1007/978-3-030-53654-1_6

Dzhafarov, E. N. (2019, November). Contextuality and probability in quantum mechanics and beyond: A preface. *Philosophical Transactions of the Royal Society A: Mathematical, Physical and Engineering Sciences, 377*(2157), 20190371. ISSN 1364-503X, 1471-2962. https://doi.org/ 10.1098/rsta.2019.0371. https://royalsocietypublishing.org/doi/10.1098/rsta.2019.0371

Dzhafarov, E. N., & Kujala, J. V. (2014). On selective influences, marginal selectivity, and bell/CHSH inequalities. *Topics in Cognitive Science, 6*(1), 121–128. ISSN 1756-8765. https:// doi.org/10.1111/tops.12060. https://onlinelibrary.wiley.com/doi/abs/10.1111/tops.12060

Dzhafarov, E. N., & Kujala, J. V. (2016). Context–content systems of random variables: The contextuality-by-default theory. *Journal of Mathematical Psychology, 23.*

Fodor, J. A. (1975). *The language of thought.* Harvard University Press.

Fodor, J. A., & Pylyshyn, Z. W. (1988, March). Connectionism and cognitive architecture: A critical analysis. *Cognition, 28*(1), 3–71. ISSN 0010-0277. https://doi.org/10.1016/0010-0277(88)90031-5. https://www.sciencedirect.com/science/article/pii/0010027788900315

Franco, R. (2009). The conjunction fallacy and interference effects. *Journal of Mathematical Psychology, 53*(5), 415–422. ISSN 1096-0880 (Electronic), 0022-2496 (Print). https://doi.org/10. 1016/j.jmp.2009.02.002

Fuss, I. G., & Navarro, D. J. (2013, September). Open parallel cooperative and competitive decision processes: A potential provenance for quantum probability decision models. *Topics in Cognitive Science*, n/a–n/a. ISSN 17568757. https://doi.org/10.1111/tops.12045

Gayler, R. W. (2003, July). Vector symbolic architectures answer Jackendoff's challenges for cognitive neuroscience. In *Joint International Conference on Cognitive Science* (p. 6). https://arxiv.org/abs/cs/0412059

Hameroff, S. (2014, January). Quantum walks in brain microtubules-a biomolecular basis for quantum cognition? *Topics in Cognitive Science, 6*(1), 91–97. ISSN 17568757. https://doi.org/10.1111/tops.12068

Hampton, J. A. (1987, January). Inheritance of attributes in natural concept conjunctions. *Memory & Cognition, 15*(1), 55–71. ISSN 1532-5946. https://doi.org/10.3758/BF03197712

Hampton, J. A. (1988a, November). Disjunction of natural concepts. *Memory & Cognition, 16*(6), 579–591. ISSN 1532-5946. https://doi.org/10.3758/BF03197059

Hampton, J. A. (1988b). Overextension of conjunctive concepts: Evidence for a unitary model of concept typicality and class inclusion. *Journal of Experimental Psychology: Learning, Memory, and Cognition, 14*(1), 12–32. ISSN 1939-1285(Electronic),0278-7393(Print). https://doi.org/10.1037/0278-7393.14.1.12

Hampton, J. A. (1997, November). Conceptual combination: Conjunction and negation of natural concepts. *Memory & Cognition, 25*(6), 888–909. ISSN 0090-502X, 1532-5946. https://doi.org/10.3758/BF03211333. http://link.springer.com/10.3758/BF03211333

Hedges, J., & Sadrzadeh, M. (2019, June). A generalised quantifier theory of natural language in categorical compositional distributional semantics with bialgebras. *Mathematical Structures in Computer Science, 29*(6), 783–809. ISSN 0960-1295, 1469-8072. https://doi.org/10.1017/S0960129518000324. https://www.cambridge.org/core/journals/mathematical-structures-in-computer-science/article/abs/generalised-quantifier-theory-of-natural-language-in-categorical-compositional-distributional-semantics-with-bialgebras/9738AE3A40B9A51AE2134D13E4FD3C5B

Hummel, J. E. (2011, June). Getting symbols out of a neural architecture. *Connection Science, 23*(2), 109–118. ISSN 0954-0091. https://doi.org/10.1080/09540091.2011.569880

Kartsaklis, D., & Sadrzadeh, M. (2016, December). Distributional inclusion hypothesis for tensor-based composition. In *Proceedings of COLING 2016, the 26th International Conference on Computational Linguistics: Technical Papers*, Osaka, Japan (pp. 2849–2860). The COLING 2016 Organizing Committee. https://www.aclweb.org/anthology/C16-1268

Khrennikov, A., Basieva, I., Dzhafarov, E. N., & Busemeyer, J. R. (2014, October) Quantum models for psychological measurements: An unsolved problem. *PLoS ONE, 9*(10), e110909. ISSN 1932-6203. https://doi.org/10.1371/journal.pone.0110909

Khrennikov, A., Basieva, I., Pothos, E. M., & Yamato, I. (2018, November). Quantum probability in decision making from quantum information representation of neuronal states. *Scientific Reports, 8*(1), 16225. ISSN 2045-2322. https://doi.org/10.1038/s41598-018-34531-3. https://www.nature.com/articles/s41598-018-34531-3

Khrennikov, A. Yu., & Haven, E. (2009, October). Quantum mechanics and violations of the sure-thing principle: The use of probability interference and other concepts. *Journal of Mathematical Psychology, 53*(5), 378–388. ISSN 0022-2496. https://doi.org/10.1016/j.jmp.2009.01.007. http://www.sciencedirect.com/science/article/pii/S0022249609000182

Lewis, M. (2019, September). Compositional hyponymy with positive operators. In *Proceedings of the International Conference on Recent Advances in Natural Language Processing (RANLP 2019)*, Varna, Bulgaria (pp. 638–647). INCOMA Ltd. https://doi.org/10.26615/978-954-452-056-4_075. https://www.aclweb.org/anthology/R19-1075

Lewis, M. (2020, September). Towards logical negation for compositional distributional semantics. *IfCoLog Journal of Applied Logics, 7*(5), 771–794.

Lorenz, R., Pearson, A., Meichanetzidis, K., Kartsaklis, D., & Coecke, B. (2021, February). QNLP in practice: Running compositional models of meaning on a quantum computer. arXiv:2102.12846 [quant-ph]. http://arxiv.org/abs/2102.12846

Martin, A. E., & Doumas, L. A. A. (2020, February). Tensors and compositionality in neural systems. *Philosophical Transactions of the Royal Society B: Biological Sciences,*

375(1791), 20190306. https://doi.org/10.1098/rstb.2019.0306. https://royalsocietypublishing. org/doi/10.1098/rstb.2019.0306

McCoy, R. T., Linzen, T., Dunbar, E., & Smolensky, P. (2019, May). RNNs implicitly implement tensor product representations. In *ICLR 2019—International Conference on Learning Representations*. https://hal.archives-ouvertes.fr/hal-02274498

Meyer, F., & Lewis, M. (2020, November). Modelling lexical ambiguity with density matrices. In *Proceedings of the 24th Conference on Computational Natural Language Learning* (pp. 276–290) (Online). Association for Computational Linguistics. https://doi.org/10.18653/v1/2020.conll-1. 21. https://www.aclweb.org/anthology/2020.conll-1.21

Moore, D. W. (2002). Measuring new types of question-order effects. *Public Opinion Quarterly, 66*(1), 80–91. ISSN 0033362X, 15375331. https://doi.org/10.1086/338631. https://academic. oup.com/poq/article-lookup/doi/10.1086/338631

Morier, D., & Bordiga, E. (1984). The conjunction fallacy: A task specific phenomenon? *Personality and Social Psychology Bulletin, 10*(2). https://doi.org/10.1177/0146167284102010. https:// journals.sagepub.com/doi/abs/10.1177/0146167284102010

Nielsen, M. A., & Chuang, I. L. (2010). *Quantum computation and quantum information: 10th Anniversary*. Cambridge University Press.

Nosofsky, R. M. (1986). Attention, similarity, and the identification–categorization relationship. *Journal of Experimental Psychology: General, 115*(1), 39–57. ISSN 1939-2222 (Electronic), 0096-3445 (Print). https://doi.org/10.1037/0096-3445.115.1.39

Penrose, R. (1990, December). The nonalgorithmic mind. *Behavioral and Brain Sciences, 13*(4), 692–705. ISSN 0140-525X, 1469-1825. https://doi.org/10.1017/S0140525X0008105X. https:// www.cambridge.org/core/product/identifier/S0140525X0008105X/type/journal_article

Piedeleu, R., Kartsaklis, D., Coecke, B., & Sadrzadeh, M. (2015). Open system categorical quantum semantics in natural language processing. In *CALCO* (p. 14).

Plate, T. A. (1995, May). Holographic reduced representations. *IEEE Transactions on Neural Networks, 6*(3), 623–641. ISSN 10459227. https://doi.org/10.1109/72.377968. http://ieeexplore. ieee.org/document/377968/

Pothos, E. M., & Busemeyer, J. R. (2009, June). A quantum probability explanation for violations of 'rational' decision theory. *Proceedings. Biological Sciences, 276*(1665), 2171–2178. ISSN 0962-8452. https://doi.org/10.1098/rspb.2009.0121

Pothos, E. M., & Busemeyer, J. R. (2013, June). Can quantum probability provide a new direction for cognitive modeling? *Behavioral and Brain Sciences, 36*(3), 255–274. ISSN 0140-525X, 1469-1825. https://doi.org/10.1017/S0140525X12001525. https://www.cambridge.org/core/product/ identifier/S0140525X12001525/type/journal_article

Pothos, E. M., & Trueblood, J. S. (2015, February). Structured representations in a quantum probability model of similarity. *Journal of Mathematical Psychology, 64–65*, 35–43. ISSN 00222496. https://doi.org/10.1016/j.jmp.2014.12.001. https://linkinghub.elsevier.com/retrieve/ pii/S0022249614000832

Pothos, E. M., Busemeyer, J. R., & Trueblood, J. S. (2013). A quantum geometric model of similarity. *Psychological Review, 120*(3), 679–696. ISSN 1939-1471, 0033-295X. https://doi.org/10.1037/ a0033142. http://doi.apa.org/getdoi.cfm?doi=10.1037/a0033142

Rosch, E. (1975). Cognitive representations of semantic categories. *Journal of Experimental Psychology: General, 104*(3), 192–233. ISSN 1939-2222 (Electronic), 0096-3445 (Print). https:// doi.org/10.1037/0096-3445.104.3.192

Sadrzadeh, M., Kartsaklis, D., & Balkır, E. (2018, April). Sentence entailment in compositional distributional semantics. *Annals of Mathematics and Artificial Intelligence, 82*(4), 189–218. ISSN 1573-7470. https://doi.org/10.1007/s10472-017-9570-x

Savage, L. J. (1954). *The foundations of statistics*. Wiley Publications in Statistics.

Smith, E. E., & Osherson, D. N. (1984, October). Conceptual combination with prototype concepts. *Cognitive Science, 8*(4), 337–361. ISSN 0364-0213. https://doi.org/10.1016/S0364-0213(84)80006-3. http://www.sciencedirect.com/science/article/pii/S0364021384800063

Smolensky, P. (1990, November). Tensor product variable binding and the representation of symbolic structures in connectionist systems. *Artificial Intelligence, 46*(1–2), 159–216. ISSN 00043702. https://doi.org/10.1016/0004-3702(90)90007-M. https://linkinghub.elsevier.com/retrieve/pii/000437029090007M

Smolensky, P., & Legendre, G. (2006, July). *The harmonic mind*. The MIT Press. https://mitpress.mit.edu/books/harmonic-mind-2-vol-set

Sozzo, S. (2014, June). A quantum probability explanation in Fock space for borderline contradictions. arXiv:1311.6050 [physics, physics:quant-ph]. http://arxiv.org/abs/1311.6050

Sozzo, S. (2015, June). Conjunction and negation of natural concepts: A quantum-theoretic modeling. *Journal of Mathematical Psychology, 66*, 83–102. ISSN 0022-2496. https://doi.org/10.1016/j.jmp.2015.01.005. https://www.sciencedirect.com/science/article/pii/S0022249615000073

Tversky, A. (1977). Features of similarity. *Psychological Review, 84*(4), 327–352. ISSN 1939-1471 (Electronic), 0033-295X (Print). https://doi.org/10.1037/0033-295X.84.4.327

Tversky, A., & Kahneman, D. (1983). Extensional versus intuitive reasoning: The conjunction fallacy in probability judgment. *Psychological Review, 90*(4), 293–315. ISSN 1939-1471 (Electronic), 0033-295X (Print). https://doi.org/10.1037/0033-295X.90.4.293

Tversky, A., & Shafir, E. (1992, September). The disjunction effect in choice under uncertainty. *Psychological Science, 3*(5), 305–310. ISSN 0956-7976. https://doi.org/10.1111/j.1467-9280.1992.tb00678.x

Wang, Z., & Busemeyer, J. R. (2013, September). A quantum question order model supported by empirical tests of an *a priori* and precise prediction. *Topics in Cognitive Science*, n/a–n/a. ISSN 17568757. https://doi.org/10.1111/tops.12040

Wang, Z., Busemeyer, J. R., Atmanspacher, H., & Pothos, E. M. (2013, September). The potential of using quantum theory to build models of cognition. *Topics in Cognitive Science*, n/a–n/a. ISSN 17568757. https://doi.org/10.1111/tops.12043

Widdows, D., & Peters, S. (2003). *Word vectors and quantum logic experiments with negation and disjunction* (p. 14). Stanford University.

Wiebe, N., Bocharov, A., Smolensky, P., Troyer, M., & Svore, K. M. (2019, February). Quantum language processing. arXiv:1902.05162 [quant-ph]. http://arxiv.org/abs/1902.05162

Wittgenstein, L. (1953). *Philosophical investigations. Philosophische Untersuchungen*. Macmillan.

The Philosophy of Quantum Computing

Michael E. Cuffaro

Abstract From the philosopher's perspective, the interest in quantum computation stems primarily from the way that it combines fundamental concepts from two distinct sciences: Physics, in particular Quantum Mechanics, and Computer Science, each long a subject of philosophical speculation and analysis in its own right. Quantum computing combines both of these more traditional areas of inquiry into one wholly new, if not quite independent, science. Over the course of this chapter we will be discussing some of the most important philosophical questions that arise from this merger and philosophical lessons to be learned.

Keywords Quantum computing · Quantum humanities · Philosophy of science · Many-worlds interpretation · Church-Turing thesis

My work on this chapter benefited significantly from my interactions with students and other audience members during and after a series of lectures I gave at the University of Urbino's twenty-third international summer school in philosophy of physics, held online in June 2020, in the midst of the first wave of the COVID-19 pandemic, as well as a further lecture I gave for Michel Janssen's "The Age of Entanglement" honors seminar at the University of Minnesota in December 2020, as the second wave of the pandemic began in earnest. Thanks to Ari Duwell, Eduardo Reck Miranda, Philip Papayannopoulos, and Lev Vaidman for comments on a previous draft of this chapter. I am also grateful for informal discussions, over the years, with Guido Bacciagaluppi, Jim Baggot, Michel Janssen, Christoph Lehner, Lev Vaidman, and David Wallace; my presentation of the Everett interpretation in Sect. 3, in particular, is significantly informed by what I take myself to have learned from these discussions, though I hold only myself responsible for any mistakes or misunderstandings in my presentation of the Everettian view. Section 2, on "Fundamental concepts" is heavily informed by my recent work on related topics with Stephan Hartmann, Michael Janas, Michel Janssen, and Markus Müller, as well as by informal correspondence with Jeffrey Bub and (the late) Bill Demopoulos; though here again, I take sole responsibility for any mistakes. Finally, I gratefully acknowledge the generous financial support of the *Alexander von Humboldt Stiftung*.

M. E. Cuffaro (✉)
Munich Center for Mathematical Philosophy, Ludwig-Maximilians-Universität München, Munich, Germany
e-mail: mike@michaelcuffaro.com

© Springer Nature Switzerland AG 2022
E. R. Miranda (ed.), *Quantum Computing in the Arts and Humanities*,
https://doi.org/10.1007/978-3-030-95538-0_3

1 Introduction

From the philosopher's perspective, the interest in quantum computation stems primarily from the way that it combines fundamental concepts from two distinct sciences: physics (especially quantum mechanics) and computer science, each long a subject of philosophical speculation and analysis in its own right. Quantum computing combines both of these more traditional areas of inquiry into one wholly new (if not quite independent) science. There are philosophical questions that arise from this merger, and philosophical lessons to be learned. Over the course of this chapter, we will be discussing what I take to be some of the most important.[1]

We begin, in Sect. 2, by introducing the fundamental concepts from physics and computation that will be essential for framing the further philosophical discussion that will follow. Section 2.1 and Sect. 2.2 introduce concepts from classical mechanics and the classical theory of computation, respectively. In Sect. 2.1, we discuss the concept of the state of a physical system as it is given in classical mechanics. We emphasize in particular the way that we are invited, in classical mechanics, to think of the state of a physical system as a compact description of the properties possessed by it. These properties determine, in advance, the answers to the experimental questions that we can pose of a system. And considering these questions and their answers leads one to an alternative representation of a system's state that is useful for representing computation, a subject we then take up in more detail in Sect. 2.2.

In Sect. 2.2, we introduce the concept of a model of computation (or computational architecture), the concept of the cost of carrying out a computation under a given computational model, and explain what it means to solve a problem just as easily under one computational model as under another. After discussing some of the more important computational complexity classes, we then introduce two theses that can be used to ground the model-independence of computational cost: the so-called universality of Turing efficiency thesis (sometimes also referred to as the "strong", or "physical", or "extended" Church-Turing thesis), and the invariance thesis. These theses are both called into question by the existence of quantum computers. And since some have taken the absolute model-independence guaranteed by them to be foundational for the science of computational complexity theory, the question arises of what to make of those foundations in light of the existence of quantum computation. We will discuss this question, in an abstract way, in Sect. 2.3, where I will argue that although the universality thesis must indeed be given up, this is not a great loss. The invariance thesis, by contrast, remains, but in a different form, transformed from a metaphysical to a methodological principle.

---

[1] Space does not permit me to exhaustively survey all of the philosophical issues brought up by quantum computing. The interested reader can find a summary of other important issues in Hagar and Cuffaro (2019).

In Sect. 2.4, we turn to the physics behind quantum computation and begin by introducing some of the more important concepts of quantum theory. Most importantly, we introduce the concept of a quantum state. We emphasize that the way that a quantum state determines what the answers to experimental questions will be is fundamentally different than the way that they are determined by a classical state. We then turn to quantum computation proper in Sect. 3, where we review the basic concepts of quantum computing and consider what we can say regarding the physical explanation of the power of quantum computers. The many-worlds explanation of quantum computing—the idea that quantum computers outperform classical computers by running their computations in exponentially many physical universes—is then introduced. We note two major problems that arise for this interpretation. The first arises from the so-called preferred basis problem. This problem is a challenge (that is arguably surmountable) for the many-worlds view in the more general context of quantum mechanics. But we will see that it is especially problematic in the context of quantum computers. The second major problem arises from the fact that there are many different models of quantum computation, but the many-worlds explanation of quantum computing only seems motivated by one of them.

In Sect. 4, we consider the role of quantum entanglement in enabling quantum computers to outperform classical computers. We summarize an unsuccessful argument to the conclusion that quantum entanglement is insufficient to enable this quantum "speedup" in Sect. 5, noting in Sect. 6 that reflecting on what it means to provide a classical computer simulation of a quantum phenomenon should convince us to reach the opposite conclusion. We continue the discussion of classically simulating quantum computers in Sect. 7, and reflect on general aspects of the computational perspective that the study of quantum computing provides on physical theory. We note that reflecting on quantum computing emphasizes that there are important differences in the methodological presuppositions that lie at the basis of physics and computer science, respectively, and that conflating these can lead to confusion. We also note the emphasis that studying quantum computation places on the fact that quantum mechanics and classical mechanics are each alternative universal languages for describing physical systems, and that the difference between quantum mechanics and classical mechanics lies, fundamentally, in the differences in the expressive power of these languages.[2] We reflect on this in Sect. 8.

---

[2] By "quantum mechanics" I mean the fundamental theoretical framework shared in common by every specific quantum-mechanical theory (quantum field theories, for instance) of a particular class of systems; see Aaronson (2013b, pp. 110–111), Janas et al. (2022, Chap. 1 and §6.3), Nielsen and Chuang (2000, p. 2), and Wallace (2019).

2   Fundamental Concepts

2.1   *Classical States*

Classical mechanics (in its various forms[3]) was, prior to quantum mechanics, our fundamental theoretical framework for describing the dynamics of physical systems (i.e., how physical systems change over time). What is meant by a physical system here is just one of the concrete objects that a particular physical theory describes. All such systems are described as evolving through time in accordance with the dynamical constraints that a theoretical framework applies universally to every physical system. In classical mechanics (and the same is true, as we will see later, in quantum mechanics), a physical system can be anything from a single particle to (in principle) the entire physical universe. Mathematically, though, what the dynamical laws of a theory actually govern are the relations between the possible state descriptions of systems. By the state description (or state specification) of a system is meant a description of the particular physical situation that it happens to be in at a given moment in time. For instance, at a given moment in time, a classical-mechanical system will have a particular kinetic energy, it will be accelerating (or not) in a particular direction, and so on.

It turns out that we do not have to explicitly specify each and every one of a system's dynamical properties to exactly and exhaustively specify the system's dynamics. In classical mechanics, given the dynamical laws, it is enough to specify values for a system's position and momentum. The value of any other dynamical property can then be calculated by relating these with each other and with the values of non-dynamical properties of the system such as, for instance, its charge or its mass.

Figure 1 illustrates the two different strategies described in the previous paragraph for specifying the state of a physical system, for instance, a system composed of a particle attached to the end of a spring constrained to move in one direction (see Hughes, 1989, p. 73). On the left, we explicitly specify values $(v_1, v_2, \ldots)$ for each and every one of the physical system's dynamical parameters. On the right, the values of momentum, p, and position, q, are specified, and all other dynamical quantities are calculated on their basis. In particular, the total energy, H, of the system is defined as the sum of the kinetic energy, T, of the particle, and the potential energy, V, stored in the spring, which are in turn defined in terms of p and q, respectively. Note that m, the system's mass, is a *non*-dynamical parameter that does not change over the history of the system, and k is the force (the first derivative of momentum) per unit distance required for the spring to be displaced from its equilibrium point (i.e., the point at which it is neither stretched nor compressed). Other forces (e.g., gravitational forces) are neglected but in principle can be included (for further discussion, see Hughes, 1989, §3.2.5).

---

[3] These include Newtonian, Lagrangian, Hamiltonian, relativistic, and classical statistical mechanics. For a recent comparison and philosophical discussion of Lagrangian and Hamiltonian mechanics, see Curiel (2014).

$$
\begin{array}{c|c}
\begin{aligned}
P_1 &= v_1 \\
P_2 &= v_2 \\
P_3 &= v_3 \\
P_4 &= v_4 \\
P_5 &= v_5 \\
&\cdots
\end{aligned}
&
\begin{aligned}
p &= v_p \\
q &= v_q \\
T &= p^2/2m \\
V &= kq^2/2 \\
H &= V + T \\
&\cdots
\end{aligned}
\end{array}
$$

Fig. 1 Two different strategies for specifying the state of a physical system, for instance, a system composed of a particle attached to the end of a spring constrained to move in one direction (see Hughes, 1989, p. 73). On the left, we explicitly specify values $(v_1, v_2, \dots)$ for each and every one of the physical system's dynamical properties. On the right, the values of momentum, p, and position, q, are specified, and all other dynamical quantities are calculated on their basis

The upshot is that specifying the position and momentum of a system provides us with enough information, in classical mechanics, to completely characterize all of the other dynamical properties of that system. Accordingly, the dynamical state, ω_t, of a system at some particular time t is given in classical mechanics by

$$
\omega_t = (\mathbf{q}_t, \mathbf{p}_t), \tag{2.1}
$$

where $\mathbf{q}_t$ and $\mathbf{p}_t$ are vectors (in three dimensions) representing the system's position and momentum, respectively, at t. Further, we can infer from the state of a system at t, and the classical-mechanical laws of motion, exactly what the state of the system will be in the next moment and at every other time both in the system's past and in its future (Hughes, 1989, §2.6).

Classical-mechanical states have another feature. Imagine all of the possible (experimental) yes-or-no questions one might want to ask about the dynamical properties of a particular system at a particular time, questions like: *Is the value of the dynamical property A within the range* Δ*?* Completely specifying a system's dynamical state, i.e., specifying its momentum, $\mathbf{p}$, and its position, $\mathbf{q}$, yields a simultaneous answer to all such questions irrespective of whether any question has actually been asked. Indeed, this is what actually justifies our having considered the values of $\mathbf{p}$, $\mathbf{q}$, and the other quantities derived from them to be (observable) *properties* of the system in the first place, properties possessed by the system whether or not, and however, we enquire concerning them (Janas et al., 2022, §6.3). The same goes for a system composed of many parts; for, after all, any observable property A_1 of a subsystem S_1 of some larger system S is also an observable property of S. Thus, given the state specification for a particular system we can construct a sort of "truth table" which could be used, in principle, to answer any conceivable yes-or-no question concerning the system and its subsystems (see Fig. 2).

In Fig. 2, I wrote Y and N (for "yes" and "no"), but I could have equally well used T (for true) and F (for false), or alternately the binary digits 0 and 1. Using binary digits, i.e., "bits", is convenient because they are useful for representing numbers, which can

$\mathbf{p}_1$	$\mathbf{q}_1$	$\mathbf{p}_2$	$\mathbf{q}_2$	A	B	...
$v^1_{p_1}$	$v^1_{q_1}$	$v^1_{p_2}$	$v^1_{q_2}$	N	N	
$v^2_{p_1}$	$v^2_{q_1}$	$v^2_{p_2}$	$v^2_{q_2}$	N	Y	
$v^3_{p_1}$	$v^3_{q_1}$	$v^3_{p_2}$	$v^3_{q_2}$	N	Y	
$v^4_{p_1}$	$v^4_{q_1}$	$v^4_{p_2}$	$v^4_{q_2}$	Y	N	
$v^5_{p_1}$	$v^5_{q_1}$	$v^5_{p_2}$	$v^5_{q_2}$	N	Y	
$v^6_{p_1}$	$v^6_{q_1}$	$v^6_{p_2}$	$v^6_{q_2}$	Y	N	
$v^7_{p_1}$	$v^7_{q_1}$	$v^7_{p_2}$	$v^7_{q_2}$	Y	N	
$v^8_{p_1}$	$v^8_{q_1}$	$v^8_{p_2}$	$v^8_{q_2}$	N	N	

...

Fig. 2 A "truth table" for a classical system composed of two subsystems whose states, respectively, are $(\mathbf{p}_1, \mathbf{q}_1)$ and $(\mathbf{p}_2, \mathbf{q}_2)$. Various combinations of values for the state parameters $\mathbf{p}_i$ and $\mathbf{q}_i$ are given on the left of the double-vertical line, with the superscript j in v^j indicating the jth combination. Relative to a given combination of values, the answers to experimental questions concerning the values of derived observable quantities A, B, ... on the right of the double-vertical line can be determined

Fig. 3 A physical system consisting of four light switches. For each switch, we can ask: "Is switch S turned on?" Using 0 to represent yes and 1 to represent no, the state of the overall system relative to this question can be represented using the bit-string 0110 (which, in base-ten, is the number 6). This is just one of the $2^4 = 16$ possible states that a system like this one can be in. More generally, for a system made up of n two-dimensional subsystems, the number of possible states for the system as a whole is 2^n

be manipulated abstractly using logico-mathematical operations.[4] In particular, if we build physical systems whose states can reliably be used to represent binary numbers (in the sense that yes-or-no questions concerning their observable properties A, B, ... can be laid out as in Fig. 2, replacing 0 for Y and 1 for N), and reliably evolve them in ways that mirror a small basic set of logico-mathematical operations, then we can (by combining these operations) use physical systems to carry out computations that can, in principle, be arbitrarily complex (see Figs. 3 and 4).[5]

---

[4] I say "logico-mathematical" because logical operations on bits can be thought of as modulo-2 arithmetical operations (see Boole, 1847).

[5] For more general accounts of how physical systems can be used to represent computations, see Fletcher (2018), Horsman et al. (2018), Maroney and Timpson (2018).

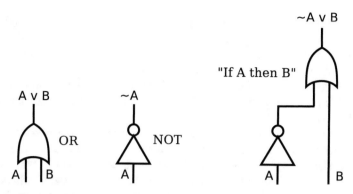

Fig. 4 Some of the logico-mathematical operations, or "logic gates", that can be used to manipulate bits. OR and NOT together constitute a universal set of gates, in the sense that any more complicated logico-mathematical operation, such as the "if-then" gate at the right, can be constructed using combinations of OR and NOT. There are other universal sets besides {OR, NOT}, for instance: {AND, NOT}, {NOR}, and {NAND}, where NAND is the "not-and" gate and NOR is the "not-or" gate

2.2 Classical Computers

When the word "computer" is used in popular discourse today, generally what is meant is a physical system like the one depicted in Fig. 3—*not*, of course, literally a collection of light switches, but a physical system composed of subsystems that, like the switches, can reliably be interpreted as being either on or off, and that can be organized into registers, random access memory, hard drives, and other peripheral systems such as keyboards, "mice", touchpads, and so on, that can be manipulated to reliably realize various logical operations like the ones depicted in Fig. 4. The laptop I am writing this chapter with is an example of such a device. We turn such devices on when we need them, and turn them off or "put them to sleep" when we do not. But the word "computer" did not always signify a kind of machine. Prior to the 1930s, "computation" generally meant an activity carried out by a human being. Even after Alan Turing's introduction of what later came to be known as the "Turing machine" in 1936—an abstract device that is itself modeled on human computation—"computers" were generally understood to be human beings until well into the 1960s.[6]

As for the Turing machine: this, as mentioned, is also not a concrete physical device (although it could, in principle, be instantiated by one) but an abstract mathematical model. Turing's primary reason for introducing it had been to address the so-called *Entscheidungsproblem* (which is the German word for 'decision problem'), an important but unsettled question that had arisen in the foundations of number theory. The *Entscheidungsproblem* concerned the existence of an effective proce-

---

[6] The role of human computers in the United States of America's space program, for instance, has been documented in Shetterly (2016).

dure for deciding whether an arbitrarily given expression in first-order logic can be proven from the axioms of first-order logic. The answer, as Turing and, independently, Alonzo Church were able to demonstrate, is that there is no such procedure. Turing's own proof relied on a penetrating philosophical analysis of the notion of effective (human) computation, and a corresponding argument that one could design an automatic machine (now called a *Turing machine*; see Fig. 5) to carry out the essential activities of a human computer.

Turing (1936–7, pp. 249–51) argued that, for a computer to carry out a computation, it is essential that she have access to a notebook from which she can read and onto which she can write various symbols in the course of her work. These symbols need to be distinguishable by her from one another, on the basis of which Turing argued that the alphabet from which she chooses her symbols must be a finite alphabet. At any given moment during a computation, a computer may find herself in one of a finite number (again, because she must be able to distinguish them) of states of mind relevant to the task at hand which summarize her memory of the actions she has performed up until that point along with her awareness of what she must now do (pp. 253–4). The actions that are available to her are characterized by a finite set of elementary operations, such as "read the next symbol" from the notebook, "write symbol a" to the notebook, and so on. In an automatic machine, a *control unit* is constructed to encode the machine's "state of mind" (i.e., a logical representation of its current state, its current input, and its transition function), which in general changes after every operation of the *read-write head*. The latter moves back and forth along a *one-dimensional tape* (the machine's "notebook"), from which it reads and onto which it writes various symbols from a finite alphabet, in conformity with a particular finite set of exactly specifiable rules. Turing was able to show that no automatic machine of this kind can be used to solve the *Entscheidungsproblem*.

Through the work of Turing, Church, and others, the modern science of computation, and in particular *computability theory*—the science of which problems can and which cannot be solved by a computer, i.e., by anything or any person that can carry out (but is restricted to) the essential activities associated with human computation—was born.[7]

Irrespective of whether it is a machine or a human being that is doing the computing, the question of how much it actually costs to compute a solution to a computable problem is also a very important one. In particular, we generally would like to know how much *time* and how much *space* are required (though time is usually regarded as the more important measure). The question of the cost of a given computation had become especially important as more and more problems came to be carried out by machines in the 1950s and 1960s. Even the earliest digital computers of the 1930s and 1940s performed their computations far faster than human beings could possibly perform them, and with improvements in design and in technology, machines could be made to run ever faster. Through the need to use these machines' resources effi-

---

[7] For more on the *Entscheidungsproblem* and the early history of computer science, see Copeland (2020), Davis (2000), Dawson Jr. (2007), Lupacchini (2018).

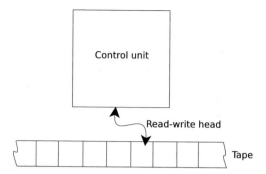

Fig. 5 A version of what is now called a 'Turing machine', with a single bi-directional tape, a control unit encoding a particular set of instructions or program, and a read-write head. Such a machine is an idealized representation of the components involved in human computation. In a *universal* Turing machine, the computer's program is read from the tape at the beginning of the computation. For a good introduction to the Turing machine and other related topics, see Martin (1997)

ciently, a sub-field of computer science began to develop whose concern was with how efficiently a given problem can be solved and on which types of machine.

In the context of the Turing machine model, time and space are quantified in terms of the number of computational 'steps' (i.e., the number of basic operations) and the number of squares of tape, respectively, that are needed to carry out a given computation. Although it is in principle possible (assuming we have enough tape) to instantiate a Turing machine physically,[8] generally it makes more sense, from a practical perspective, to design a mechanical computer in accordance with some other model of computation. Particularly important is the von Neumann architecture, a type of stored-program computer,[9] that involves a central processing unit (CPU), an instruction register and program counter, internal memory used to store data (including the program being run), external memory used to store less frequently needed data, and finally an input and an output mechanism. A von Neumann machine can in principle compute all of the same problems as a universal Turing machine (see Fig. 5), but a time step for a von Neumann machine is not concretely interpreted in the

---

[8] A Turing machine's tape does not need to be actually infinite in length. What is required is only that the tape be *indefinitely* long, so that, for a given (finite) computation, the machine can be supplied with enough tape to carry it out. To put it another way: What constitutes 'enough tape' to carry out a computation is not part of the general definition of a Turing machine. It is, rather, assumed, for the purposes of that definition, that enough tape to carry out a given computation can be supplied. That said, certain variations on the Turing machine model restrict the ways in which tape can be read by the control unit in various ways. For instance, some variants employ separate tape(s) for the machine to write "rough work" on in addition to an output tape, some variants only allow the read-write head to move in one direction along the tape, and so on.

[9] Although the von Neumann architecture, or 'von Neumann machine' is only one of a number of various types of stored-program computer, the terms have today (inappropriately, from a historical point of view) come to be understood synonymously (Copeland, 2017).

same way as it is for a Turing machine, just as 'space' for a von Neumann machine is not measured in squares of tape.

There are many other universal models of computation besides von Neumann machines and Turing machines. Nevertheless, at least when it comes to reasonable classical computational models, it is possible to talk about the computational cost of solving a particular problem (like, for instance, the problem of finding the prime factors of a given integer), in a model-independent way; i.e., without having to specify which machine model that cost is defined with respect to. There are two things involved in explicating what it means for the computational cost of solving a given problem to be model-independent in this sense. The first, mentioned just above is the idea of a reasonable machine architecture.

Roughly, a *reasonable* (universal) machine architecture or model is one that it would be possible to physically instantiate. Turing machines like the one described in Fig. 5 and von Neumann machines are examples of reasonable machine models. A computational model that employs an unbounded amount of parallel processing, by contrast, is an example of an unreasonable model, or what van Emde Boas (1990, p. 5) calls a model of the "second machine class". There are legitimate theoretical reasons for studying the complexity-theoretic properties of unreasonable models. But no finite physical system could possibly instantiate an unreasonable architecture. There are thus good reasons to want to distinguish the reasonable from the unreasonable ones (see also, Dean, 2016).

The second thing we need to do, in order to explicate what it means for the computational cost of solving a given problem to be model-independent, is to make sense of what it means to solve a given problem *just as easily* with one computational architecture as with some other one. But before we can make sense of solving something just as easily, we need to first try and make sense of what it means to solve something *easily* under one computational model. Consider, to begin with, a human computer working alone. For a human computer, the difference between a problem that could in principle require up to a thousand steps to solve, and one that could in principle require up to a million steps, is very great indeed. With some justification, this person might call the former problem easy to solve, and the latter problem hard. Any criterion based on a fixed number of computational steps, however, would not be very useful in the context of machines, at any rate certainly not any criterion based on numbers this low. Even by the 1950s, a mechanical computer could carry out a computation far faster than a human being. As of 2020, a typical laptop computer can perform hundreds of thousands of Millions of Instructions per Second (MIPS) (Wikipedia contributors, 2020), and computers are becoming more and more performant every year. In the context of machine computation, it is far more useful to ask how a given solution to a computational problem *scales*; i.e., how the number of resources required to carry out a given solution to that problem grows with the size of the input to the problem. Take, for instance, the problem to sort a given set of numbers. All else equal, the more numbers there are to sort, the longer it will take. How much longer? The fastest sorting algorithms (one of which is `MergeSort`) will take on the order of $n \log n$ steps to sort n numbers in the worst case (Mehlhorn & Sanders, 2008).

In modern computational complexity theory, we say that a problem is *easy* to solve, or alternately that it is *tractable*, if there exists an efficient algorithm to solve it. If no efficient algorithm exists, then we say that the problem is *hard*, or *intractable*. This definition, of course, merely trades in one informal notion for another, so let us state more precisely what we mean by 'efficient'. By an *efficient algorithm* for solving a given problem, we mean an algorithm that will solve it in *polynomial time*; i.e., in a number of time steps that are no more than a polynomial function of the size of its input. Stated in mathematical terms, this means that for input size n, such an algorithm will take on the order of n^k steps, where k is some constant.[10] As for an algorithm that takes more than a polynomial number of steps, we consider it to be *inefficient*, even if it might be the most efficient algorithm known for a particular problem. An example of an inefficient algorithm is one that requires *exponential time*; i.e., on the order of k^n steps for input size n and some constant k.[11] Even harder problems take *factorial time*, i.e., on the order of $n!$ steps, and so on.

The idea of using polynomial time as our criterion for distinguishing between efficient and inefficient algorithms—what we will henceforth call the *polynomial principle*—was introduced independently by Alan Cobham and Jack Edmonds in 1965.[12] There are a number of good reasons for adopting the polynomial principle. One reason is simply that the set of problems picked out by it has so far tended to correspond with those that we would (so to speak) pre-theoretically regard as efficiently solvable (Cuffaro, 2018b, §11.6).[13] Perhaps the most useful feature of the polynomial principle stems from the fact that polynomial functions compose (see Fig. 6). That is, given an algorithm that runs in polynomial time, if we add to it any number of calls to polynomial-time subroutines, the total running time of the algorithm will still be polynomial (Arora & Barak, 2009, p. 27).[14]

Now that we have a handle on what it means for a problem to be solvable easily, we can state what it means for a problem to be solvable just as easily on one type of computational architecture as on another. We say that a problem is solvable *just as easily* on machine model $\mathfrak{M}_1$ as on machine model $\mathfrak{M}_2$ if and only if there is an algorithm for solving the problem on $\mathfrak{M}_1$ that requires no more than a polynomial

---

[10] "On the order of" is a technical term, usually symbolized in "big-oh notation" as $O(T(n))$. An algorithm is $O(T(n))$ for some function $T(n)$ if for every sufficiently large n, its actual running time $t(n) \leq c \cdot T(n)$ for some constant c. For instance, an algorithm that never takes more than $5n^3$ steps is $O(n^3)$.

[11] A famous example of a problem for which only exponential-time solutions are known is the Traveling Salesman Problem (Cook, 2012).

[12] In some literature it is referred to as the *Cobham-Edmonds thesis*. Kurt Gödel anticipated the principle, to some extent, in a private letter he wrote to John von Neumann in 1956. For further discussion, see Cuffaro (2018b).

[13] This correspondence is not perfect, but the usefulness of the polynomial principle is such that we appeal to it despite this (Cuffaro, 2018b, §11.6).

[14] It is easy to see this: Consider a program that consists of n^k calls of a subroutine that takes n^l steps, where n is the number of bits used to represent the input, and k and l are finite constants. The total number of steps needed to carry out this program will be n^{k+l}. If k and l are finite constants then so is $k + l$. In other words, n^{k+l} is still a polynomial.

```
// A polynomial-time algorithm:          // Another polynomial-time algorithm:
myProgram1(arg1, arg2) {                 myProgram2(arg1, arg2) {
   statement1;                              statement1;
   statement2;                              statement2;
   ...                                      ...
}                                           myPolynomialTimeSubroutine(arg1);
                                            ...
                                            myPolynomialTimeSubroutine(arg2);
                                         }
```

Fig. 6 Inserting any number of calls to polynomial-time subroutines within the body of a polynomial-time algorithm results in a polynomial-time algorithm

number of extra time steps as compared to $\mathfrak{M}_2$.[15] Another way of putting this is that anything that $\mathfrak{M}_2$ can do is *efficiently simulable* by $\mathfrak{M}_1$.

Any problem that is solvable in polynomial time by a Turing machine is said to belong to the class **PTIME** (for "polynomial time"), though it is usually referred to simply as **P**. **P** is, of course, not the only complexity class. If a given problem is such that, for a given solution to it, there is a Turing machine that will *verify* that the solution is indeed a solution, then the problem is said to belong to the class **NP**, for "nondeterministic polynomial time". The reason for the name is that, equivalently, this class of problems can be defined as those that can be solved in polynomial time on a nondeterministic Turing machine. A *nondeterministic Turing machine*, or *choice machine*, is such that at a given step of a given computation, an imagined external operator of the machine can choose to have it transition in one way rather than another. This is unlike a standard (deterministic) Turing machine for which every computational step is completely determined in advance given a particular input. Choice machines are interesting from a theoretical point of view, but given that they involve an external operator they are not really automatic machines in the sense that is relevant to our discussion here.

A variant of the choice machine that *is* automatic is a *probabilistic Turing machine*. Such a machine's choices are not made by an external operator but by the machine itself, which we can imagine as equipped with the equivalent of a fair coin that it can flip a number of times to decide whether to transition one way or another whenever such a choice is open to it.[16] The class of problems solvable in polynomial time on a probabilistic Turing machine is called **BPP** (for "bounded error probabilistic polynomial"). Note that what it means to be able to solve a given problem is not the same for a probabilistic Turing machine as it is for a deterministic Turing machine. In particular, a given "solution" output by a probabilistic Turing machine is allowed to be wrong. We only demand that it be right with high enough probability so that,

---

[15] Note that it makes sense to talk about solving a given problem just as easily on $\mathfrak{M}_1$ as on $\mathfrak{M}_2$ even when the problem under consideration is actually intractable for both. For instance, if some problem requires 2^n steps to solve on $\mathfrak{M}_1$ and $2^n + n^3$ steps to solve on $\mathfrak{M}_2$ then it is no harder, from the point of view of the polynomial principle, to solve it on $\mathfrak{M}_2$ than on $\mathfrak{M}_1$.

[16] For more on probabilistic and nondeterministic Turing machines and how they compare to their deterministic counterparts, see Cuffaro (2018b, §11.3).

if we re-run the machine for the given input on the order of a polynomial number of further times, our confidence in the majority answer will approach certainty.

There are very many more complexity classes besides these ones (see Aaronson, 2016). But we will stop here as we now have the basic ingredients with which to state a couple of related theses that can be used to explicate what it means for the cost of solving a given problem to be model-independent. As we will see a little later, there are good reasons to prefer the second thesis, but for now, we simply introduce both. First, the *universality of Turing efficiency thesis* (sometimes also called the "strong", "physical", or "extended" Church-Turing thesis[17]) asserts that any problem that can be efficiently solved on *any* reasonable machine model $\mathfrak{M}$ (von Neumann architecture, Harvard architecture, or whatever) can be efficiently solved on a probabilistic Turing machine, or more formally

$$\bigcup \text{Poly}_{\mathfrak{M}} = \mathbf{BPP}. \tag{2.2}$$

In other words, the thesis implies that the set of problems solvable in polynomial time does not grow beyond **BPP** if we allow ourselves to vary the underlying model. Assuming the thesis is true, we do not need to worry about what model an algorithm is implemented on when discussing the computational complexity of various problems; we can simply focus on the abstract probabilistic Turing machine model and carry out our analyses in relation to it.[18]

The second thesis we will introduce is called the *invariance thesis* (van Emde Boas, 1990, p. 5),[19] which asserts that given any two reasonable machine models $\mathfrak{M}_i$ and $\mathfrak{M}_j$, $\mathfrak{M}_i$ can efficiently simulate $\mathfrak{M}_j$; i.e., $\mathfrak{M}_i$ can solve any problem just as easily (in the sense explained above) as $\mathfrak{M}_j$ can. More formally

$$\forall_{i,j} \ \mathfrak{M}_i \overset{poly}{\sim} \mathfrak{M}_j. \tag{2.3}$$

Note that the invariance thesis implies the universality thesis, but not vice versa.

---

[17] See Timpson (2013, Chap. 6) for discussion of a different, more general thesis, that is only indirectly relevant to computational complexity theory. For a discussion of how these theses relate, see Cuffaro (2018b, §11.4).

[18] We could have also expressed the thesis in terms of **P** rather than **BPP**. Although it was thought, for many years, that there are more problems efficiently solvable on a probabilistic Turing machine than on a standard Turing machine, a number of recent results have pointed in the opposite direction (e.g., Agrawal et al., 2004), and it is now generally believed that classical probabilistic computation does not offer any performance advantage over classical deterministic computation (Arora Barak, 2009, Chap. 20). In other words, it is now widely believed that **P** = **BPP**, or that it is just as easy to solve a given problem on a deterministic Turing machine as it is on a probabilistic one. We have nevertheless stated the universality thesis in terms of **BPP** because this will prove convenient when it comes time to discuss the differences between classical and quantum computation. A quantum computer is, from one point of view, just another kind of probabilistic computer (that calculates probabilities differently), and it has the same success criterion as a classical probabilistic computer, i.e., we only demand that a given "solution" be correct with "high enough" probability.

[19] See also: Goldreich (2008, p. 33), who names it differently.

2.3 Physical Perspectives on Computer Science

Neither the universality of Turing efficiency thesis, nor the invariance thesis, is a mathematical theorem. These statements can be true or false, and for a long time they were thought to be true, for none of the reasonable (classical) universal models of computation that had been developed since the time of Turing were found to be more efficient than the Turing model by more than a polynomial factor, and all of them had been shown to be able to simulate one another efficiently (van Emde Boas, 1990). Over the last three decades, however, evidence has been mounting against universality and invariance, primarily as a result of the advent of quantum computing (Aaronson, 2013b, Chaps. 10 and 15).

Quantum mechanics, as we will see in the next section, is an irreducibly probabilistic theory. Thus, a quantum computer is a type of probabilistic machine. Analogously to the way we defined **BPP** as the class of problems solvable (in the probabilistic sense) in polynomial time on a probabilistic Turing machine, we can also define the complexity class **BQP** (for "bounded error quantum polynomial") as the class of problems solvable (in the same sense) in polynomial time on a quantum computer. It is easy for a quantum computer to simulate a probabilistic Turing machine. Thus

$$\mathbf{BPP} \subseteq \mathbf{BQP}, \tag{2.4}$$

i.e., the class of problems efficiently solvable on a quantum computer is at least as large as the class of problems efficiently solvable on a probabilistic Turing machine. What is still unknown, though it would be surprising if it were not the case, is whether

$$\mathbf{BPP} \subsetneq \mathbf{BQP}, \tag{2.5}$$

i.e., whether the class of problems efficiently solvable on a quantum computer (a physically realizable computational architecture) is *strictly larger than* the class of problems efficiently solvable on a probabilistic Turing machine. Note that if Eq. (2.5) is true, then both the universality and invariance theses are false.

There are some authors who view the consequences of the falsification of the universality thesis, in particular, to be profound. Bernstein and Vazirani (1997), for example, take it that computational complexity theory "rests upon" this thesis (p. 1411), and that the advent of quantum computers forces us to "re-examine the foundations" (p. 1412) of the theory. The sense in which complexity theory rests upon universality is expressed by Nielsen and Chuang (2000), who write that the falsity of the thesis implies that complexity theory cannot achieve an "elegant, model independent form" (p. 140). For Hagar (2007), the failure of model-independence shakes, not only the foundations of complexity theory, but certain views in the philosophy of mind that depend on the model-independence of computational kinds (pp. 244–245).

If we actually examine what the universality thesis is saying, however, then it is not really clear, at least not *prima facie*, how it can ground the model-independence of complexity-theoretic concepts. The statement of the thesis is that any efficiently solvable problem is solvable efficiently by a probabilistic Turing machine. A probabilistic Turing machine is a particular model of computation, though. How can a thesis whose very definition makes reference to a particular computational model ground the model-independence of computational complexity theory? In fact, there is a weak notion of model-independence being alluded to here. The point (see Nielsen & Chuang, 2000, p. 140) is that, for any assertion of the form: "problem P is efficiently solvable under computational model $\mathfrak{M}$", the qualification "under computational model $\mathfrak{M}$" can *always* (given the truth of the universality thesis) be substituted with "by a probabilistic Turing machine" without changing the truth value of that sentence. Further, to show that such a sentence is true in general, it suffices to show that it is true for a probabilistic Turing machine. Finally, because "by a probabilistic Turing machine" qualifies every such sentence we can leave it off and still expect to be understood. "Problem P is efficiently solvable by a probabilistic Turing machine" is thus abbreviated to "problem P is efficiently solvable". That we can do this is not insignificant, but this is arguably not a particularly deep sense of model-independence (see Cuffaro, 2018b, §11.6).

Far more interesting in relation to the question of model-independence is the invariance thesis. Unlike the universality thesis, model-independence is built right into the very statement of invariance. For after all, it amounts to the quite direct claim that the details of any particular reasonable machine model, since these can be efficiently simulated by any other reasonable model, are irrelevant for the purposes of characterizing the complexity of a given problem.

Invariance, if taken to be true without qualification, clearly brings with it an absolute notion of model-independence (at least with respect to physically reasonable models). And if taken to be false (as it seems we should, given the existence of quantum computation) it clearly precludes such a notion. Arguably, however, what is (and always was) most valuable about the idea of invariance is not the absolute model-independence that is implied when it is taken to hold without qualification. As we will see later, the term "quantum computer" does not refer to some single particular model of computation but is, rather, an umbrella term for a number of distinct computational models,[20] all of which have been shown to be computationally equivalent in the sense that they are all efficiently simulable by one another.[21] In other words, what we have learned from the study of quantum computing is that, in addition to the existence of an equivalence class of reasonable classical models that satisfy the invariance thesis with respect to one another, there is a further class of reasonable computational models, based on the principles of quantum mechanics, that satisfy the invariance thesis with respect to one another. Thus, there are *two* distinct equivalence classes of physically reasonable computational models from the viewpoint of computational complexity theory. This is a discovery.

---

[20] For instance, the quantum Turing model (Deutsch, 1985), the quantum circuit model (Deutsch, 1989), the cluster-state model (Briegel et al., 2009), the adiabatic model (Farhi et al., 2000), and so on.

[21] See, for instance, Aharonov et al. (2007), Nishimura and Ozawa (2009), Raussendorf and Briegel (2002).

Invariance, thought of as a guiding rule or methodological principle, rather than as an absolute thesis, can be understood as grounding these investigations, and arguably this was the point all along (Cuffaro, 2018b, §11.6). Through the search for equivalence classes, we carve out the structure of the space of computational models, yielding a notion of *relative* model-independence among the machine models comprising a particular equivalence class. And the fact that relative model-independence exists within the space of computational models at all arguably tells us something deep about how computer science connects up with the world, for the differences in computational power between these two reasonable classes of computational model are best understood by considering the *physics* needed to describe them. We discussed the physics of classical computers in Sect. 2.1. In the next section, we turn to the physics of quantum computers.

2.4 Quantum States and Operations

We saw earlier that in classical mechanics, assigning a particular state, $\omega = (\mathbf{q}, \mathbf{p})$, to a system fixes the answer to every conceivable yes-or-no experimental question that we can ask about it in advance, irrespective of whether we actually ask any questions or not. And we saw how to arrange the possible questions we can ask about a system, and their possible answers, into a truth-table-like structure like the one given in Fig. 2. Note that specifying the values of the answers to the questions on the right-hand side of the double-vertical line in Fig. 2 is another way of representing the state of a system like the one depicted in Fig. 3. We saw how one can manipulate such a system in ways that can be represented abstractly as logico-mathematical operations on binary digits, where the binary digits are abstract representations of the properties of the system's (two-level) subsystems.

The subject matter of *quantum* mechanics, just as it is for classical mechanics, is physical systems, where these can be anything from a single particle to the entire physical universe. In other words, quantum mechanics is not just a theory of the small. It is, rather, a new universal language for describing physical systems,[22] and it describes our experience with those systems better and more accurately than classical mechanics describes it. In quantum mechanics, just as in classical mechanics, a system can be, at a given moment in time, in any one of a number of possible physical states. These states are represented by *state vectors* (or "wave functions"). The state vector for a two-dimensional quantum system or "qubit", for instance, is given in general by

$$|\psi\rangle = \alpha|0\rangle + \beta|1\rangle. \tag{2.6}$$

When the complex numbers α and β are both non-zero, a qubit is said to be in a *superposition* of the *basis vectors* $|0\rangle$ and $|1\rangle$. Just as an ordinary vector in the

---

[22] See note 2 above.

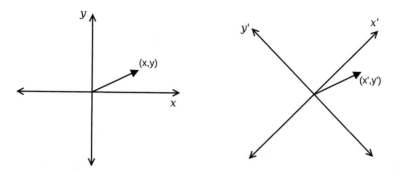

Fig. 7 An ordinary vector in the Cartesian plane. It can be decomposed into x and y coordinates, as on the left, or into the alternative coordinates x' and y' as on the right. Similarly, a state vector, in quantum mechanics, can be decomposed into more than one Hilbert space basis. Two common bases (but not the only two) with which to decompose the state vector for a qubit are the $\{|0\rangle, |1\rangle\}$ basis (known as the computational basis) and the $\{|+\rangle, |-\rangle\}$ basis

Cartesian plane can be decomposed, mathematically, into x and y coordinates, a state vector for a two-dimensional quantum system can be decomposed, mathematically, into two basis vectors for the (Hilbert) space that it is situated in (see Fig. 7). Two common bases (but not the only two) with which to decompose the state vector for a qubit are the $\{|0\rangle, |1\rangle\}$ basis (known as the computational basis) and the $\{|+\rangle, |-\rangle\}$ basis.

Associated with a given basis is a particular experimental question; for instance, "Is the qubit in the state $|0\rangle$ (as opposed to the state $|1\rangle$)?" in the case of the computational basis, and in the case of the $\{|+\rangle, |-\rangle\}$ basis, "Is the qubit in the state $|+\rangle$ (as opposed to the state $|-\rangle$)?" When a system in the state $\alpha|0\rangle + \beta|1\rangle$ is 'asked' whether it is in state $|0\rangle$ as opposed to the state $|1\rangle$—i.e., when it is *measured* in the computational basis—then with probability $|\alpha|^2$ the answer will come back as $|0\rangle$ and with probability $|\beta|^2$ it will come back as $|1\rangle$; i.e.

$$M\left(\alpha|0\rangle + \beta|1\rangle\right) \xrightarrow{\text{Pr} = |\alpha|^2} |0\rangle \tag{2.7}$$

$$M\left(\alpha|0\rangle + \beta|1\rangle\right) \xrightarrow{\text{Pr} = |\beta|^2} |1\rangle \tag{2.8}$$

where M is a *measurement operator* used to represent a computational basis measurement. Note that there is also a basis corresponding to the question: "Is the qubit in the state $|\psi\rangle$ (as opposed to the state $|\phi\rangle$)?" Why not just ask this question? One reason is that measuring a quantum-mechanical system involves using a concrete physical device, and it turns out that it is technologically more feasible to construct measurement devices corresponding to some experimental questions than to others (Janas et al., 2022, §6.5).

In terms of their dynamics, states of quantum systems evolve linearly

$$U\big(\alpha|0\rangle + \beta|1\rangle\big) = \alpha U|0\rangle + \beta U|1\rangle \tag{2.9}$$

and unitarily:

$$UU^\dagger|\psi\rangle = U^\dagger U|\psi\rangle = |\psi\rangle \tag{2.10}$$

over time, where U is a *unitary operator* and $U^\dagger$ is its *adjoint*.

Classical and quantum mechanics differ in the way that answers to experimental questions are determined. In contrast to classical mechanics, specifying a quantum system's state at a given moment in time does not fix in advance the answers to the experimental questions that can be asked of that system, or as Bub and Pitowsky (2010) put it, the quantum state is not a *truthmaker* in relation to those questions. It fails to be a truthmaker in two senses: First, a given state specification yields, in general, only the probability that the answer to a given experimental question will take on one or another value when asked. This in itself is not as much of a departure from classical mechanics as one might think, however, because conditional upon the selection of an experimental question, one can, in quantum mechanics, describe the observed probabilities as stemming from a prior classical probability distribution over the dynamical properties of the system that *is* determined in advance by the quantum state.[23]

This brings us to the second, more important, sense in which the state of a quantum-mechanical system fails to be a truthmaker in relation to questions that can be asked about that system's dynamical properties: The probability distributions associated with the answers to individual experimental questions cannot be embedded into a global prior probability distribution over all of the answers, as they can be for a classically describable system (see Fig. 2). In quantum mechanics, one can only say that *conditional* upon our inquiring about the observable A, there will be a particular probability for the answer to that question to take on a particular value. Thus, in quantum mechanics, unlike in classical mechanics, we actually have to ask the system a question in order to get an answer from it.[24]

When quantum-mechanical systems are combined, they can sometimes become *entangled* with one another. Consider, by way of contrast, the simple illustration in Fig. 3 of a classically describable system composed of many subsystems. This system is in a *product state*, which means that the state of the overall system can be expressed as a product of the states of each individual subsystem, i.e., $0_A 1_B 1_C 0_D$, where the individual states of the subsystems are 0_A, 1_B, 1_C, and 0_D. In classical mechanics, there is no other way to describe the combined state of a number of subsystems. Even

---

[23] This is analogous to the way we interpret probabilities in classical *statistical* mechanics.

[24] See also Janas et al. (2022, Chaps. 1 and 6), who call this (ironically) the "small" measurement problem in contrast to the "big" (this label is also ironic) problem described in the previous paragraph. These labels are originally due to Bub and Pitowsky (2010), who used them ironically too.

when these subsystems are correlated with one another (as the state of one side of a coin is correlated with the state of the other side, for instance), as long as we include facts about those correlations in our description of the overall system, then its overall state can be factored into the individual states of its subsystems (Bell, 2004 [1981]). This is true even if the subsystems happen to be far apart, for instance, if I print out two copies of a document and mail each of them to distant parts of the globe.

In quantum mechanics, in contrast, a system composed of multiple subsystems can sometimes be in a state like the following:

$$\frac{1}{\sqrt{2}}\big(|0\rangle_A|1\rangle_B - |1\rangle_A|0\rangle_B\big). \tag{2.11}$$

This state, describing a system composed of two subsystems A and B, is *not* a product of the individual states of its subsystems, for there is no way to factorize it into a state of the form: $|\psi\rangle_A|\phi\rangle_B$ (the reader is invited to try). It is sometimes said that in an entangled quantum system the whole is somehow prior to the parts (Howard, 1989), or that the systems are *nonlocally correlated* (Maudlin, 2011), for if Alice measures her qubit in the computational basis and receives the result $|0\rangle_A$, then a measurement by Bob on his qubit (which in principle he may have taken to the opposite side of the universe) is thereby instantaneously determined to be $|1\rangle_B$, although there is no way for Bob to take advantage of this instantaneous determination for the purpose of predicting the outcome of a measurement on his subsystem, short of waiting for Alice to send him a classical signal (a text message or a phone call, for instance).[25]

Note that the correlation between the results of computational (i.e., $\{|0\rangle, |1\rangle\}$) basis measurements on an entangled system in this state do not, in themselves, represent as much of a departure as one might think from what one can describe using classical language. For with respect to this specific correlation, Bell (2004 [1964]) showed that it can be seen as arising from a classically describable system that has been prepared in a particular way. It is only when we consider other experiments on such a system that we see a departure from the possibilities inherent in classical description. In particular, because

$$|+\rangle = \frac{|0\rangle + |1\rangle}{\sqrt{2}}, \qquad |-\rangle = \frac{|0\rangle - |1\rangle}{\sqrt{2}}, \tag{2.12}$$

Equation (2.11) can be rewritten, in the $\{|+\rangle, |-\rangle\}$ basis, as

$$\frac{1}{\sqrt{2}}\big(|+\rangle_A|-\rangle_B - |-\rangle_A|+\rangle_B\big). \tag{2.13}$$

Indeed, we obtain the *same form* for the state of the system *regardless* of which basis we choose to express it in:

---

[25] This restriction on the use that can be made of nonlocal correlations in quantum mechanics is called "no signaling". For discussion, see Bub (2016), Cuffaro (2020).

$$\frac{1}{\sqrt{2}}\left(|b_1\rangle_A|b_2\rangle_B - |b_2\rangle_A|b_1\rangle_B\right). \tag{2.14}$$

Equations (2.11), (2.13), and (2.14) all yield the same probabilities for the results of experiments.

If we collect a number of pairs of subsystems prepared in this entangled state, and then ask of each pair: "Is this pair in the state $|0\rangle|1\rangle$ (as opposed to $|1\rangle|0\rangle$)?," then we will find that the answers "yes" and "no" will be obtained with equal probability. Conditional upon this question, we can imagine these answers as arising from a prior classical probability distribution over the properties of the subsystem pairs, half of which we imagine to be in the state $|0\rangle|1\rangle$, and half of which we imagine to be in the state $|1\rangle|0\rangle$. This prior probability distribution is *incompatible*, however, with the prior classical probability distribution that we might imagine to be responsible for the answers yielded from repeatedly asking the question: "Is this pair in the state $|+\rangle|-\rangle$ (as opposed to $|-\rangle|+\rangle$)?" In the context of the latter question, we can imagine the answers as arising from a prior classical probability distribution over the properties of the subsystem pairs, half of which we imagine to be in the state $|+\rangle|-\rangle$, and half of which we imagine to be in the state $|-\rangle|+\rangle$. Asking a question of an imagined ensemble of systems, half of which are in the state $|0\rangle|1\rangle$, and half of which are in the state $|1\rangle|0\rangle$, will yield different statistics, however, than the ones that would be yielded by asking the same question of an imagined ensemble of systems, half of which are in the state $|+\rangle|-\rangle$, and half of which are in the state $|-\rangle|+\rangle$. These probability distributions are, in this sense, incompatible. And yet *all* of these statistics are predicted by one and the same quantum state, the one expressed in the unitarily equivalent forms given in Eqs. (2.11–2.14). In quantum mechanics, unlike in classical mechanics, the questions we ask or do not ask actually matter for the ways that we can conceive of the underlying properties of a physical system, even when the state of that system is fully specified (Janas et al., 2022, §6.5).

3 Quantum Computation and Parallel Universes

The are problems that are hard for a classical computer, but that a quantum computer can solve easily (where "easy" and "hard" are meant in the computational sense defined above in Sect. 2.2).[26] The most famous example of such a problem is that of factoring large integers into primes. The best-known classical algorithm for factoring is the number field sieve (Lenstra et al., 1990), which takes on the order of $2^{(\log N)^{1/3}}$ steps to factor a given integer N.[27] No one has yet proven that this is the best that a

---

[26] There are also problems for which a quantum computer, despite being unable to solve them easily, can nevertheless solve them significantly *more* easily than a classical computer can. An example is the problem to search an unstructured database, for which a quantum ("Grover's") algorithm can reduce the number of steps required by a quadratic factor over any known classical algorithm. See: Bennett et al. (1997), Grover (1996), and for further discussion see Cuffaro (2018b, p. 269).

[27] For the meaning of 'on the order of' see fn. 10.

classical computer can do. All the same, encryption algorithms such as RSA (named for the paper by Rivest, Shamir, and Adleman in which it was introduced), that are today widely used for secure transactions on the internet, rely on the assumption that it is; i.e., that one cannot factor in polynomial time. In 1994, Peter Shor discovered a *quantum* algorithm—now known as Shor's algorithm—that can factor integers in on the order of $\log N$ steps for a given integer N—an exponential "speedup" over the number field sieve.

What explains this phenomenon? What is it, exactly, that allows quantum computers to compute certain problems more efficiently than classical computers can? Surprising as it may seem, there is still no consensus regarding the answer, neither among the researchers working in the field, nor among the philosophers commenting on this work. By contrast, in the popular literature on the topic, one answer has tended to dominate all the others. This is the idea that quantum computers are actually *parallel* devices that perform exponentially many classical computations simultaneously, with each computation taking place in a parallel physical universe (or "world"), different in certain ways from our own but just as real as it is.

Perhaps the strongest proponent of this view, which we will call the *many-worlds explanation* of the power of quantum computers, is one of the fathers of the field, David Deutsch, discoverer of the first quantum algorithm and of the universal quantum Turing machine (Deutsch, 1985). The many-worlds explanation of quantum computing is, for Deutsch, not just a speculative conjecture. For Deutsch, many-worlds are the only plausible explanation for how quantum computers work. As he puts it in his 1997 book, *The Fabric of Reality*: "[t]o those who still cling to a single-universe world-view, I issue this challenge: Explain how Shor's algorithm works" (1997, p. 217).

Before we discuss what those who, like Deutsch, defend this view of quantum computation are really saying, let us review briefly what the view amounts to in more general terms in the context of quantum mechanics, for what the many-worlds explanation of quantum computing amounts to is an application of this more general philosophical interpretation of quantum mechanics to the case of quantum-mechanical computers.[28] Thus far, our examples of quantum-mechanical systems have been of qubits, either considered singly or in combination with other qubits. But quantum mechanics is not restricted, of course, to describing qubits. Quantum mechanics is, rather, a universal language for describing any physical system, be it a single particle or (in principle) the entire physical universe. From the point of view of quantum mechanics, you and I, the laptop I am writing this chapter with, the table

---

[28] The interpretation of quantum mechanics that we will be discussing in this section is one of a number of related interpretations of quantum mechanics that are collectively referred to as the "Everett interpretation". These include, but are not limited to Hugh Everett III's original formulation (Everett III, 1956), the "Berlin Everettianism" of Lehner (1997), Lev Vaidman's version of Everett (Vaidman, 1998), so-called "many minds" variants (Albert & Loewer, 1988), and finally the "many-worlds" variants that are the direct inspiration for the many-worlds explanation of quantum computing. Belonging to the last group are DeWitt's (1973 [1971]) view, as well as the "Oxford Everett" interpretation (Deutsch, 1997; Saunders, 1995; Wallace, 2003, 2012) with which we will be mostly concerned here.

it is sitting on, and everything else in the world is a quantum-mechanical system. And just as a qubit can be in a superposition of its basis states (see Eq. (2.6) and Fig. 7), so too can *any* physical system. As Schrödinger famously remarked, according to quantum mechanics there are even ways of preparing a cat in a superposition of "dead" and "alive" states (Schrödinger, 1935a).

When we actually come to pose a question to the system; i.e., to measure the qubit, the cat, or what have you, we never actually find that it is in a superposition state. Quantum mechanics predicts that the cat will be found to be either dead or alive, the qubit to be either $|0\rangle$ or $|1\rangle$, and so on, with a certain probability. Thus, unlike the physical state of a system as described by classical mechanics, we cannot simply read off from the quantum state description of a system what the outcome of an experiment on that system will be (except in special cases). This raises the question of how to interpret the quantum state. Are we to take superpositions literally, despite the fact that we never observe them? What would it mean if we did? The proponent of the many-worlds interpretation of quantum mechanics answers yes to the first question. As for the second question, the view of the many-worlds advocate is that each branch of a superposition represents a distinct, classically describable (from the point of view of that branch), physical universe.

Just as it is with a cat, or with any physical system, so it is with a quantum computer. What is different, arguably, about a quantum computer (according to the defender of the many-worlds explanation of quantum computing) is that in a quantum computer the resources made available by these multiple physical universes are harnessed to perform computations. This is a striking claim. But it can be said, in favor of the many-worlds explanation, that some quantum algorithms certainly do give us this impression. The following evolution is representative of a typical step in many quantum algorithms (note that normalization factors have been omitted for simplicity):

$$\sum_{x=0}^{2^n-1} |x\rangle^n |0\rangle \rightarrow \sum_{x=0}^{2^n-1} |x\rangle^n |f(x)\rangle, \tag{3.1}$$

What is being depicted here is the state of a quantum computer, composed of $n + 1$ qubits, where the first n qubits are called the *input qubits* of the computer, and the last qubit is called the computer's *output qubit*. The computer begins in the state described on the left-hand side of the equation, and then transitions to the state described on the right-hand side.

This notation is very compact, so let us unpack it. Considering, first, the left-hand side: If we take the n input qubits as comprising, together, one subsystem of the system, and the output qubit as comprising another subsystem, then we can say that the input and output subsystems of the computer begin in a product state. In other words, we can write the left-hand side of the equation in product form as

$$\text{LHS} = \left(\sum_{x=0}^{2^n-1} |x\rangle^n \right) |0\rangle. \tag{3.2}$$

Note that, in Eq. (3.2), the n input qubits are together in a superposition of all of the possible computational basis states for a system of n qubits. This is obscured somewhat, both by the summation notation as well as by the shorthand, $|x\rangle^n$, being used to represent each superposition term. To clarify this, recall that for a two-dimensional system, i.e., a single qubit, the computational basis states are, as we pointed out earlier (see Eq. (2.6) and Fig. 7), the states $|0\rangle$ and $|1\rangle$. As for a system composed of *two* qubits, it has four computational basis states, namely, $|0\rangle|0\rangle$, $|0\rangle|1\rangle$, $|1\rangle|0\rangle$, and $|1\rangle|1\rangle$. In general, for a system of n qubits there are 2^n basis states.

$$|0\rangle|0\rangle \ldots |0\rangle|0\rangle,$$
$$|0\rangle|0\rangle \ldots |0\rangle|1\rangle,$$
$$|0\rangle|0\rangle \ldots |1\rangle|0\rangle,$$
$$\ldots$$
$$|1\rangle|1\rangle \ldots |1\rangle|1\rangle, \tag{3.3}$$

each of which can be thought of as representing a binary number. For short, we can represent these in decimal as $\{|\mathbf{0}\rangle, |\mathbf{1}\rangle, |\mathbf{2}\rangle, |\mathbf{3}\rangle, \ldots |\mathbf{2^{n-1}}\rangle\}$. A superposition of all of these basis states is then given by

$$|\mathbf{0}\rangle + |\mathbf{1}\rangle + |\mathbf{2}\rangle + |\mathbf{3}\rangle + \ldots + |\mathbf{2^{n-1}}\rangle = \sum_{x=0}^{2^n-1} |x\rangle^n, \tag{3.4}$$

exactly as we see in Eq. (3.2).

On the right-hand side of Eq. (3.1), the $n + 1$ qubits of the quantum computer are no longer in a product state. The state of the computer has now transitioned to an *entangled* state: a superposition in which the state obtained for the output qubit, conditional upon measuring it, is correlated with the state that will be obtained for the input qubits, conditional upon measuring them. For instance, if we measure the input qubits and get the state $|\mathbf{1}\rangle$, then the state of the output qubit will be $|f(\mathbf{1})\rangle$, for the given function f, and similarly for every other value of x. What is the function f? That will depend on the particular problem that the algorithm is designed to solve. More importantly, from the point of view of our current discussion, notice that the right-hand side of the equation encodes exponentially many evaluations of that function in the state of the computer! We can, or so it seems, appeal to the results of all of these evaluations to help us with whatever problem we are trying to solve.

In reality, things are not so easy, for when we actually come to read off the result of the computation, we will only ever find the computer's output qubit to be in *one* of its exponentially many superposition terms; i.e., in some state $|f(a)\rangle$ for some one particular a in the domain of f. If we are tempted to read the presence of these exponentially many function evaluations in the description of the state of the computer literally, then the fact that only one of them can ever be accessed in a particular run should give us some pause (cf. Mermin, 2007, p. 38). That said, in any given run, there will be an, in general, non-zero probability of obtaining any one of them, and this, perhaps, speaks in favor of viewing them all as somehow literally there. As for the goal (which we should not lose sight of) of actually solving the problem at hand, what is just as important as achieving the form of the state of the computer on the right-hand side of Eq. (3.1) is the *next* step in a computation like this one, which requires that we manipulate the system cleverly enough so that one of the desired solutions is found with higher probability than the other, undesirable, solutions (Pitowsky, 2002, §3).

The many-worlds explanation of a quantum computational process enjoins us to take an evolution like the one given in Eq. (3.1) ontologically seriously. It affirms, in other words, that things are indeed as they seem: The computer most definitely *is* performing many simultaneous function evaluations in parallel when it is in a state like the one on the right-hand side of Eq. (3.1). Moreover, the many-worlds explanation directly answers the question of *where* this parallel computation occurs, namely, in distinct physical universes. For this reason, it is also, arguably, one of the more intuitive of the physical explanations of quantum speedup. And it is certainly thought-provoking. It is no wonder, then, that it is the one most often mentioned in the popular literature on quantum computation. And it has advocates in the more serious literature as well.[29]

So far we have highlighted one potential problem for the many-worlds explanation: Despite the appearance of parallel processing in an equation like Eq. (3.1), only one of these function evaluations is ever accessible on a given run of the computer. This is not to say that no story can be told from the point of view of the many-worlds explanation about why this is so, but at the very least this is enough to cast doubt on the claim that an evolution like the one given in Eq. (3.1) constitutes, all by itself, evidence for the many-worlds explanation of the power of quantum computers. We will return to this issue later. But for now, we need to consider a somewhat deeper problem faced by the advocate of the many-worlds explanation. Recall that decomposing a qubit in the state $|\psi\rangle$ into the computational basis states $|0\rangle$ and $|1\rangle$ (see Eq. (2.6) and Fig. 7) is only one way to express this state. The same state can also be expressed in other bases besides the computational basis. This means that a system that is in the state

$$\frac{1}{\sqrt{2}}|0\rangle + \frac{1}{\sqrt{2}}|1\rangle, \tag{3.5}$$

from the point of view of the computational basis, is simply in the state:

$$|+\rangle \tag{3.6}$$

---

[29] In addition to Deutsch's 1997 book, see Deutsch (2010), and see also Vaidman (2018 [2002], §7) and Wallace (2012, Chap. 10). The strongest and most in-depth defence of the many-worlds explanation of quantum computing that I am aware of is the one given by Hewitt-Horsman (2009).

from the point of view of the $\{|+\rangle, |-\rangle\}$ basis (see Eq. (2.12)). But now it is no longer clear just how many universes we should take this qubit to exist in. If we decompose its state in the computational basis, then it would seem that it exists in two worlds, and if we decompose it into the $\{|+\rangle, |-\rangle\}$ basis then it would seem that it only exists in one. Which is it? If we decide that one of these decompositions is to be preferred to the other, then the challenge for the advocate of the many-worlds explanation is to give a compelling reason why, for *prima facie* there does not appear to be any reason for preferring one of these decompositions over the other.

This is known as the *preferred basis problem*, and it is a problem for the many-worlds view more generally, i.e., not just in the context of quantum computing. In this more general context, advocates of the many-worlds picture (see, for instance, Wallace, 2003) attempt to solve the preferred basis problem by appealing to the dynamical process of *decoherence*. In quantum mechanics, as we saw (see Eq. (2.6) and Fig. 7), the state of a system at a particular moment in time is described by a state vector that evolves linearly and unitarily. However, the evolution thereby described is the evolution of a system that is completely isolated from its external environment, which is an idealization; in reality, it is actually never possible to completely isolate a system from its environment,[30] unless, perhaps, we take our system of interest to be the universe in its entirety.[31] But barring the universe as a whole, when a system interacts with an external environment—for example, with our measurement apparatus as we ask the system an experimental question—then the terms in the superposition describing its state begin to decohere (Zurek, 2003 [1991]) and come to achieve a kind of independence from one another (although they never decohere completely). On the many-worlds picture, we are to think of such (approximately) decoherent terms as existing in independently evolving physical universes. And in each of these independently evolving physical universes, there is, in addition, a different version of ourselves, all of whom receive a different answer to the experimental question that was asked. And with every additional question, the universe as we know it is branching, spawning exponentially more and more versions of the system, and more and more versions of ourselves along with it, and so on and on and on.

Fundamentally, however, decoherence is an approximate phenomenon, and some small amount of residual interference between worlds always remains despite it. Nevertheless, decoherence tells us that, when the environment and system happen to be correlated in a particular way, then *a particular basis will emerge with respect to which* we can effectively describe the superposition terms expressed in the state of the system as evolving independently of one another. As Wallace (2003, p. 90) puts it: "the basic idea is that dynamical processes cause a preferred basis to emerge rather than having to be specified a priori". In this way, superposition terms that for all practical purposes evolve stably and independently over time with respect to the decoherence basis can be identified with different copies of measurement pointers,

---

[30] At the very least, the gravitational effects of other distant systems will not be able to be neglected.

[31] Some philosophers have questioned whether we should think of even the universe as a whole as a closed system (see, for instance, Cuffaro & Hartmann, 2021).

cats, experimenters, and whatever else is theoretically useful for us to include in our ontology: "the existence of a pattern as a real thing depends on the usefulness—in particular, the explanatory power and predictive reliability—of theories which admit that pattern in their ontology" (Wallace, 2003, p. 93). Whatever else one may think of the many-worlds view, decoherence, at least, does provide a principled way to identify worlds in the wave-function.

For the advocate of the many-worlds explanation of *quantum computation*, however, there is still a problem. Appealing to decoherence may solve the preferred basis problem for the purposes of describing the world of our everyday experience,[32] but the inner workings of a quantum computer are not part of that everyday experience. The problem is not just that qubits are too small to see. The problem is that the superpositions characteristic of quantum algorithms are *coherent* superpositions (Nielsen Chuang, 2000, p. 278; see also Aaronson, 2013c; Cuffaro, 2012). Thus, the terms in the wave-function of a quantum computer do not seem to meet the criterion for world-identification advocated for by the many-worlds view.

Now, to be fair, a similar thing can be said with regard to our everyday experience. So-called decoherent superpositions are (as I mentioned) not *really* decoherent, after all. But they are decoherent enough, and for long enough, that it is useful (or so says the many-worlds advocate) to think of the terms in such a superposition as independent, and the worlds that they describe as ontologically real. Likewise, although the superposition on the right-hand side of Eq. (3.1) is actually a coherent superposition, it may nevertheless be useful to think of the terms in that superposition as independent, at least for the short time that the quantum computer is in that state (Hewitt-Horsman, 2009, p. 876). The problem, however, is that it is the very fact that they are coherent that allows us to "cleverly" extract desirable solutions from these superpositions with higher probability than undesirable solutions (Bub, 2010; Duwell, 2007).

Even if we grant that there is some heuristic value in describing a quantum computer in a superposition state as evaluating functions in exponentially many parallel worlds (I do not doubt that this was of some heuristic use to Deutsch, for instance, even if I question whether it is *necessary* to think of such superpositions in this way), it does not follow that this is enough to licence granting ontological status to those worlds. Wallace (2003, p. 93) mentions (as we saw) explanatory power and predictive reliability, for instance, and discusses the way that these and other ideas are applied in contemporary physics to support the many-worlds view outside of the context of quantum computing. It is not at all clear that these criteria are met in the context of quantum computing, however, and even Wallace admits that they, for the most part, are not: "There is no particular reason to assume that *all* or even *most* interesting quantum algorithms operate by any sort of 'quantum parallelism'" (Wallace, 2010, p. 70, n. 17). Wallace goes on: "Shor's algorithm, at least, does seem to operate in

---

[32] The preferred basis problem is not the only challenge that needs to be met by an advocate of the Everett interpretation of quantum mechanics. Another issue that has been much discussed in recent literature is the problem of how to account for probabilities on the Everettian view. For more on this issue, see Adlam (2014), Dawid and Thébault (2015), Greaves and Myrvold (2010), Vaidman (1998, 2012), and Wallace (2007).

this way" (ibid.), but he does not describe how. Yet there are very plausible accounts of how Shor's algorithm works that do not appeal to massive parallelism at all (see Bub, 2010). Far from it, on Jeffrey Bub's account of Shor's algorithm, the quantum algorithm is more efficient than known classical algorithms because it performs *fewer*, not more, computations (see also Bub, 2008).

The final reason that I will mention for being skeptical of the many-worlds explanation of quantum computing is that it only really seems to be useful in the context of one particular model of quantum computing. This is the so-called *quantum circuit model*, the model for which many of the first quantum algorithms were designed (Deutsch, 1989). This model is useful for abstract theoretical purposes, as well as for pedagogical purposes, as it borrows many familiar ideas from the classical circuit model of computation (see Fig. 4). In the quantum circuit model, similarly to the classical circuit model, logical circuits are constructed out of various "quantum logic gates". These instantiate unitary transformations of one or more qubits that are prepared beforehand in computational basis states (typically qubits begin a computation in the state $|0\rangle$). The unitary gates transform the qubits' states into various superpositions of computational basis states, and at the end of the computation a measurement is performed, again in the computational basis, on (some of) the qubits and the results are read out.

Figure 8a depicts a number of important one- and two-qubit quantum gates. The X gate implements a qubit-flip operation; i.e., it takes $|0\rangle \rightarrow |1\rangle$ and vice versa. The Y gate takes $|0\rangle \rightarrow i|1\rangle$ and $|1\rangle \rightarrow -i|0\rangle$. The Z gate takes $|0\rangle \rightarrow |0\rangle$ and $|1\rangle \rightarrow -|1\rangle$. The R gate takes $|0\rangle \rightarrow |0\rangle$ and $|1\rangle \rightarrow i|1\rangle$. The H (or Hadamard) gate takes $|0\rangle \rightarrow (|0\rangle + |1\rangle)/\sqrt{2}$ and $|1\rangle \rightarrow (|0\rangle - |1\rangle)/\sqrt{2}$. The S gate takes $|0\rangle \rightarrow |0\rangle$ and $|1\rangle \rightarrow e^{i\pi/4}|1\rangle$. At the extreme right is the two-qubit CNOT (or controlled-not) gate. It leaves the topmost qubit unchanged. The bottom qubit is then assigned the output of taking the exclusive-or of both, i.e., this gate takes $|0\rangle|0\rangle \rightarrow |0\rangle|0\rangle, |0\rangle|1\rangle \rightarrow |0\rangle|1\rangle$, $|1\rangle|0\rangle \rightarrow |1\rangle|1\rangle$, and $|1\rangle|1\rangle \rightarrow |1\rangle|0\rangle$. The X, Y, Z, R, H, and CNOT gates together form the *Clifford group* of gates, which we will have more to say about later. If we add the S gate to the Clifford group, they together form a *universal set* of gates, i.e., any quantum circuit implementing any series of unitary transformations can be simulated to arbitrary accuracy using combinations of these seven gates.

Figure 8b depicts a quantum circuit diagram for Deutsch's Algorithm, which determines whether a given function f on one bit is constant ($f(0) = f(1)$) or balanced ($f(0) \neq f(1)$): Two qubits are prepared in the product state $|0\rangle|0\rangle$ and are each sent through an X-gate and a Hadamard gate, after which they are together input to the two-qubit entangling unitary gate U_f. The first qubit is then sent through a further Hadamard gate and finally measured to yield the answer (see Deutsch, 1989).

Just as there are various models of universal classical computation (for instance, the various versions of the Turing machine, as well as the von Neumann architecture, and so on, that I mentioned above), there are various models of universal quantum computation. One computational model that presents a particularly difficult problem for those who would advocate for the many-worlds explanation is the *cluster-state* model of quantum computing, also known as *measurement-based* and *one-way* quantum computing (Raussendorf & Briegel, 2002; Raussendorf et al., 2003; Nielsen,

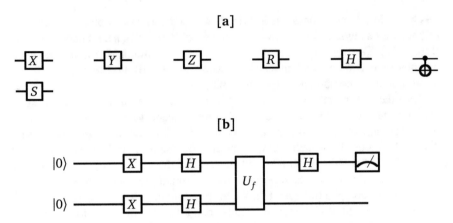

Fig. 8 **a** A number of important one- and two-qubit quantum gates. The X, Y, Z, R, H, and CNOT gates together form the *Clifford group*. If we add the S gate to the Clifford group, they together form a *universal set*. **b** A quantum circuit diagram for Deutsch's Algorithm, which determines whether a given function f on one bit is constant ($f(0) = f(1)$) or balanced ($f(0) \neq f(1)$)

2006).[33] In the cluster-state model, the computer begins, not in a product state but in a highly entangled state, and measurements are performed not only at the end of the computation but throughout it. These measurements are *adaptive*, in the sense that each measurement is performed in a *different basis*, which depends on the random outcomes of whatever previous measurements have been performed.

Why is this a problem for the advocate of the many-worlds explanation of quantum computing? Because the fact that measurements are adaptive means, on the one hand, that there is no principled way to select a preferred basis *a priori* in the context of a given computation (Cuffaro, 2012, §4). Whichever basis we choose, few qubits will actually be measured in that basis in the course of the computation. On the other hand, there is no sense in which we can say that a preferred basis "emerges" from the computational process. There is, therefore, no way to identify the worlds that the computation as a whole is supposed to be happening in.

As I alluded to above, both the cluster-state model and the circuit model are universal models of quantum computing. Thus, anything that the circuit model can do can be done (and, as it happens, just as efficiently) in the cluster-state model and vice versa. Perhaps, then, it could be argued that understanding the "true nature" of algorithms in the cluster-state model requires that we first translate them into the language of the circuit model, though I can think of no reason why one should think so other than a desire to hold on to the many-worlds idea come what may. The proper response to anyone who would put forward such an argument is that a large part of what motivates those who adhere to the many-worlds explanation of quantum computing in the first place is that it is useful for algorithm analysis and design to

---

[33] For introductions to cluster-state quantum computing aimed at philosophers, see Cuffaro (2012, §4) and Duwell (2018, §4.5).

believe that a quantum computer is carrying out its computations in parallel worlds. This does not seem to be so for the cluster-state model. On the contrary, dogmatically holding on to the view that many-worlds are, at root, physically responsible for the speedup evinced in the cluster-state model is at best useless, for it is of no help in designing algorithms for the cluster-state model. At worst, dogmatically holding on to the many-worlds idea could prove positively detrimental if it prevents us from exploiting the power of the cluster-state model or discovering other quantum computational models in the future.

4 Quantum Computation and Entanglement

The many-worlds explanation of the power of quantum computers is not the only explanation that has been proposed by philosophers. I have already mentioned the upshot of Bub's analysis of Shor's algorithm in the previous section. Bub's view, more generally (2006, 2010), is that the key to understanding the power of quantum computers lies in understanding the way that they exploit the novel logico-probabilistic structure of quantum mechanics (Pitowsky, 1989). We saw above that in quantum mechanics, the (classical) probability distribution over the answers to a particular experimental question—the one that we infer, conditional upon our asking that question—cannot be embedded into a global prior probability distribution over all of the answers to all of the questions we might want to ask. But although these individual probability distributions do not logically fit together as they do for a classical system, there are logical relations between them nonetheless, that are compactly described by the quantum state.

For instance, as a general rule (which admits exceptions) one cannot use a quantum system to compute a logical disjunction by computing the individual values of each disjunct. This is simply because in the general case both disjuncts will not be globally defined. However, quantum mechanics' logical structure provides other ways to compute a logical disjunction, and these other ways (which are not available to a classical computer) are exploited by a quantum computer to compute certain problems more efficiently than a classical computer can compute them.

Another view is Duwell's (2018, 2021), who in contrast, agrees with the proponent of the many-worlds explanation in identifying quantum parallelism as at least part of the source of the power of quantum computers. Duwell, however, resists the temptation to make the a metaphysical inference from parallelism to many computational worlds. For Duwell, the source of the power of a quantum computer lies in the way that it can efficiently correlate multiple values of a function and use these correlations to efficiently extract global information about the function. The disagreement over whether a quantum computer performs more, or fewer, computations than a classical computer is one that Duwell views as arising from conflicting intuitions about how to appropriately describe the quantum systems that perform computational tasks.

In these and other candidate explanations of the power of quantum computers that one encounters in the philosophical literature on the topic, the fact that quantum-mechanical systems exhibit *entanglement* (see Sect. 2.4) invariably plays an important role. For Bub entanglement is absolutely central, as entanglement is a (direct) manifestation of the fact that the logical structures of quantum and classical mechanics differ. As for Duwell, he takes his quantum parallelism thesis to be completely compatible with the idea that entanglement plays a central role in quantum computing, even if his explanation emphasizes the correlations between the values of the function being evaluated by a system rather than the logical structure of its underlying state space (Duwell, 2018, p. 101). The many-worlds advocate, as well, views quantum entanglement to be indispensable in the analysis of the power of quantum computing (Hewitt-Horsman, 2009, p. 889), even if for the many-worlds advocate it does not, by itself, suffice as a philosophical explanation for it.

The debate over the interpretation of the phenomenon of quantum entanglement has historically been one of the central debates in the controversy over quantum theory's conceptual foundations. First emphasized by Albert Einstein, Boris Podolsky, and Nathan Rosen in their 1935 criticism of the orthodox interpretation of quantum mechanics, Erwin Schrödinger called it "*the* characteristic trait of quantum mechanics, the one that enforces its entire departure from classical lines of thought" (Schrödinger, 1935b, p. 555, emphasis in the original). It was only with the work of Bell (2004 [1964], 2004 [1966]), however, that its significance for our understanding of the break that quantum mechanics makes with classical physics was first made fully clear.

Consider Louise and Robert, two friends who live in the 15th and 11th arrondissements, on the left bank and the right bank of the Seine, respectively, in the city of Paris. Every morning, Louise and Robert each receive a letter in their respective mailboxes that consists of a single sheet of paper inscribed with either a large bass clef:

𝄢

or a large treble clef

After awhile, Louise determines that it is equally likely, on any given day, that she will receive a bass clef as it is that she will receive a treble clef. After awhile, Robert determines the same. But when they compare their notes, they find that whenever Robert receives a bass clef, so does Louise. Similarly, every time he receives a treble clef, she does too. In other words, there is a one-in-two chance, on any given day, that they both receive a bass clef, and a one-in-two chance that they both receive a treble clef. No other combinations are ever observed. Thus, their outcomes are *correlated*, such that the probability distribution over the possible combinations of outcomes is given by[34]:

$$\frac{1}{2}[\text{𝄢}]_L[\text{𝄢}]_R + \frac{1}{2}[\text{𝄞}]_L[\text{𝄞}]_R.$$

(4.1)

What explains this correlation? Well in this case it turns out that Louise and Robert both play in a jazz ensemble. The band's leader, Carsten, lives in the center of the city (on the Île Saint-Louis). Every afternoon he flips a fair coin. Depending on whether the coin lands heads or tails, he either writes a large bass clef or a large treble clef on a sheet of paper, photocopies it, and sends one copy each to Robert and Louise. If Robert receives a treble clef from Carsten, then he will know to take his tenor horn to the jazz club that night. Otherwise, he will take his trombone. As for Linda, if she receives a treble clef, she will know to bring her soprano clarinet. Otherwise, she will bring her bassoon.

The result of Carsten's coin flip is called the *common cause* of Louise's and Robert's outcomes, and the story we tell about how Carsten's coin flip determines Louise's and Robert's outcomes is called a *causal model* for the correlations that they see (see Pearl, 2009). If Louise and Robert do not know how Carsten is determining what to send them, then they will wonder about the *hidden-variable* that explains their correlation. In this case, the result of Carsten's coin flip is actually a *local* hidden-variable, since the process by which the outcome of the coin flip determines what is written on the letters is confined to the small localized region in the vicinity of the desk in Carsten's central office. He flips the coin, takes note of the outcome (the local hidden-variable), and writes the corresponding symbol.

Instead of simply flipping a coin while sitting at his desk, we can imagine a more complicated, spatially distributed, process by which Carsten determines what to write. For instance, Carsten might begin by flipping his coin, and then, corresponding to heads or tails he might telephone Tilde or Bjarne, who live in the northern and southern suburbs of the city, respectively, and ask whichever of them he calls to roll a six-sided die at exactly seven-o'clock in the evening, and to afterward call him back at either 7:13 PM, if it is Bjarne, or 7:18 PM, if it is Tilde, to tell him the outcome. Then, if the result of the die roll is one, three, or six, Carsten will write a treble clef on the sheet of paper before photocopying it and sending copies to Robert and Louise, while if it is two, four, or five, he will write a bass clef. All of these actions together constitute, just like the coin flip in our simpler scenario, a locally causal model to explain Louise's and Robert's correlation. Why do we call it local even though it is spatially distributed? Because the physical processes (the coin flips, die rolls, and

---

[34] If the outcomes were completely uncorrelated, the probability distribution would be

$$\frac{1}{4}[\text{𝄢}]_L[\text{𝄢}]_R + \frac{1}{4}[\text{𝄢}]_L[\text{𝄞}]_R + \frac{1}{4}[\text{𝄞}]_L[\text{𝄢}]_R + \frac{1}{4}[\text{𝄞}]_L[\text{𝄞}]_R.$$

telephone calls) by which Carsten determines what to write propagate locally in the *physical sense*, i.e., at a finite speed that is less than the speed of light.

Bell (2004 [1964]) showed, first, that there are certain probabilistic constraints— what are now called *Bell inequalities*—on the statistics arising from measurements on *any* locally correlated system. He then showed that these constraints are violated by statistics that are predicted to arise from certain experiments on a system composed of two qubits in the quantum-mechanical state[35]

$$\frac{1}{\sqrt{2}}\left(|0\rangle_A|1\rangle_B - |1\rangle_A|0\rangle_B\right), \tag{4.2}$$

where the A and B subsystems can be as far apart as one likes. The proof of this violation is known as *Bell's theorem* and it, and its variants, have since been experi- mentally confirmed many times over (Genovese, 2016).

The predicted violation only occurs for certain measurements. If we measure both the A and B qubits in the computational basis (see Eq. (2.6) and Fig. 7), then the predicted statistics will actually be compatible with the constraints imposed by local hidden-variable theories, as Bell himself showed (2004 [1964], p. 16). But as we rotate the measurement basis (see Fig. 7) that we use for B away from the measurement basis that we use for A, quantum mechanics predicts that we will see a violation, that it will reach a maximum at a certain point, and then decrease again as the measurement basis for B begins to line up again with the measurement basis for A.

Another name for the computational basis is the Z-basis. We call it the Z-basis because the two computational basis vectors $|0\rangle$ and $|1\rangle$ correspond to the two possible outcomes of a *Z-basis measurement* on a qubit, where "measuring a qubit in the Z- basis" means sending it through a Z-gate (see Fig. 8) and then recording its state (Janas et al., 2022, §6.5). Similarly, the basis vectors $|+\rangle$ and $|-\rangle$ are the two possible outcomes of an X-basis measurement on a qubit, and $|y^+\rangle$ and $|y^-\rangle$ are the two possible outcomes of a Y-basis measurement. The X, Y, and Z gates, together with the trivial I gate that leaves a qubit's state unchanged, are known as the *Pauli* gates.

I will have more to say about this family of gates later. For now, I want to point out that, for a system of qubits in the state given by Eq. (4.2), as long as both the A and B qubit are measured in one of the Pauli bases, the predicted statistics arising from those measurements will not violate the constraints that Bell's inequality imposes on local hidden-variable theories. In other words, if all that we have access to are measurement devices that can measure a qubit in one of the Pauli bases, then there will be no way to disprove some local hidden-variable story a skeptic might cook up to explain the observed statistics. To experimentally disprove such a story, we will need to have measurement devices that can measure in other measurement bases besides these ones, the ones for which Bell showed that a violation of the Bell inequalities will occur.

---

[35] This state is identical to the one given in Eq. (2.11) but we repeat it here for convenience.

Bell's theorem should convince us that no local hidden-variable theory can reproduce the correlations arising from such measurements (i.e., from measurements in
bases other than X, Y, and Z), but this does not bar a skeptic from positing a *nonlocal*
hidden-variable theory to recover them instead. An example of such a theory is one
in which the outcome of a measurement on A depends on the measurement basis
used to measure B. But since A and B could in principle be far apart in space, the
causal influence from B to A in such a theory will have to be propagated faster than
the speed of light. This is a hard pill to swallow, but one might be inclined to swallow
it anyway in order to avoid the other alternatives.[36]

Coming back to our discussion of quantum computers, recall that when we discussed the many-worlds explanation of quantum computing, we noted that the computer's state as given on the right-hand side of Eq. (3.1) is entangled. The question
arises, then, as to what role is played by entanglement more generally in enabling
quantum computers to outperform their classical competitors. There are two ways to
think of this question. We might ask, on the one hand, whether realizing an entangled
state is *necessary* to enable a quantum speedup. On the other hand, we might ask
whether it is *sufficient* to enable it.

There has been some debate surrounding the first question. On the one hand, Jozsa
and Linden (2003) have proven that realizing an entangled state is necessary to enable
speedup if a quantum computer can be assumed to be in a so-called "pure state", i.e.,
in a state that represents a maximally specific description of the system from the point
of view of the quantum-mechanical formalism. This is the case when we represent
a system with a state vector as we have been doing up until now (and will continue
to do). On the other hand, it has been argued (Biham et al., 2004) that quantum
computers that are in "mixed states"—i.e., states which describe the computer in
a less than maximally specific way, either because we are ignorant of the actual
state of the computer, or because the computer is coupled to its environment—are in
some cases capable of achieving a quantum speedup over their classical competitors
without ever being entangled. There is insufficient space to discuss this (largely
technical) debate here, but in Cuffaro (2013), I argue that the purported counter-
examples to what I there call the "necessity of entanglement thesis" do not actually
demonstrate what they purport to show, but instead clarify the necessary role that
entanglement does play in quantum computation.

From the philosopher's point of view, the more interesting question in relation to
the role of entanglement in quantum computing is the question regarding sufficiency,
for as we will see, reflecting on this question sheds light on the tension between
physical and computational ways of thinking that is at the heart of the science of
quantum computing, the tension that is the primary source of the insight this new
science brings into both physics and computation.

---

[36] For further discussion, see Myrvold et al. (2020).

5   The Gottesman–Knill Theorem

The main reason for being skeptical of the idea that entanglement suffices to enable quantum speedup is the Gottesman–Knill theorem (Gottesman, 1999). This theorem states that any quantum algorithm that employs (exclusively) some combination of the following operations (which together form what we will call the *Gottesman–Knill set*) is efficiently simulable by a classical computer: (i) The Clifford group of gates (see Fig. 8), i.e., the X, Y, Z, R, H, and CNOT gates; (ii) Clifford group gates conditioned on the values of classical bits (indicating, e.g., the results of previous measurements); (iii) state preparation of a qubit in the computational basis (as one typically does for each qubit prior to the beginning of a computation); (iv) measurements in one of the Pauli bases (as one does at the end of a computation) (Nielsen & Chuang, 2000, §10.5.4).

The reason the theorem represents a challenge, to the view that entanglement suffices to enable a quantum computer to outperform a classical computer, is that by using only the operations given in (i)–(iv), which according to the theorem are all efficiently classically simulable, it is possible to generate an entangled state. Figure 9 depicts a quantum circuit to realize an entangled state using only operations from the Gottesman–Knill set. It begins with the preparation of two qubits in the computational basis state $|0\rangle$, then subjects them each to an X-gate, as a result of which they will both be in the state $|1\rangle$. The first qubit is then run through a Hadamard gate which transforms it into the superposition state $(|0\rangle - |1\rangle)/\sqrt{2}$, and following that both qubits are run through a CNOT gate. The resulting state of the computer is an entangled state.

It would be wrong to conclude (cf. Jozsa & Linden, 2003, pp. 2029–30), on the basis of this, that entanglement is not sufficient to enable a quantum computer to outperform a classical computer. Why? Well, let us reflect on what the Gottesman–Knill theorem is saying. The Gottesman–Knill theorem shows that there is a certain set of quantum operations that can be efficiently simulated on a classical computer. Let us then consider what we mean when we say that we have "efficiently classically simulated" something. We can say, to start with, that in a classical computer simulation (efficient or otherwise), whatever computer is doing the simulating will be such that

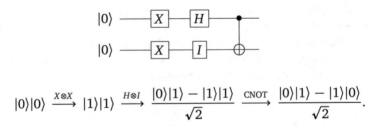

$$|0\rangle|0\rangle \xrightarrow{X \otimes X} |1\rangle|1\rangle \xrightarrow{H \otimes I} \frac{|0\rangle|1\rangle - |1\rangle|1\rangle}{\sqrt{2}} \xrightarrow{\text{CNOT}} \frac{|0\rangle|1\rangle - |1\rangle|0\rangle}{\sqrt{2}}.$$

Fig. 9 A quantum circuit (above), written out explicitly in terms of the unitary transformations required to realize it (below), involving only operations from the Gottesman–Knill set. At the end of this sequence of operations, the computer will be in an entangled state

it can be given a classical description. As we saw in Sect. 2.1, a complete description of a system, in classical mechanics, is always factorisable into complete descriptions of all of its subsystems (see Fig. 3). Besides this, classically describable systems and processes are locally causal, just like the convoluted procedure that Carsten uses to determine what to stuff in the envelopes he sends to Louise and Robert. The classical language of classical mechanics *constrains* all the descriptions of systems that can be given in that language to be this way. The behavior of a classically describable system, in other words, is *always* expressible in terms of local causes and effects, and correlations between effects manifested by such a system's subsystems can always be explained by appealing either to common or direct causes, whose influences propagate no faster than the speed of light. This is true no matter how large a system is conceived to be. For classical mechanics aims to be a universal language for describing *any* physical system, be it a single particle or the entire physical universe. The local causes of an effect will not always be known to us, but classical mechanics tells us that they are there, and that we can find them in principle.

Let us return to Carsten, our band leader who lives on the Île Saint-Louis in the center of Paris. Carsten has lately become interested in quantum computers as a way of helping with composition,[37] and has acquired a shiny silver box, now sitting on his desk, which he claims is one. Quadeisha, a friend and fellow musician who has happened by, is skeptical, and asks him to show her how it works. Obligingly, Carsten turns on his computer and has it run what he tells her is a calculation whose result depends on the prior generation of an entangled state. Quadeisha, still not satisfied, asks Carsten to describe the algorithm by which the computer arrives at its result. Upon discovering (after looking through the owner's manual for the device) that the algorithm employs no operations outside of the Gottesman–Knill set, she refuses to believe that Carsten's computer is quantum.

Quadeisha is right to be skeptical, *even if* the box on Carsten's desk (which, let us assume, he does not know how to open) *actually is* a quantum computer.[38] Moreover, she need not have heard of the Gottesman–Knill theorem to be justified in her skepticism. She need only be familiar with the work of John Bell. For it can be shown that the combined effect of any sequence of Gottesman–Knill operations, for any subsystem of the system to which they are applied, is equivalent to a measurement in one of the Pauli bases, X, Y, Z, on a system whose state is given by one of the basis vectors of a Pauli basis (Cuffaro, 2017, p. 115). Bell already showed us how to provide a locally causal model to reproduce the statistics of such measurements on a system in the state given in Eq. (4.2). Moreover, his technique is straightforwardly extendable to other similarly entangled two-party systems (Cuffaro, 2017 §A.1), and further techniques have been devised for constructing efficient locally causal models to recover the statistics arising from Pauli-basis measurements on entangled

---

[37] For a review of some of the uses envisioned for quantum computers in music, see Miranda (2021).

[38] This is actually the attitude (and for similar reasons) that many theorists take toward certain claims by private corporations to have built a working quantum computer (see, e.g., Aaronson (2013a), Shin et al. (2014)).

systems composed of three or more parties as well (see Cuffaro, 2017, §A.2, which summarizes a technique first described in Tessier, 2004).[39]

The upshot of the Gottesman–Knill theorem is that a certain number of quantum operations are efficiently classically simulable despite being capable of generating entangled states. And we have just seen that to say that some phenomenon is "efficiently classically simulable" is to say that it can be efficiently redescribed in a locally causal way. Since any sequence of Gottesman–Knill operations is equivalent to a measurement in one of the Pauli bases, X, Y, Z, on a system whose state is given by one of the basis vectors of a Pauli basis, the Gottesman–Knill theorem is essentially telling us that there are some statistics associated with measurements on systems in entangled states that admit of a locally causal description.

We do not really need the Gottesman–Knill theorem to tell us this, for Bell's and related inequalities amount to general constraints that any locally causal model of combined measurement statistics must satisfy, and we already know that in some cases, i.e., for measurements in one of the Pauli bases, the predictions of a locally causal model are compatible with quantum-mechanical predictions. It is therefore misleading to conclude, on the basis of the Gottesman–Knill theorem, that entanglement is not sufficient to enable quantum computational speedup. What the Gottesman–Knill theorem shows us is that one also needs to *use* this entanglement to its full potential, in particular, by not restricting a quantum computer to use only a small proportion of the operations of which it is capable (i.e., by not restricting it to the Gottesman–Knill set).

In the context of more general discussions of quantum mechanics, there are those who claim that entanglement is *the* one and only distinguishing feature of the theory (Schrödinger, 1935b). This is a controversial claim, but it is not proved false by the fact that Bell (2004 [1964]) showed that some operations on some entangled states can be redescribed in a locally causal way. Likewise, pointing out essentially the same thing in the context of quantum computation is not an objection to the claim that quantum entanglement constitutes a sufficient physical resource for realizing speedup over classical computation. Indeed, the Gottesman–Knill theorem serves to highlight the role that is actually played by entanglement in a quantum computer and to clarify why and in what sense it is sufficient to preclude the computer's evolution from plausibly being classically simulated (for more on this, see Cuffaro, 2017).

6 Computer Simulations and Locally Causal Models of Quantum Phenomena

Above we mentioned that techniques have been developed for providing classical computer simulations, i.e., locally causal descriptions, to recover the statistics arising from Pauli-basis measurements on entangled systems whose subsystems are in superpositions of eigenstates of Pauli observables, for any number of parties. The

---

[39] The n-party case, for $n \geq 3$, introduces subtleties which we will discuss in the next section.

classical computer simulations of three or more party systems are especially interesting in that they necessarily involve a small amount of communication. For three parties, A, B, and C, one bit must be communicated between B and A in order to make the protocol work. For n parties, $n - 2$ bits are required, i.e., the number of bits required scales linearly with n (Tessier, 2004). Note that this counts as being "easy" in the complexity-theoretic sense we discussed in Sect. 2.2.

From a certain point of view, however, it may seem quite odd to call a classical computer simulation of quantum correlations, that involves a certain amount of communication between parties, a locally causal model of those correlations. Recall that, earlier, when we introduced Bell's theorem, we claimed that communication between distant parties in the context of a Bell experiment constitutes a *nonlocal* influence, i.e., one that would have to be communicated at a speed exceeding that of light.

We can start to resolve the tension here if we consider, first, just what a locally causal redescription of quantum measurement statistics is required to do. If a pair of qubits, A and B, are prepared in an entangled state and then spatially separated, quantum mechanics tells us that when, for instance, a Z-basis measurement is performed on B, A's state changes "instantaneously", so that a Z-basis measurement on A will yield an outcome that is correlated with the result of the experiment on B in a particular way. Our challenge, however, is not to reproduce these apparent dynamics. Our challenge is to come up with an alternative description of the statistics that we *actually observe* in our experiments.

In order to assert that one has actually observed the results of a *combined* experiment on two or more spatially separated subsystems, it must be assumed that the results associated with the individual experiments on those subsystems have actually been combined in some way. That is, we must somehow gather together the results registered locally at the sites of the experiments on A and B. If Alice (a bassist) is assigned to measure A, for instance, and Bob (who plays the piano) to measure B, then once they record their individual results, they need to communicate them, either to one another, or perhaps to Carsten who is sitting in his office on the Île Saint-Louis, where he is in the process of showing Quadeisha how his quantum computer is but one part of a network of spatially distributed quantum computers that are together performing a computation that depends on the prior generation of a system in an entangled state. While Alice and Bob are making their way toward Carsten and Quadeisha with their results, however, there is time for Alice and Bob to, surreptitiously, exchange further information with one another by local means (Bob might call Alice on his cell phone, for instance, as they are making their way toward Carsten's office) in order to coordinate the outcomes of their individual measurements and "correct" them if necessary, unbeknownst to Quadeisha. And since this further information is propagated locally, i.e., at a speed no greater than that of light, it can be considered as *part of the common cause* of the actually observed (combined) measurement event, and thus as part of a classical, locally causal, description of that measurement event (see Cuffaro, 2017, §4, Fig. 1).

Now, given a particular combined calculational outcome at a time t, one can try to explain it in one of two ways: (i) Carsten's way, i.e., as the result of a *bona fide* quantum-mechanical process manifesting nonlocal behavior, or (ii) Quadeisha's more skeptical way, i.e., as the result of a locally causal mechanism whose different computational components are communicating with one another at less than the speed of light. Without further evidence—*if Carsten cannot or will not open his silver box*—it is perfectly legitimate to side with Quadeisha here, and be a skeptic with respect to Carsten's quantum-mechanical description of the box's observed behavior. Quadeisha's alternative account of the observed correlations is a perfectly good one in such a scenario, for her proposed classical alternative can solve the problem just as easily, in the complexity-theoretic sense, as Carsten's purported quantum computer.

Bell's and related inequalities specify *constraints* on what a locally causal description that accounts for a combined probability distribution must be like, but in and of themselves they are little more than formal statements. Bell's (2004 [1964]) original inequality, for instance, is essentially just a theorem of probability (Bell, 2004 [1981]; Janas et al., 2022, §3.4, Pitowsky, 1989, 1994). This is no fault. But if we are to make a meaningful distinction between what is and is not ruled out by them, then we must consider the context within which they are being applied. Normally this does not need to be made explicit, for normally the context is what I have elsewhere called the *theoretical* context (Cuffaro, 2017, p. 107), wherein we are considering alternative theories of the natural world. In this context, Bell's inequalities help us to answer the question of whether there may be some deeper hidden-variable theory of the natural world underlying quantum mechanics, and what such a theory needs to be like.

They do not answer that question all by themselves, however. Any alternative hidden-variable theory of the natural world must do more than merely satisfy the constraints imposed by the Bell inequalities. It must also be *plausible*. Thus, besides reproducing the well-confirmed statistical predictions of quantum mechanics, any deeper candidate theory should also, for instance, be consistent with our other well-confirmed theories of physics such as special and general relativity (and if not, then a compelling reason will need to be given for why we should not worry about such contradictions). It is on the basis of such plausibility constraints, and not on the basis of the Bell inequalities, that we rule out many of the so-called "loopholes" to Bell's theorem.

Quadeisha's alternative account of the observed correlations in Carsten's office counts as a particularly implausible example of such a loophole in the theoretical context.[40] But the theoretical context is not the only context that the Bell inequalities are relevant to. Nor is it the context appropriate to a discussion of the respective capabilities of quantum and classical computers. The context that is appropriate here is what I have elsewhere called the *practical* context (Cuffaro, 2017, p. 107). In the practical context, what concerns us is not alternative theories of the natural world. What concerns us is what *we* are capable of *building* with the aim of reproducing

---

[40] Quadeisha's loophole is actually conceptually similar to the "collapse locality" loophole at the heart of Adrian Kent's *causal quantum theory* (Kent, 2005). For discussion see Cuffaro (2017, pp. 104–106). For a more general discussion of the methodology of no-go theorems, see Dardashti (2021).

the statistical predictions of quantum mechanics. The plausibility constraints that restrict us in the practical context are not the same as the ones that restrict us in the theoretical context. But there are plausibility constraints in the practical context nevertheless.

In particular, in the practical context, we should rule out, as implausible, alternative locally causal descriptions of systems that would be *too hard* (in the complexity-theoretic sense) for us to build. And we should rule in alternative locally causal descriptions of systems that can be built *easily* (again, in the complexity-theoretic sense). The complexity involved in the specification of such a system, in other words, needs to be *tractable*. If, for instance, Quadeisha's basis for being skeptical of Carsten's claim to own a quantum computer is that she can specify a locally causal model for his computer's calculation that requires a number of additional resources that scales exponentially, then Carsten will be justified in dismissing her skepticism, for he can with justice say that it is wildly implausible that anyone can have built a machine according to Beatrice's specifications, because simulating the quantum calculation would amount, for such a machine, to an intractable task (Cuffaro, 2017, §5).[41]

7   Computational Perspectives on Physics

In the previous section we discussed classical computer simulations of quantum-mechanical phenomena, and we made the point that such simulations can be thought of as locally causal alternative descriptions of those phenomena. This is not the way that they are normally framed in the literature on quantum information and computation. The way that it is normally put is that such simulations quantify the extent to which quantum mechanics differs from classical mechanics, by enumerating the number of resources required to account for quantum-mechanical phenomena using classical means (see, for instance, Brassard, 1999; Rosset et al., 2013; Tessier, 2004; Tessier et al., 2005; Toner & Bacon, 2003).

Framing their significance in this way is certainly not incorrect, and it can be very theoretically useful.[42] But from the philosopher's perspective there is value in, so to speak, calling a spade a spade. A description of a plausible classical computational model for efficiently simulating some of the operations that can be carried out by a quantum computer is the same kind of thing as a plausible alternative local hidden-variable account of some particular class of quantum phenomena, in the simple sense that in both cases these constitute plausible locally causal models of quantum phenomena. The difference between them lies in the way that one interprets the word

---

[41] Note that I am taking 'tractable' here in a relative sense. That is, the resources required by a classical computer to reproduce a particular effect should differ tractably from those required by a quantum computer. Or in other words: it should *not be essentially harder* for the classical system to produce the same effect as the quantum system.

[42] The general subject of classical simulations of quantum systems is an important and burgeoning area of modern physics (see, for example, Borges et al., 2010; Lee & Thomas, 2002).

'plausible' in each case. In the practical context we assume that a system has been built by a rational agent to achieve a particular purpose. We do not presuppose this in the theoretical context.[43] Clearly, certain assumptions that would be regarded as plausible in the former context will not be in the latter and it is important to be on guard against conflating them. As we saw in the previous section, practical investigators attempting to isolate and/or quantify the computational resources provided by physical systems may be in danger of conceptual confusion if they are not cognisant of the differences between the practical and theoretical contexts. More positively, it must not be assumed by practical investigators that every 'no-go result', formulated in the theoretical context, will constrain what they can accomplish in the practical context (Cuffaro, 2018a). More generally it is important to think seriously about what one means by a 'plausible locally causal model' regardless of the context of investigation.

Calling a spade a spade gives us, moreover, insight into how the practical context can illuminate the investigations of the traditional philosopher of physics in the context of quantum mechanics. The silly story we told in the previous section about how Carsten, Alice, and Bob manage to reproduce, by communicating just a few classical bits, all of the correlational phenomena associated with Pauli-basis measurements on an n-party quantum system in an entangled state, besides representing just another silly classical story, emphasizes perhaps the most fundamental difference that we can point to between classical and quantum mechanics. Classical and quantum mechanics are, fundamentally, alternative universal languages for describing physical systems (Janas et al., 2022, §6.3). And by investigating the computational power inherent in those languages we gain insight into the respective logical structures of classical and quantum theory, into the logical structure of quantum mechanics that enables the efficient representation of correlational phenomena by quantum-mechanical systems, and the logical structure of classical mechanics which precludes this (Janas et al., 2022, §4.2).

8 Concluding Remarks

Reflecting on the lessons of quantum mechanics, Niels Bohr wrote that:

> In representing a generalization of classical mechanics suited to allow for the existence of the quantum of action, Quantum mechanics offers a frame *sufficiently wide* to account for empirical regularities *which cannot* be comprised in the classical way of description (Bohr, 1948, p. 316, my emphasis).

This is a lesson with which the artist will be familiar.[44] From time to time, in literature, music, and in other forms of art, new methods of writing and composing emerge that allow one to express the subtleties and nuances of experience more

---

[43] This statement is not meant to express any sort of theological opinion. It is merely a statement about how science operates, at least in this century.

[44] A similar point is made in Janas et al. (2022, Sect. 6.3, note 22).

easily than before. E. M. Forster (1942, p. 28) once said, about Virginia Woolf, that "She pushed the light of the English language a little further against the darkness." In music, the (re-)introduction, into the Western musical tradition, of modality and chromaticism through Jazz, Blues, and other forms of popular music, and also through art music (see Vincent, 1951), has enabled modern composers in the Western tradition to explore musical landscapes and express aspects of our emotional and intellectual experience that composers limited to the major and minor scales, so dominant during Western music's classical period, cannot easily capture. In his musings about quantum mechanics, Bohr himself was wont to appeal to an analogy with poetry. Werner Heisenberg recalls, in his memoirs, a conversation he had with Bohr in which the latter stated that:

> We must be clear that, when it comes to atoms, language can be used only as poetry. The poet, too, is not nearly so concerned with describing facts as with creating images and establishing mental connections" (Heisenberg, 1971, p. 41).

These and like analogies were not lost on actually practising poets and other artists of the same period (see Mairhofer, 2021). The subject of quantum mechanics has even been found to lend itself very naturally to the format of the comic book (see Bub & Bub, 2018, and the review by Cuffaro & Doyle, 2021, especially §4). As for the science of quantum computing: The science of quantum computing takes advantage, as we have seen throughout the course of this chapter, of the increased expressive power of the quantum-mechanical language. It shows us that using the new language of quantum mechanics, we can interact with what we take to be quantum-mechanical systems in ways that cannot be easily replicated classically.

In the course of this chapter we have reflected on the fundamentals of classical and quantum physics and computation, on the resources used and on the explanation of the power of quantum computers, and finally on the broader insights that can be gleaned from the science of quantum computing both for physics and for computation. The philosophical issues addressed in this chapter are not the only issues brought up by the science of quantum computing. But I hope to have convinced the reader, with this sample of some of the more central ones, of quantum computing's potential for illuminating the world that we are all participants in.

References

Aaronson, S. (2013a). D-Wave: Truth finally starts to emerge. Posted: June 05, 2013. Retrieved August 11, 2014, from
www.scottaaronson.com/blog/?p=1400

Aaronson, S. (2013b). *Quantum computing since Democritus*. Cambridge University Press.

Aaronson, S. (2013c). Why philosophers should care about computational complexity. In B. J. Copeland, C. J. Posy, & O. Shagrir (Eds.), *Computability: Turing, Gödel, Church, and beyond* (pp. 261–327). MIT Press.

Aaronson, S. (2016). Complexity zoo. complexityzoo.uwaterloo.ca/Complexity_Zoo.

Adlam, E. (2014). The problem of confirmation in the Everett interpretation. *Studies in History and Philosophy of Modern Physics, 47*, 21–32.

Agrawal, M., Kayal, N., & Saxena, N. (2004). PRIMES is in P. *Annals of Mathematics, 160*, 781–793.

Aharonov, D., van Dam, W., Kempe, J., Landau, Z., Lloyd, S., & Regev, O. (2007). Adiabatic quantum computation is equivalent to standard quantum computation. *SIAM Journal on Computing, 37*, 166–194.

Albert, D., & Loewer, B. (1988). Interpreting the many worlds interpretation. *Synthese, 77*, 195–213.

Arora, S., & Barak, B. (2009). *Computational complexity: A modern approach*. Cambridge University Press.

Bell, J. S. (2004 [1964]). On the Einstein-Podolsky-Rosen paradox. In *Speakable and unspeakable in quantum mechanics* (pp. 14–21). Cambridge University Press.

Bell, J. S. (2004 [1966]). On the problem of hidden variables in quantum mechanics. In *Speakable and unspeakable in quantum mechanics* (pp. 1–13). Cambridge University Press.

Bell, J. S. (2004 [1981]). Bertlmann's socks and the nature of reality. In *Speakable and unspeakable in quantum mechanics* (pp. 139–158). Cambridge University Press.

Bennett, C. H., Bernstein, E., Brassard, G., & Vazirani, U. (1997). Strengths and weaknesses of quantum computing. *SIAM Journal on Computing, 26*, 1510–1523.

Bernstein, E., & Vazirani, U. (1997). Quantum complexity theory. *SIAM Journal on Computing, 26*, 1411–1473.

Biham, E., Brassard, G., Kenigsberg, D., & Mor, T. (2004). Quantum computing without entanglement. *Theoretical Computer Science, 320*, 15–33.

Bohr, N. (1948). On the notions of causality and complementarity. *Dialectica, 2*, 312–319.

Boole, G. (1847). *The mathematical analysis of logic*. Philosophical Library. Reprinted: Thoemmes Press (1998).

Borges, C. V. S., Hor-Meyll, M., Huguenin, J. A. O., & Khoury, A. Z. (2010). Bell-like inequality for the spin-orbit separability of a laser beam. *Physical Review A, 82*, 033833.

Brassard, G., Cleve, R., & Tapp, A. (1999). Cost of exactly simulating quantum entanglement with classical communication. *Physical Review Letters, 83*, 1874–1877.

Briegel, H. J., Browne, D. E., Dür, W., Raussendorf, R., & den Nest, M. V. (2009). Measurement-based quantum computation. *Nature Physics, 5*, 19–26.

Bub, J. (2006). Quantum computation from a quantum logical perspective. arXiv:quant-ph/0605243v2

Bub, J. (2008). Quantum computation and pseudotelepathic games. *Philosophy of Science, 75*, 458–472.

Bub, J. (2010). Quantum computation: Where does the speed-up come from? In A. Bokulich & G. Jaeger (Eds.), *Philosophy of quantum information and entanglement* (pp. 231–246). Cambridge University Press.

Bub, J. (2016). *Bananaworld, quantum mechanics for primates*. Oxford University Press.

Bub, J., & Pitowsky, I. (2010). Two dogmas about quantum mechanics. In S. Saunders, J. Barrett, A. Kent, & D. Wallace (Eds.), *Many worlds? Everett, quantum theory, and reality* (pp. 433–459). Oxford University Press.

Bub, T., & Bub, J. (2018). *Totally random: Why nobody understands quantum mechanics*. Princeton University Press.

Church, A. (1936). An unsolvable problem of elementary number theory. *American Journal of Mathematics, 58*, 345–363.

Cobham, A. (1965). The intrinsic computational difficulty of functions. In Y. Bar-Hillel (Ed.), *Logic, methodology and philosophy of science: Proceedings of the 1964 international congress* (pp. 24–30). North-Holland.

Cook, W. J. (2012). *In pursuit of the traveling salesman: Mathematics at the limits of computation*. Princeton University Press.

Copeland, J. B. (2017). The modern history of computing. In E. N. Zalta (Ed.), *The Stanford encyclopedia of philosophy* (Winter 2017 ed.). Metaphysics Research Lab, Stanford University.

Copeland, J. B. (2020). The Church-Turing thesis. In E. N. Zalta (Ed.), *The Stanford encyclopedia of philosophy* (Summer 2020 ed.). Metaphysics Research Lab, Stanford University.

Cuffaro, M. E. (2012). Many worlds, the cluster-state quantum computer, and the problem of the preferred basis. *Studies in History and Philosophy of Modern Physics, 43*, 35–42.

Cuffaro, M. E. (2013). On the necessity of entanglement for the explanation of quantum speedup. arXiv:1112.1347v5

Cuffaro, M. E. (2017). On the significance of the Gottesman-Knill theorem. *The British Journal for the Philosophy of Science, 68*, 91–121.

Cuffaro, M. E. (2018). Reconsidering no-go-theorems from a practical perspective. *The British Journal for the Philosophy of Science, 69*, 633–655.

Cuffaro, M. E. (2018). Universality, invariance, and the foundations of computational complexity in the light of the quantum computer. In S. O. Hansson (Ed.), *Technology and mathematics: Philosophical and historical investigations.* Springer.

Cuffaro, M. E. (2020). Information causality, the Tsirelson bound, and the 'being-thus' of things. *Studies in History and Philosophy of Modern Physics, 72*, 266–277.

Cuffaro, M. E., & Doyle, E. P. (2021). Essay review of Tanya and Jeffrey Bub's *Totally random: Why nobody understands quantum mechanics: A serious comic on entanglement. Foundations of Physics, 51*, 28:1–28:16.

Cuffaro, M. E., & Hartmann, S. (2021). The open systems view. arXiv:2112.11095v1

Curiel, E. (2014). Classical mechanics is Lagrangian; it is not Hamiltonian. *The British Journal for Philosophy of Science, 65*, 269–321.

Dardashti, R. (2021). No-go theorems: What are they good for? *Studies in History and Philosophy of Science, 86*, 47–55.

Davis, M. (2000). *The universal computer: The road from Leibniz to Turing.* W. W. Norton and Company.

Dawid, R., & Thébault, K. P. Y. (2015). Many worlds: Decoherent or incoherent? *Synthese, 192*, 1559–1580.

Dawson, J. W., Jr. (2007). Classical logic's coming of age. In D. Jacquette (Ed.), *Philosophy of logic* (pp. 497–522). Elsevier.

Dean, W. (2016). Squeezing feasibility. In A. Beckmann, L. Bienvenu, & N. Jonoska (Eds.) *Pursuit of the Universal: Proceedings of the 12th Conference on Computability in Europe* (pp. 78–88). Springer International Publishing.

Deutsch, D. (1985). Quantum theory, the Church-Turing principle and the universal quantum computer. *Proceedings of the Royal Society of London. Series A. Mathematical and Physical Sciences, 400*, 97–117.

Deutsch, D. (1989). Quantum computational networks. *Proceedings of the Royal Society of London. Series A. Mathematical and Physical Sciences, 425*, 73–90.

Deutsch, D. (1997). *The fabric of reality.* Penguin.

Deutsch, D. (2010). Apart from universes. In S. Saunders, J. Barrett, A. Kent, & D. Wallace (Eds.), *Many worlds? Everett, quantum theory, and reality* (pp. 542–552). Oxford University Press.

DeWitt, B. S. (1973 [1971]). The many-universes interpretation of quantum mechanics. In DeWitt & Graham (1973, pp. 167–218).

Duwell, A. (2007). The many-worlds interpretation and quantum computation. *Philosophy of Science, 74*, 1007–1018.

Duwell, A. (2018). How to make orthogonal positions parallel: Revisiting the quantum parallelism thesis. In M. E. Cuffaro & S. C. Fletcher (Eds.), *Physical perspectives on computation, computational perspectives on physics* (pp. 83–102). Cambridge University Press.

Duwell, A. (2021). *Computation and physics.* Cambridge University Press.

Edmonds, J. (1965). Paths, trees, and flowers. *Canadian Journal of Mathematics, 17*, 449–467.

Einstein, A., Podolsky, B., & Rosen, N. (1935). Can quantum-mechanical description of physical reality be considered complete? *Physical Review, 47*, 777–780.

Everett III, H. (1956). The theory of the universal wave function. In DeWitt & Graham (1973, pp. 3–140).

Farhi, E., Goldstone, J., Gutmann, S., & Sipser, M. (2000). Quantum computation by adiabatic evolution. Tech. Rep. MIT-CTP-2936, MIT. arXiv:quant-ph/0001106

Fletcher, S. C. (2018). Computers in abstraction/representation theory. *Minds & Machines, 28*, 445–463.

Forster, E. M. (1942). *Virginia Woolf. The Rede Lecture 1941*. Cambridge University Press.

Genovese, M. (2016). Experimental tests of Bell's inequalities. In Bell & Gao (2016, pp. 124–140).

Gödel, K. (1956). Private letter to John von Neumann, 20 March 1956. Translated by Wensinger in: Sipser (1992).

Goldreich, O. (2008). *Computational complexity: A conceptual perspective*. Cambridge University Press.

Gottesman, D. (1999). The Heisenberg representation of quantum computers. In S. P. Corney, R. Delbourgo, & P. D. Jarvis (Eds.), *Group22: Proceedings of the XXII International Colloquium on Group Theoretical Methods in Physics* (pp. 32–43). International Press. Longer version available at: arXiv:quant-ph/9807006v1

Greaves, H., & Myrvold, W. (2010). Everett and evidence. In S. Saunders, J. Barrett, A. Kent, & D. Wallace (Eds.), *Many worlds? Everett, quantum theory, and reality* (pp. 181–205). Oxford University Press.

Grover, L. K. (1996). A fast quantum mechanical algorithm for database search. In *Proceedings of the Twenty-eighth Annual ACM Symposium on Theory of Computing*, STOC '96 (pp. 212–219). Association for Computing Machinery.

Hagar, A. (2007). Quantum algorithms: Philosophical lessons. *Minds & Machines, 17*, 233–247.

Hagar, A., & Cuffaro, M. (2019). Quantum computing. In E. N. Zalta (Ed.), *The Stanford encyclopedia of philosophy* (Winter 2019 ed.). Metaphysics Research Lab, Stanford University.

Heisenberg, W. (1971). *Physics and beyond*. Harper & Row.

Hewitt-Horsman, C. (2009). An introduction to many worlds in quantum computation. *Foundations of Physics, 39*, 869–902.

Horsman, D., Kendon, V., & Stepney, S. (2018). Abstraction/representation theory and the natural science of computation. In M. E. Cuffaro & S. C. Fletcher (Eds.), *Physical perspectives on computation, computational perspectives on physics* (pp. 127–152). Cambridge University Press.

Howard, D. (1989). Holism, separability, and the metaphysical implications of the Bell experiments. In J. T. Cushing & E. McMullin (Eds.), *Philosophical consequences of quantum theory* (pp. 224–253). University of Notre Dame Press.

Hughes, R. I. G. (1989). *The structure and interpretation of quantum mechanics*. Harvard University Press.

Janas, M., Cuffaro, M. E., & Janssen, M. (2022). *Understanding quantum Raffles: Quantum mechanics on an informational approach: Structure and interpretation*. Springer.

Jozsa, R., & Linden, N. (2003). On the role of entanglement in quantum-computational speed-up. *Proceedings of the Royal Society of London. Series A. Mathematical, Physical and Engineering Sciences, 459*, 2011–2032.

Kent, A. (2005). Causal quantum theory and the collapse locality loophole. *Physical Review A, 72*, 012107.

Lee, K. F., & Thomas, J. E. (2002). Experimental simulation of two-particle quantum entanglement using classical fields. *Physical Review Letters, 88*, 097902.

Lehner, C. (1997). What it feels like to be in a superposition. And why. *Synthese, 110*, 191–216.

Lenstra, A. K., Lenstra Jr., H. W., Manasse, M. S., & Pollard, J. M. (1990). The number field sieve. In *Proceedings of the Twenty-second Annual ACM Symposium on Theory of Computing*, STOC '90 (pp. 564–572). Association for Computing Machinery.

Lupacchini, R. (2018). Church's thesis, Turing's limits, and Deutsch's principle. In M. E. Cuffaro & S. C. Fletcher (Eds.), *Physical perspectives on computation, computational perspectives on physics* (pp. 60–82). Cambridge University Press.

Mairhofer, L. (2021). *Atom und Individuum: Bertolt Brechts Interferenz mit der Quantenphysik*. De Gruyter.

Maroney, O. J. E., & Timpson, C. G. (2018). How is there a physics of information? On characterising physical evolution as information processing. In M. E. Cuffaro & S. C. Fletcher (Eds.), *Physical perspectives on computation, computational perspectives on physics* (pp. 103–126). Cambridge University Press.

Martin, J. C. (1997). *Introduction to languages and the theory of computation* (2nd ed.). McGraw-Hill.

Maudlin, T. (2011). *Quantum non-locality and relativity* (3rd ed.). Wiley-Blackwell.

Mehlhorn, K., & Sanders, P. (2008). *Algorithms and data structures*. Springer.

Mermin, D. N. (2007). *Quantum computer science: An introduction*. Cambridge University Press.

Miranda, E. R. (2021). Quantum computer: Hello, music! In E. R. Miranda (Ed.) *Handbook of artificial intelligence for music: Foundations, advanced approaches, and developments for creativity*. Springer.

Myrvold, W., Genovese, M., & Shimony, A. (2020). Bell's theorem. In E. N. Zalta (Ed.) *The Stanford encyclopedia of philosophy* (Fall 2020 ed.). Metaphysics Research Lab, Stanford University.

Nielsen, M. A. (2006). Cluster-state quantum computation. *Reports on Mathematical Physics, 57*, 147–161.

Nielsen, M. A., & Chuang, I. L. (2000). *Quantum computation and quantum information*. Cambridge University Press.

Nishimura, H., & Ozawa, M. (2009). Perfect computational equivalence between quantum Turing machines and finitely generated uniform quantum circuit families. *Quantum Information Processing, 8*, 13–24.

Pearl, J. (2009). *Causality: Models, reasoning, and inference* (2nd ed.). Cambridge University Press.

Pitowsky, I. (1989). *Quantum probability—Quantum logic*. Springer.

Pitowsky, I. (1994). George Boole's 'conditions of possible experience' and the quantum puzzle. *British Journal for the Philosophy of Science, 45*, 99–125.

Pitowsky, I. (2002). Quantum speed-up of computations. *Philosophy of Science, 69*, S168–S177.

Raussendorf, R., & Briegel, H. J. (2002). Computational model underlying the one-way quantum computer. *Quantum Information and Computation, 2*, 443–486.

Raussendorf, R., Browne, D. E., & Briegel, H. J. (2003). Measurement-based quantum computation on cluster states. *Physical Review A, 68*, 022312.

Rivest, R. L., Shamir, A., & Adleman, L. (1978). A method for obtaining digital signatures and public-key cryptosystems. *Communications of the ACM, 21*, 120–126.

Rosset, D., Branciard, C., Gisin, N., & Liang, Y.-C. (2013). Entangled states cannot be classically simulated in generalized Bell experiments with quantum inputs. *New Journal of Physics, 15*, 053025.

Saunders, S. (1995). Time, quantum mechanics, and decoherence. *Synthese, 102*, 235–266.

Schrödinger, E. (1935a). Die gegenwärtige situation in der quantenmechanik. *Naturwissenschaften, 23*, 807–812; 823–828; 844–849. Translated in: Trimmer, J. D. (1980) *Proceedings of the American Philosophical Society, 124*, 323–338.

Schrödinger, E. (1935b). Discussion of probability relations between separated systems. *Mathematical Proceedings of the Cambridge Philosophical Society, 31*, 555–563.

Shetterly, M. L. (2016). *Hidden figures*. HarperCollins.

Shin, S. W., Smith, G., Smolin, J. A., & Vazirani, U. (2014). How "quantum" is the D-Wave machine? arXiv:1401.7087v2

Shor, P. W. (1994). Algorithms for quantum computation: Discrete logarithms and factoring. In *35th Annual Symposium on Foundations of Computer Science, 1994 Proceedings* (pp. 124–134).

Tessier, T. E. (2004). *Complementarity and entanglement in quantum information theory*. Ph.D. thesis, The University of New Mexico, Albuquerque, New Mexico.

Tessier, T. E., Caves, C. M., Deutsch, I. H., & Eastin, B. (2005). Optimal classical-communication-assisted local model of *n*-qubit Greenberger-Horne-Zeilinger correlations. *Physical Review A, 72*, 032305.

Timpson, C. G. (2013). *Quantum information theory & the foundations of quantum mechanics*. Oxford University Press.

Toner, B. F., & Bacon, D. (2003). Communication cost of simulating Bell correlations. *Physical Review Letters, 91*, 187904.

Turing, A. M. (1936–7). On computable numbers, with an application to the Entscheidungsproblem. *Proceedings of the London Mathematical Society. Second Series, s2-42*, 230–265.

Vaidman, L. (1998). On schizophrenic experiences of the neutron or why we should believe in the many-worlds interpretation of quantum theory. *International Studies in the Philosophy of Science, 12*, 245–261.

Vaidman, L. (2012). Probability in the many-worlds interpretation of quantum mechanics. In Y. Ben-Menahem & M. Hemmo (Eds.), *Probability in physics* (pp. 299–311). Springer.

Vaidman, L. (2018 [2002]). Many-worlds interpretation of quantum mechanics. In E. N. Zalta (Ed.) *The Stanford encyclopedia of philosophy* (Fall 2018 ed.). Metaphysics Research Lab, Stanford University. First published: 2002.

van Emde Boas, P. (1990). Machine models and simulations. Algorithms and Complexity. In J. van Leeuwen (Ed.), *Handbook of theoretical computer science* (Vol. A, pp. 1–66). MIT Press/Elsevier.

Vincent, J. (1951). *The diatonic modes in modern music*. Mills Music. Page references to second edition (1974) published by Curlew Music Publishers.

Wallace, D. (2003). Everett and structure. *Studies in History and Philosophy of Modern Physics, 34*, 87–105.

Wallace, D. (2007). Quantum probability from subjective likelihood: Improving on Deutsch's proof of the probability rule. *Studies in History and Philosophy of Modern Physics, 38*, 311–332.

Wallace, D. (2010). Decoherence and ontology. In S. Saunders, J. Barrett, A. Kent, & D. Wallace (Eds.), *Many worlds? Everett, quantum theory, and reality* (pp. 53–72). Oxford University Press.

Wallace, D. (2012). *The emergent multiverse*. Oxford University Press.

Wallace, D. (2019). On the plurality of quantum theories: Quantum theory as a framework, and its implications for the quantum measurement problem. In S. French & J. Saatsi (Eds.), *Realism and the quantum* (pp. 78–102). Oxford University Press.

Wikipedia contributors. (2020). Instructions per second. In *Wikipedia, The Free Encyclopedia*. Posted on 15 September, 2020, 11:15.

Zurek, W. H. (2003 [1991]). Decoherence and the transition from quantum to classical—revisited. arXiv:quant-ph/0306072v1

Quantum-Like Cognition and Rationality: Biological and Artificial Intelligence Systems

Andrei Khrennikov

Abstract This is a short introductory review on quantum-like modeling of cognition with applications to decision-making and rationality. The aim of the review is twofold: (a) to present briefly the apparatus of quantum information and probability theory useful for such modeling; (b) to motivate applications of this apparatus in cognitive studies and artificial intelligence, psychology, decision-making, social and political sciences. We define quantum rationality as decision-making that is based on quantum information processing. Quantumly and classically rational agents behaves differently. A quantum-like agent can violate the Savage Sure Thing Principle, the Aumann theorem on impossibility of agreeing to disagree. Such an agent violates the basic laws of classical probability, e.g., the law of total probability and the Bayesian probability inference. In some contexts, "irrational behavior" (from the viewpoint of classical theory of rationality) can be profitable, especially for agents who are over-loaded by a variety of information flows. Quantumly rational agents can save a lot of information processing resources. At the same time, this sort of rationality is the basis for quantum-like socio-political engineering, e.g., social laser. This rationality plays the important role in the process of decision-making not only by biosystems, but even by AI-systems. The latter equipped with quantum(-like) information processors would behave irrationally, from the classical viewpoint. As for biosystems, quantum rational behavior of AI-systems has its advantages and disadvantages. Finally, we point out that quantum-like information processing in AI-systems can be based on classical physical devices, e.g., classical digital or analog computers.

Keywords Mathematical formalism · Quantum mechanics · Open quantum systems · Quantum instruments · Quantum Markov dynamics · Psychological effects · Cognition · Decision-making · Classical rationality · Quantum rationality · Information overload

A. Khrennikov (✉)
International Center for Mathematical Modeling in Physics and Cognitive Sciences, Linnaeus University, 351 95 Växjö, Sweden
e-mail: andrei.khrennikov@lnu.se

© Springer Nature Switzerland AG 2022
E. R. Miranda (ed.), *Quantum Computing in the Arts and Humanities*,
https://doi.org/10.1007/978-3-030-95538-0_4

1   Introduction

'Information' is the new paradigm reflecting the main character of modern human society as an information society. The tremendous development of information technologies of the last 20 years has dramatically changed our lifestyle through the World Wide Web of the Internet and the mobile-connectivity web that have led to the creation of virtual social networks. Recent information technologies caused the digital transformation of human communications and, consequently, of the whole society. Unavoidably, challenges of the new information culture become the focus of intensive studies and a meeting point for researchers not only from physics, information engineering, and artificial intelligence, but also from biology, psychology, decision-making, cognitive, social and political sciences, economics and finance. The aim of this note to review briefly the recent applications of the quantum information and probability theories outside of physics. This line of research is known as quantum-like modeling (Asano et al., 2011, 2012, 2014, 2015a, 2015b, 2017a, 2017b; Basieva & Khrennikov, 2015; Bagarello & Oliveri, 2013; Busemeyer & Bruza, 2012; Bagarello et al., 2019; 2018; Basieva et al., 2017; Busemeyer et al., 2006; Boyer-Kassem et al., 2015;Busemeyer et al., 2014; Dzhafarov & Kujala, 2012; Dzhafarov et al., 2015; Haven, 2005; Haven & Khrennikov, 2013; Haven et al., 2017; Khrennikov, 1999, 2003, 2004a, 2004b, 2006, 2010, 2014, 2015a, 2015b, 2016; Khrennikov et al., 2014; Khrennikova, 2016, 2017; Khrennikov & Basieva, 2014b,2014a; Khrennikov et al., 2018; Ozawa & Khrennikov, 2020a, 2020b; Pothos & Busemeyer, 2009; Surov et al., 2019; Wang & Busemeyer, 2013, 2014; White et al., 2014). This is not consideration of genuine quantum physical processes in cognitive systems, but modeling behavior of generally macroscopic cognitive or AI agents processing information on the basis of quantum laws.

We want to present in more detail consequences of such information processing for rationality. In classical decision-making, rational agents are mathematically modeled as probabilistic information processors using Bayesian update of probabilities: rational = Bayesian. Quantum state update is generally non-Bayesian (Ozawa & Khrennikov, 2020b). We define quantum rationality as decision-making that is based on quantum state update. Quantum and classical rational agents behave differently. For instance, a quantum(-like) agent can violate the Savage Sure Thing Principle (Savage, 1954) (see Busemeyer & Bruza, 2012; Haven & Khrennikov, 2013; Haven et al., 2017) and the Aumann theorem (Aumann, 1976) on impossibility of agreeing to disagree (see Haven et al., 2017; Khrennikov & Basieva, 2015b; Khrennikov, 2014b).

In quantum-like modeling, the brain is treated as a black box in which information processing cannot be described by classical probability (Kolmogorov, 1933) (CP) (cf. with genuine quantum physical models of brain's functioning (Arndt et al., 2009; Bernroider, 2017; Bernroider & Summhammer, 2012; Hameroff, 1994; Penrose, 1989; Umezawa, 1993; Vitiello, 1995, 2001)). And there is a plenty of such nonclassical statistical data—in cognitive psychology, game theory, decision-making, social science, economics, finances, and politics. In decision theory, such data was coupled to probability fallacies and irrational behavior of agents. We propose to apply the most well-developed nonclassical theory of probability, namely, based on the mathematical formalism of quantum theory.

One may think that the appeal to quantum probability (Khrennikov, 2016) (QP) and information to model decision-making by humans is too exotic. However, we recall that as early as the 1970s, Tversky (one of the most cited psychologists of all time) and Kahneman (Nobel prize in economics in 2002, for prospect theory, which he co-developed with Tversky) have been demonstrating cases where CP-prescription and human behavior persistently diverge (Kahneman, 2003; Kahneman & Tversky, 1972, 1979, 1984; Kahneman & Thaler, 2006). Today, we are at the theoretical cross-roads, with huge divisions across conflicting, entrenched theoretical positions.

Should scientists continue use CP as the basis for descriptive and normative predictions in decision making and accept continuous generation of paradoxes?

Should we abandon probability theory completely and instead pursue explanations based on heuristics, as Tversky and Kahneman proposed?

The use of the probabilistic and statistical methods is really the cornerstone of the modern scientific methodology. Thus, although the heuristic approach to decision-making cannot be discarded completely, it seems more natural to search novel probabilistic models for decision-making. Our suggestion is to use QP and more generally quantum information, instead of heuristics of Tversky and Kahneman.

We stress that quantum-like modeling does not appeal to genuine quantum physical processes in biosystems. Quantum-like information processing can be performed by macroscopic systems, as cells, animals, or humans. Even artificial intelligence need not be based on quantum physical processors, as, e.g., quantum computers. Quantum-like modeling opens the door to creation of AI-systems processing information by respecting the laws of quantum theory, but equipped with classical physical processors. Some step in this direction was done within the recent studies on quantum(-like) information retrieval (see, e.g., Aerts et al., 2019; Melucci, 2015; van Rijsbergen, 2004): algorithms based on the complex Hilbert state representation of information, but driven on classical computers, demonstrate superiority comparing with the traditional ("classical") algorithms. Of course, creation of successfully working genuine quantum computers and simulators would give the possibility for creation of the genuine quantum AI-systems. However, since behavior of quantum-like and genuine quantum AI-systems is based on the same formalism and methodology, the theory of quantum-like cognition, rationality, and artificial intelligence would be useful even for quantum physical AI-systems.

For reader's convenience, this paper contains basics of CP and QP. We present them in two steps. Section 2 is the informal comparative introduction to CP versus QP; then Sects. 7, 9 present briefly, but mathematically rigorously the CP and QP formalisms (the later is presented as a part of the quantum formalism: quantum states and observables, the Born rule for probability of outcome of a measurement, the projection postulate and quantum state update). However, these formal presentations of CP and QP are used only in the last part of this work devoted to comparison of classical and quantum versions of the Aumann theorem. The preceding part devoted to classical vs. quantum rationality is written at the heuristic level.

2   Brief Comparison: Classical Versus Quantum Probability

CP was mathematically formalized by Kolmogorov (1933) (see Sect. 7 for details). This is the calculus of probability measures, where a nonnegative weight $p(A)$ is assigned to any event A. The main property of CP is its additivity: if two events O_1, O_2 are disjoint, then the probability of disjunction of these events equals the sum of probabilities:

$$P(O_1 \vee O_2) = P(O_1) + P(O_2).$$

In fact, powerful integration theory that is needed for calculation of averages demands σ-additivity:

$$P(\cup_j O_j) = \sum_j P(O_j), \tag{1}$$

where $O_j \cap O_i = \emptyset, i \neq j$.

QP is the calculus of complex amplitudes or in the abstract formalism complex vectors. Thus, instead of operations on probability measures one operates with vectors. We can say that QP is a *vector model of probabilistic reasoning*. Each complex amplitude ψ gives the probability by Born's rule: *Probability is obtained as the square of the absolute value of the complex amplitude.*

$$p = |\psi|^2 \tag{2}$$

(for the Hilbert space formalization, see Sect. 9, Formula (17). By operating with complex probability amplitudes, instead of the direct operation with probabilities, one can violate the basic laws of CP.

In CP, the *law of total probability* (LTP) is derived by using additivity of probability and *the Bayes formula*, the definition of conditional probability,

$$P(O_2|O_1) = \frac{P(O_2 \cap O_1)}{P(O_1)}, \ \ P(O_1) > 0. \tag{3}$$

Consider the pair, A and B, of discrete classical random variables. Then

$$P(B = \beta) = \sum_\alpha P(A = \alpha) P(B = \beta | A = \alpha).$$

Thus, in CP the B-probability distribution can be calculated from the A-probability and the conditional probabilities $P(B = \beta | A = \alpha)$.

In QP (Khrennikov, 2016), classical LTP is perturbed by the interference term (Khrennikov, 2010, 2016); for dichotomous quantum observables A and B of the von Neumann-type, i.e., given by Hermitian operators $\hat{A}$ and $\hat{B}$, the quantum version of LTP has the form:

$$P(B = \beta) = \sum_{\alpha} P(A = \alpha)P(B = \beta|a = \alpha) \tag{4}$$

$$+ 2 \sum_{\alpha_1 < \alpha_2} \cos \theta_{\alpha_1 \alpha_2} \sqrt{P(A = \alpha_1)P(B = \beta|A = \alpha_1)P(A = \alpha_2)P(B = \beta|a = \alpha_2)}$$
$$\tag{5}$$

If *the interference term* is positive, then the QP-calculus would generate a probability that is larger than its CP-counterpart given by the classical LTP (2). In particular, this probability amplification is the basis of the quantum computing supremacy.

3 Classical (Bayesian) Versus Quantum (Generally Non-Bayesian) Rationality

In classical theory of decision-making, rational behavior of agents is formalized with the *Savage Sure Thing Principle* (STP) (Savage, 1954):

If you prefer prospect b_+ to prospect b_- if a possible future event A happens ($a = +1$); and you prefer prospect b_+ still if future event A does not happen ($a = -1$); then you should prefer prospect b_+, despite having no knowledge of whether or not event A will happen.

Savage's illustration refers to a person deciding whether or not to buy a certain property shortly before a presidential election, the outcome of which could radically affect the property market:

"Seeing that he would buy in either event, he decides that he should buy, even though he does not know which event will obtain".

STP is considered as the axiom of rationality of decision makers (Savage, 1954). It plays the important role in decision-making and economics in the framework of Savage's subjective utility theory. In the latter, probability is formalized in the classical probabilistic framework (Kolmogorov, 1933) and it is endowed with the subjective interpretation.

We remark that STP is a simple consequence of the law of total probability—LTP (see (5)). Violation of LTP implies violation of STP. Thus, the degree of satisfaction of LTP can be used as a statistical test of classical (STP-type) rationality.

In cognitive psychology, violation of STP is known as the *disjunction effect*. A plenty of statistical data was collected in cognitive psychology in experiments demonstrating disjunction effect. For example, in experiments of the Prisoners' Dilemma type (Kahneman, 2003; Kahneman & Thaler, 2006; Kahneman & Tversky, 1972, 1979, 1984). Such data violate LTP. The latter implies irrationality (from classical viewpoint) of agents participating in experiments (mainly students).

We recall that LTP is derived from two assumptions that are firmly incorporated into the Kolmogorov axiomatics:

1. Additive law for probability.
2. Baeys formula for conditional probability.

Therefore, violation of LTP and, hence, of STP (and classical rationality) is generated either by violation of additivity of probability or the Bayes law for conditional probability or by the combination of these factors. Generally, this leads to the impossibility to use in decision-making Bayesian inference. Quantum(-like) agents proceed with more general inference machinery based on the quantum state update.

Hence, classical rationality is Bayesian inference rationality and quantum rationality is non-Bayesian inference rationality.[1] In the light of above considerations, one can ask:

Are quantum agents irrational?

As was discussed, by using QP it is possible to violate LTP and hence STP. Therefore, generally quantum-like agents are (classically) irrational. However, we can question the classical probabilistic approach to mathematical formalization of decision making and, consequently, the corresponding notion of rationality. We define quantum(-like) rationality as respecting the quantum calculus of probabilities and the quantum formula for interference of probabilities, LTP with the interference term (5).

4   Advantages, Disadvantages, and Roots of Quantum Rationality

4.1   Liberalization of Decision-Making

Quantum rationality means more freedom for decision-making, liberalization of this process. Generally, liberalization has its advantages and disadvantages. The main advantage of quantum rationality is that such agents can come to essentially larger spectrum of possible decisions than classically rational agents. Some quantum decisions are classically unapproachable. One of the main disadvantage is that decisions—events of decisions—can belong to incompatible decision algebras (σ-algebras of events in Kolmogorov axiomatics of CP, see Sect. 7). In such a case, it is impossible to come to consensus. We recall that in CP all possible decision events are unified in one common event-algebra. They can always be joined consistently with operations of conjunction, disjunction, and negation.

We remark that the CP-QP interplay is closely connected to the interplay of classical and quantum logic. Classical logic operates in a single Boolean algebra. Quantum logic operates in the lattice of orthogonal projectors in complex Hilbert space. This lattice can be represented as union of partially overlapping Boolean algebras representing compatible events given by commuting projectors. We recall that measure-theoretic realization of Boolean algebras are precisely σ-algebras of CP (Sect. 7). However, quantum logic is not just a collection of Boolean algebras.

---

[1] Of course, non-Bayesian probability updates are not reduced to quantum, given by state transformations in the complex Hilbert space. One may expect that human decision-making violates not only classical, but even quantum rationality.

It is based on the consistent transition from one algebra to another—through unitary transformations. The latter can be treated as change of logic coordinates.

By solving a problem, a cognitive system selects one of Boolean subalgebras of quantum logic; typically as simple as possible. Why? We shall discuss this question below.

4.2 Dysfunctional Disagreement and Information Overload

We point out that recently humans started to use widely the quantum decision-making technique, especially in social and political decision-making. In particular, this can explain the recent trend *in increasing dysfunctional disagreement in, e.g., political debates and generally in social life of the modern society.* This disagreement cannot be explained by insufficient information supply. There is a lot of information. The problem is in opposite—in *information overload* (Khrennikov, 2020). The information flows generated by mass-media and internet are so powerful, that by making a decision on the concrete complex problem humans are not able to construct the joint algebra for all possible events related to the problem under consideration. Different issues involved in a problem are treated in different algebras. Moreover, these algebras can differ essentially depending on agent's social network and his information environment (or it would better to say "a part of the information environment that is selected by this agent").

In the situation when an agent is overloaded with variety of information and he or she should make rapidly the decision on a complex problem, the quantum information processing (without attempting to unify all received information within a single Boolean algebra) demonstrates its superiority over classical one, at least from the viewpoint of minimization of computational resources and speed up of decision-making. Thus, in the modern society quantum rationality beats classical rationality, with "just one" casualty—inconsistency of some decisions. However, this is inconsistency w.r.t. classical Boolean logic. If all agents behave quantumly, then such inconsistencies become invisible.

The above discussion is applicable not only to biological systems, as say humans, but also to AI-systems. The latter when operating with powerful information flows would also prefer to use the quantum logic and QP. For each problem, a quantum(-like) AI-system, say robot, selects a proper Boolean algebra, but at the same time it keeps the possibility to use other Boolean algebras, corresponding to selection of different orthonormal bases in the Hilbert state space. However, transition from one Boolean algebra to another realized with a unitary operator U in the state space demands a lot of computational resources, because state space has big dimension; for n-qubit state space, it is 2^n. In situation of information overload and temporal constraints, an agent (biological or AI) has no possibility to perform such transition. Moreover, to get more or less complete image of the situation, an agent has to make transitions to a variety of Boolean algebras corresponding to incompatible variables which are represented by noncommuting Hermitian operators.

4.3 Social Laser

One of the consequences of information overload is that information loses its content. A human has no possibility to analyze deeply the content of communications delivered by mass-media and social networks. People process information without even attempting to construct an extended Boolean algebra of events. They operate with labels such as say COVID-19, vaccination, pandemy without trying to go deeper beyond this labels. Contentless information behaves as a bosonic quantum field which is similar to the quantum electromagnetic field. Interaction of humans with such quantum information field can generate a variety of quantum-like behavioral effects. One of them is social lasing, stimulated amplification of social actions (SASA) (Khrennikov, 2015a, 2016, 2018; Khrennikov et al., 2018, 2019; Tsarev et al., 2019). In social laser theory, humans play the role of atoms, social atoms (s-atoms). Interaction of the information field composed of indistinguishable (up to some parameters, as say social energy) excitations with gain medium composed of s-atoms generate the cascade type process of emission of social actions. SASA describes well, e.g., color revolutions and other types of mass protests (see Khrennikov, 2020 for detailed presentation).

5 Classical Versus Quantum Approach to the Problem of Agreement on Disagree

Aumann's approach (Aumann, 1976) to common knowledge and his theorem that rational agents cannot agree on disagree play the crucial role in theory of rationality. This theorem implies that rational agents acting under very natural conditions would never agreeing on disagree. Hence, it couples rationality with consistency of actions of decision makers. The main puzzle raised by Aumann's theorem is that people often "agree on disagree".

How can we explain this contradiction between the statement which was mathematically proved in the well-established framework of decision theory and the real behavior of humans?

The typical solution is that one of the two basic assumptions of the Aumann theorem is violated in the real processes of decision-making. We recall the Aumann's assumptions:

1. Common knowledge.
2. Common priors.

However, in modern information society these assumptions are very natural. Society is homogeneous and the majority of people have common priors. Information is openly distributed via mass-media and internet. Of course, there are attempts to use insider information. But, such attempts, e.g., at the financial market, are subject for punishment. Violations of the Aumann theorem cannot be reduced to insiders' activity.

In my paper (Khrennikov & Basieva, 2014b) (see also Haven et al., 2017; Khren-nikov, 2015b), it was pointed out that the Aumann theorem (Aumann, 1976) is also based on the third, so to say hidden, assumption, namely, rationality of agents. As was discussed in Sect. 3, classical notion of rationality is based on Bayesian infer-ence in decision-making which is CP-formalized in scientific theory. If real agents are not classically, but quantumly rational, they can violate the Aumann theorem, even under the assumptions of common knowledge and common priors.

The basic QP-departure from the classical Aumann's model is the existence of *incompatible information representations of the world* by different agents. To model this situation, instead of the set-theoretical (Boolean) partitions of the space of the states of the world Ω, we have to consider partitions of the unit operator in com-plex Hilbert space H (space of the quantum states of the world) consisting of the mutually orthogonal projectors. In general these partitions can be incompatible, i.e., the corresponding question-operators of different agents need not commute. Here, we proceed with the simplest mathematical model of quantum measurements based on the projection operators, measurements of the von Neumann-Lüders type (Von Neumann-Luders, 1955). Generalization to measurements represented as quantum instruments (Davies, 1976; Davies & Lewis, 1970; Okamura & Ozawa, 2016; Ozawa, 1984, 1997, 2004, Yuen, 1987) seems to be possible, but technically nontrivial.

In short, the main reason for this is that the basic quantum element introducing violations into Aumann's theorem is usage of a more general rule of updating of probabilities, the quantum analog of the Bayesian updating, see Sect. 3 for the gen-eral discussion. A few different sources of incompatibility are combined in this rule. Besides the most evident source, namely, possible incompatibility of information representations of agents (decision makers), two other sources also play important roles. These are possible incompatibilities of information representations with quan-tum events and common prior states. They both can contribute non-trivially into the interference term perturbing the matching of posterior probabilities; even in contexts with common prior states and nontrivial common knowledge (see Haven et al., 2017; Khrennikov, 2015b for details).

6 Common Knowledge: Illustrative Examples

The notion of *common knowledge* plays the crucial role in various problems of coordi-nation of actions and approaching conventions—in philosophy, economics (including accounting and capital market research), game theory, statistics, computer science, artificial intelligence. This notion is not simple and we shall devote this section to the informal discussion, irrelevant to the use of CP or QP. For illustration, we shall present the examples of the crucial impact of common knowledge in the problems of social coordination. The first one is based on author's personal experience and the second one is commonly used in literature on common knowledge.

We start with the remark that common knowledge is generalization of a simpler notion *mutual knowledge:* everybody in a group of people is aware about some fact or event. Now we present the first illustrative example for the notion of common knowledge:

Example 1 (*Earthquake, Japan, 11 March 2011* (Haven et al., 2017)) At that time I was participating in Quantum Bio-information workshop at the Noda-city campus of Tokyo University of Science, the fourth floor of the conference building. Around 15.00 the building started to shake, strongly shake.[2] This shaking of the building and that this is a sign of a strong earthquake was the mutual knowledge for the workshop participants, everybody seen this. However, people did not try to escape from the building until somebody loudly said: "this is a very strong earthquake." Immediately we sprung outside the building. This announcement of the fact known to everybody in the conference room made the mutual knowledge common and this changed our behavior crucially. After this announcement each workshop's participant knew: "each participant knows that each participant knows that earthquake is very strong." and so on, ad infinitum.

However, this was not the end of this story. After a half-hour staying at some distance from the conference building and seeing that there were no more signs of earthquake, we decided to return to the conference room and continue workshop. In the middle of the session the building started to shake again and sufficiently strongly. However, nobody said publicly that shaking is strong. Session continued. (May be participants expected such an announcement from session's chairman who may be expected it from conference's organizers.) My collaborator who was the last speaker of this session told that she was really scared during her talk and very angry when people started to ask questions after she finished.

Thus common knowledge is an essentially stronger assumption than simply mutual knowledge. To have *common knowledge means not only that everybody knows some information E, but even that everybody knows that everybody knows E and that everybody knows that everybody knows that everybody knows E and so on, ad infinitum.*

Remark. (Ad infinitum) The definition of common knowledge is based on the *infinite* hierarchy of levels of knowing. This presence of infinity might make the impression that this notion is not useful for concrete applications in which the infinite level of commonality is inapproachable. However, this is not the case. Of course, common knowledge is an ideal notion, but its role in science is similar to the role of other ideal notions, such as, e.g., a point, straight line, irrational number. We cannot proceed mathematically without such ideal notions. This is the exhibition of the transcendental structure of human reasoning.

Example 2 (*Blue eyes paradox* (Friedell, 1969) There is an island populated by people with blue and green eyes, say k people have blue eyes, others have green

---

[2] Comment: earthquakes nearby Tokyo happen often; in some periods things in room shake practically everyday. Amplitudes vary day to day; of course each time I estimated their strength. But this is difficult to do subjectively....

eyes. For island's inhabitants, the number k is unknown. At the beginning nobody knows the color of her/his eyes. There is the very strict rule:

If a person finds that she/he has blue eyes, that person must move from the island before sunrise.

At this island everybody knows eye colors of others. But, it is forbidden to discuss eye colors and there are no mirrors or similar devices. Once a stranger comes to the island and announced to all the people:

"At least one of you has blue eyes".

Thus it became common knowledge. The problem: what is the eventual outcome of this public announcement?

Consider first the simplest case, $k = 1$. Some person will recognize that she/he alone has blue eyes (by seeing only green eyes in the others) and leave at the first sunrise. Let now $k = 2$. At the first sunrise nobody leaves the island. Then two people having blue eyes by seeing only one person with blue eyes, and that no one left on the first sunrise understand that $k > 1$. They leave at the second sunrise. And so on by using the inductive argument. The paradox is that if $k > 1$ then the stranger told to people at this island what they already have known: there are blue-eyed people. However, without stranger's announcement this fact was not common knowledge. Its becoming common knowledge had dramatic consequences for inhabitants of the island.

For k = 2, it is first-order knowledge. Each person having blue eyes knows that there is someone with blue eyes, but she/he does not know that the other blue-eyed person has this same knowledge. For k = 3, it is second order knowledge. After 2 days, each person having blue eyes knows that a second blue-eyed person knows that a third person has blue eyes, but no one knows that there is a third blue-eyed person with that knowledge, until the third day arrives. And so on...

7   Kolmogorov's Model of Classical Probability Theory

Since CP-formalization of common knowledge is performed in the measure-theoretic framework, it is useful to recall the Kolmogorov's model (Kolmogorov, 1933) of CP.

The *Kolmogorov probability space* (Kolmogorov, 1933) is any triple

$$(\Omega, \mathcal{F}, P),$$

where Ω is a set of any origin and $\mathcal{F}$ is a σ-algebra of its subsets, P is a probability measure on $\mathcal{F}$. The set Ω represents random parameters of the model. Kolmogorov called elements of Ω *elementary events*. This terminology is standard in mathematical literature. Sets of elementary events are regarded as *events*.

The essence of Kolmogorov's approach is that not any subset of Ω can be treated as an event. For any stochastic model, the system of events $\mathcal{F}$ is selected from the very beginning. The main mathematical point is that $\mathcal{F}$ has to be a σ-*algebra*.

We remind that a σ-algebra is a system of sets which contains Ω and empty set, it is closed with respect to the operations of countable union and intersection and to the operation of taking the complement of a set. For example, the collection of all subsets of Ω is a σ-algebra. This σ-algebra is used in the case of finite or countable sets:

$$\Omega = \{\omega_1, ..., \omega_n, ...\}. \tag{6}$$

However, for "continuous sets", e.g., $\Omega = [a, b] \subset \mathbf{R}$, the collection of all possible subsets is too large to have applications. Typically it is impossible to describe a σ-algebra in the direct terms. To define a σ-algebra, one starts with a simple system of subsets of Ω and then consider the σ-algebra which is generated from this simple system with the aid of aforementioned operations. In particular, one of the most important for applications σ-algebras, the so-called σBorel-algebra, is constructed in this way by staring with the system consisting of all open and closed intervals of the real line. In a metric space (in particular, in a Hilbert space), the Borel σ-algebra is constructed by starting with the system of all open and closed balls.

Finally, we remark that in American literature the term σ-field is typically used, instead of σ-algebra.

The probability is defined as a *measure*, i.e., a map from $\mathcal{F}$ to nonnegative real numbers which is σ-additive:

$$P(\cup_j A_j) = \sum_j P(A_j), \tag{7}$$

where $A_j \in \mathcal{F}$ and $A_i \cap A_j = \emptyset, i \neq j$. The probability measure is always normalized by one:

$$P(\Omega) = 1. \tag{8}$$

In the case of a discrete probability space, see (6), the probability measures have the form

$$P(A) = \sum_{\omega_j \in A} p_j, \quad p_j = P(\{\omega_j\}).$$

In fact, any finite measure μ, i.e, $\mu(\Omega) < \infty$, can be transformed into the probability measure by normalization:

$$P(A) = \frac{\mu(A)}{\mu(\Omega)}, A \in \mathcal{F}. \tag{9}$$

A (real) random variable is a map $\xi : \Omega \to \mathbf{R}$ which is measurable with respect to the Borel σ-algebra $\mathcal{B}$ of $\mathbf{R}$ and the σ-algebra $\mathcal{F}$ of Ω. The latter means that, for any set $B \in \mathcal{B}$, its preimage $\xi^{-1}(B) = \{\omega \in \Omega : \xi(\omega) \in B\}$ belongs to $\mathcal{F}$. This condition provides the possibility to assign the probability to the events of the type "values of ξ belong to a (Borel) subset of the real line." The probability distribution of ξ is defined as

$$P_\xi(B) = P(\xi^{-1}(B)). \tag{10}$$

In the same way we define the real (and complex) vector-valued random variables, $\xi : \Omega \to \mathbf{R}^n$ and $\xi : \Omega \to \mathbf{C}^n$.

Let $\xi_1, ..., \xi_k$ be real-valued random variables. Their join probability distribution $P_{\xi_1,...,\xi_k}$ is defined as the probability distribution of the vector-valued random variable $\xi = (\xi_1, ..., \xi_k)$. To determine this probability measure, it is sufficient to define probabilities

$$P_{\xi_1,...,\xi_k}(\Gamma_1 \times ... \times \Gamma_k) = P(\omega \in \Omega : \xi_1(\omega) \in \Gamma_1, ..., \xi_k(\omega) \in \Gamma_k)$$

where Γ_j, $j = 1, ..., k$, are intervals (open, closed, half-open) of the real line.

We remark once again that LTP (5) is a theorem within Kolmogorov probability theory (Kolmogorov, 1933). We also recall that LTP plays the basic role in Bayesian probability inference.

8 Classical Formalization of Common Knowledge and Aumann Theorem

8.1 States of the World

Aumann's considerations are applicable to a finite number of *agents*, call them $i = 1, 2, ..., N$. These individuals are about to learn the answers to various multi-choice *questions*, to make observations.

Mathematically the situation is represented with the aid of classical probability space (based on the Kolmogorov axiomatics, 1933). Typically it is assumed that the state space Ω representing all possible states of the world is finite.

8.2 Agents' Information Representations

Each agent creates its *information representation* for possible states of the world based on its own possibilities to perform measurements, "to ask questions to the world." Mathematically these representations are given by partitions of $\Omega : \mathcal{P}^{(i)} = (P_j^{(i)})$, where, for each agent i,

$$\cup_j P_j^{(i)} = \Omega \text{ and } P_j^{(i)} \cap P_k^{(i)} = \emptyset, j \neq k.$$

Thus an agent cannot get to know the state of the world ω precisely; she/he can only get to know to which element of its information partition $P_j^{(i)} \equiv P^{(i)}(\omega)$ this ω belongs. In this set-theoretic model of knowledge, by definition the agent i knows an event E in the state of the world ω if the element of his information partition containing this ω is contained in E :

$$P^{(i)}(\omega) \subset E. \tag{11}$$

In logical terms, this can be written as $P^{(i)}(\omega) \Rightarrow E$, the event $P^{(i)}(\omega)$ implies the event E; we also remark that $\{\omega\} \Rightarrow P^{(i)}(\omega)$.

8.3 Common Prior

It is assumed that on Ω there is defined a probability p, *the common prior* of all agents. In the accordance with the measure-theoretic model of probability theory (Kolmogorov, 1933) there is given a σ-algebra, say $\mathcal{F}$, of subsets of Ω, its elements represent events ("propositions" in some interpretations), and there is given a probability measure p defined on $\mathcal{F}$. In the knowledge models it is typically assumed that $\mathcal{F}$ is generated by agents' partitions, i.e., this is the minimal σ-algebra containing all systems of set $\mathcal{P}^{(i)}$, $i = 1, ..., N$. It is important to point out that, in particular, such a σ-algebra contains all subsets of the form $P_{j_1}^{(1)} \cap ... \cap P_{j_N}^{(N)}$. Hence, in the classical knowledge model the prior probability is assigned not only to the individual elements of agents' information representations, i.e., $P_j^{(i)} \to p(P_j^{(i)})$, but even to more complex events

$$P_{j_1}^{(1)} \cap ... \cap P_{j_N}^{(N)} \to p_{j_1...j_N} \equiv p(P_{j_1}^{(1)} \cap ... \cap P_{j_N}^{(N)}). \tag{12}$$

Thus by agreeing on the prior the agents have to agree on numerous conjunctive probabilities.

8.4 CP-Formalization of the Notion of Common Knowledge

We consider the systems of sets $\tilde{\mathcal{P}}^{(i)} = \{\cup_m P_{j_m}^{(i)}\}$ consisting of finite unions of the elements of the systems $\mathcal{P}^{(i)}$ and the system $\tilde{\mathcal{P}} = \cap_i \tilde{\mathcal{P}}^{(i)}$. A set O belongs to the system $\tilde{\mathcal{P}}$ if it belongs to any $\tilde{\mathcal{P}}^{(i)}$. Thus, for each i, it can be represented as

$$O = \cup_m P_{j_m}^{(i)}, \tag{13}$$

for some finite set of indexes (depending on i).

We now repeat the definition of common knowledge for two agents (and we continue to proceed with two agents):

ACN *An event E is common knowledge at the state of the world ω if 1 knows E, 2 knows E, 1 knows 2 knows E, 2 knows 1 knows E, and so on.*

In theory of common knowledge, the basic role is played by the set of all states of the world for which E is common knowledge; it is denoted by the symbol κE. As was shown by Aumann (1976), this set of states of the world belongs to $\tilde{P}$ and, hence, for each i, it can be represented (in the case $\kappa E \neq \emptyset$) in the form (see (13)):

$$\kappa E = \cup_m P_{j_m}^{(i)}. \tag{14}$$

Let E be an event. For a state of the world w, each agent i updates the common prior $p(E)$ on the basis of the observation the element $P^{(i)}(w)$ of its information partition. (For this agent, it means that the state of the world w is contained in $P^{(i)}(w)$.) This update is given by the conditional probability

$$\mathbf{q}_i(w) = p(E \cap P^{(i)}(w))/p(P^{(i)}(w)).$$

We remark that the conditional probability $\mathbf{q}_i(w)$ is defined to be the same for all states of the world w in a given element of partition. Thus, in fact,

$$\mathbf{q}_i(w) \equiv \mathbf{q}_{ik},$$

where $w \in P_k^{(i)} = P^{(i)}(w)$.

8.5 Aumann Theorem

Now, Aumann's theorem states that if both

$$\mathbf{q}_1(w) = q_1 \text{ and } \mathbf{q}_2(w) = q_2 \tag{15}$$

are common knowledge and prior probabilities are the same, then necessarily $q_1 = q_2$—simply because

$$q_i = p(E|\kappa C_{q_1 q_2}) = p(E \cap \kappa C_{q_1 q_2})/p(\kappa C_{q_1 q_2}), \tag{16}$$

where $C_{q_1 q_2}$ is the event (15): "the first agent by updating the prior probability of the event E assigns the value q_1 and the second agent the value q_2."

9 Basics of Quantum Formalism

9.1 States

Denote by H a complex Hilbert space. For simplicity, we assume that it is finite dimensional. Pure states of a system S are given by normalized vectors of H and mixed states by density operators (positive semi-definite operators with unit trace).

9.2 Observables, Born Rule for Probability, and Projection Postulate

In the original quantum formalism (Von Neumann, 1955), physical observable A is represented by a Hermitian operator $\hat{A}$. We consider only operators with discrete spectra: $\hat{A} = \sum_x x \, \hat{E}^A(x)$, where $\hat{E}^A(x)$ is the projector onto the subspace of H corresponding to the eigenvalue x. Suppose that system's state is mathematically represented by a density operator ρ. Then the probability to get the answer x is given by *the Born rule*:

$$\Pr\{A = x \| \rho\} = Tr[\hat{E}^A(x)\rho] = Tr[\hat{E}^A(x)\rho\hat{E}^A(x)] \tag{17}$$

and according to the projection postulate the post-measurement state is obtained via the state transformation:

$$\rho \to \rho_x = \frac{\hat{E}^A(x)\rho\hat{E}^A(x)}{Tr\hat{E}^A(x)\rho\hat{E}^A(x)}. \tag{18}$$

For reader's convenience, we present these formulas for a pure initial state $\psi \in H$. The Born's rule has the form:

$$\Pr\{A = x \| \rho\} = \|\hat{E}^A(x)\psi\|^2 = \langle\psi|\hat{E}^A(x)\psi\rangle. \tag{19}$$

The state transformation is given by the projection postulate:

$$\psi \to \psi_x = \hat{E}^A(x)\psi/\|\hat{E}^A(x)\psi\|. \tag{20}$$

Here the observable-operator $\hat{A}$ (its spectral decomposition) uniquely determines the feedback state transformations $\mathcal{I}_A(x)$ for outcomes x

$$\rho \to \mathcal{I}_A(x)\rho = \hat{E}^A(x)\rho\hat{E}^A(x). \tag{21}$$

The map $x \to \mathcal{I}_A(x)$ given by (21) is the simplest (but very important) example of quantum instrument (Davies, 1976; Davies & Lewis, 1970; Ozawa, 1984, 1997; Yuen, 1987).

9.3   Quantum Logic

Following von Neumann (1955) and Birkhoff and von Neumann (1936) we represent *events, propositions,* as orthogonal projectors in complex Hilbert space H.

For an orthogonal projector P, we set $H_P = P(H)$, its image, and vice versa, for subspace L of H, the corresponding orthogonal projector is denoted by the symbol P_L.

The set of orthogonal projectors is a *lattice* with the order structure: $P \leq Q$ iff $H_P \subset H_Q$ or equivalently, for any $\Psi \in H$, $\langle \Psi | P\Psi \rangle \leq \langle \Psi | Q\Psi \rangle$. This lattice is known as *quantum logic*. Thus in classical Boolean logic events are represented by sets and in quantum logic events are represented by orthogonal projectors.

We recall that the lattice of projectors is endowed with operations "and" ($\wedge$), conjunction, and "or" ($\vee$), disjunction. For two projectors P_1, P_2, the projector $R = P_1 \wedge P_2$ is defined as the projector onto the subspace $H_R = H_{P_1} \cap H_{P_2}$ and the projector $S = P_1 \vee P_2$ is defined as the projector onto the subspace H_R defined as the minimal linear subspace containing the set-theoretic union $H_{P_1} \cup H_{P_2}$ of subspaces H_{P_1}, H_{P_2} : this is the space of all linear combinations of vectors belonging these subspaces. The operation of negation is defined as the orthogonal complement:

$$P^{\perp} = \{ y \in H : \langle y | x \rangle = 0 : \text{for all } x \in H_P \}.$$

In the language of subspaces the operation "and", conjunction, coincides with the usual set-theoretic intersection, but the operations "or", disjunction, and "not", negation, are nontrivial deformations of the corresponding set-theoretic operations. It is natural to expect that such deformations can induce deviations from classical Boolean logic.

Consider the following simple example. Let H be two-dimensional Hilbert space with the orthonormal basis (e_1, e_2) and let $v = (e_1 + e_2)/\sqrt{2}$. Then $P_v \wedge P_{e_1} = 0$ and $P_v \wedge P_{e_2} = 0$, but $P_v \wedge (P_{e_1} \vee P_{e_2}) = P_v$. Hence, for quantum events, in general *the distributivity law is violated*:

$$P \wedge (P_1 \vee P_2) \neq (P \wedge P_1) \vee (P \wedge P_2) \tag{22}$$

As can be seen from our example, even mutual orthogonality of the events P_1 and P_2 does not help to save the Boolean laws.

Thus quantum logic relaxes some constraints set by classical Boolean logic, in particular, the distributivity between the operations of conjunction and disjunction.

10   Quantum Formalization of Common Knowledge and Aumann Theorem with Interference Term

10.1   Quantum States of the World

In our quantum-like model, the *"states of the world"* are given by pure states. Thus the unit sphere $S_1(H)$ in a complex Hilbert space H represents (up to phase factors) all possible states of the world.

10.2   Agents' Quantum Information Representations

Questions posed by agents are mathematically described by Hermitian operators, say $A^{(i)}$. We state again that events (propositions) are identified with orthogonal projectors. For the state of the world Ψ, an event P *occurs* (takes place with probability 1) if Ψ belongs to H_P.

To simplify considerations, we proceed in the case of the finite dimensional state space of the world, $m = \dim H < \infty$. Each Hermitian operator can be represented as a linear combination of orthogonal projectors to its eigen-subspaces; the questions of agents can be expressed as

$$A^{(i)} = \sum_j a_j^{(i)} P_j^{(i)}, \qquad (23)$$

where $(a_j^{(i)})$ are real numbers, all different eigenvalues of $A^{(i)}$, and $(P_j^{(i)})$ are the orthogonal projectors onto the corresponding eigen-subspaces. Here (a_j) encode possible answers to the question of the ith agent.[3] The system of projectors $\mathcal{P}^{(i)} = (P_j^{(i)})$ is the spectral family of $A^{(i)}$. Hence, for any agent i, it is a "disjoint partition of unity":

$$\bigvee_k P_k^{(i)} = I, \quad P_k^{(i)} \wedge P_m^{(i)} = 0, k \neq m. \qquad (24)$$

We remark that (24) is simply the lattice-theoretical expression of the following operator equalities:

$$\sum_k P_k^{(i)} = I, \quad P_k^{(i)} P_m^{(i)} = 0, k \neq m. \qquad (25)$$

---

[3] Although in quantum physics the magnitudes of these numbers play an important role, in quantum information theory the eigenvalues are merely formal labels encoding information which can be extracted from a state with the aid of an observable. In the case of dichotomous answers, we can simply use zero to encode "no" and one to encode "yes".

This spectral family can be considered as *information representation* of the world by the ith agent. In particular, "getting the answer $a_j^{(i)}$" is the event which is mathematically described by the projector $P_j^{(i)}$.

If the state of the world is represented by Ψ and, for some k_0, $P_\Psi \leq P_{k_0}^{(i)}$, then

$$p_\Psi(P_{k_0}^{(i)}) = \text{Tr} P_\Psi P_{k_0}^{(i)} = 1 \text{ and, for } k \neq k_0, \ p_\Psi(P_k^{(i)}) = \text{Tr} P_\Psi P_k^{(i)} = 0.$$

Thus, in this case, the event $P_{k_0}^{(i)}$ happens with the probability one and other events from information representation of the world by the ith agent have zero probability.

However, opposite to the classical case, in general Ψ need not belong to any concrete subspace[4] $H_{P_k^{(i)}}$. Nevertheless, for any pure state Ψ, there exists the minimal projector $Q_\Psi^{(i)}$ of the form $\sum_m P_{j_m}^{(i)}$ such that $P_\Psi \leq Q_\Psi^{(i)}$. This projector can be constructed in the following way. Each state Ψ determines the set of indexes

$$O_\Psi^{(i)} = \{j : P_j^{(i)} \Psi \neq 0\}. \tag{26}$$

Then the minimal projector majorating the one-dimensional projector corresponding to the state Ψ has the form:

$$Q_\Psi^{(i)} = \sum_{j \in O_\Psi^{(i)}} P_j^{(i)}. \tag{27}$$

The projector $Q_\Psi^{(i)}$ represents the ith agent's knowledge about the Ψ-world. We remark that $p_\Psi(Q_\Psi^{(i)}) = 1$.

Consider the system of projectors $\tilde{\mathcal{P}}^{(i)}$ consisting of sums of the projectors from $\mathcal{P}^{(i)}$:

$$\tilde{\mathcal{P}}^{(i)} = \{P = \sum_m P_{j_m}^{(i)}\}. \tag{28}$$

Then

$$Q_\Psi^{(i)} = \min\{P \in \tilde{\mathcal{P}}^{(i)} : P_\Psi \leq P\}, \tag{29}$$

see (26), (27) for the constructive definition.

Thus in general, for the ith agent, the picture of the world is not only fuzzy (i.e., based on her partition of unity), but also probabilistic corresponding to the set of probabilities, $(p_\Psi(P_k^{(i)}), k = 1, 2, ...)$.

---

[4] We state again that in the classical probability model the states of the world are encoded by points of Ω. Take one fixed state ω. Since information representation of each agent is a partition of Ω, for each i there exists an element of partition, say $P_j^{(i)}$, containing this ω. For this state of the world, the ith agent should definitely get the answer $a_j^{(i)}$ corresponding the element $P_j^{(i)}$. Thus any agent is able to resolve uncertainty at least for her/his information representation (although she/he is not able to completely resolve uncertainty about the state of the world). In the quantum case, an agent is not able to resolve uncertainty even at the level of her/his information representation. And the prior probability is updated in this uncertainty context.

The ith agent picture of the world can be represented by the density operator (mixed quantum state)

$$\rho^{(i)} = \sum_k P_k^{(i)} P_\Psi P_k^{(i)} = \sum_k p_\Psi(P_k^{(i)})\rho_k^{(i)},$$

where $\rho_k^{(i)} = \frac{P_k^{(i)} P_\Psi P_k^{(i)}}{\mathrm{Tr} P_k^{(i)} P_\Psi P_k^{(i)}}$. Since each $\rho_k^{(i)}$ is a pure state, the ith picture of the world is given by the mixture of pure states, corresponding to "cells" $P_k^{(i)}$.

10.3 Quantum Way of Common Knowledge Formalization

Definition 1 For the Ψ-state of the world and the event E, the ith agent knows E if

$$Q_\Psi^{(i)} \le E. \tag{30}$$

It is evident that if, for the state of the world Ψ, the ith agent knows E, then $\Psi \in H_E$. In general the latter does not imply that E is known (for the state Ψ).[5] However, if $\Psi \in E = P_j^{(i)}$, then this event is known for i. The same is valid for any event of the form $E = P_{j_1}^{(i)} \vee \ldots \vee P_{j_k}^{(i)} (= P_{j_1}^{(i)} + \ldots + P_{j_k}^{(i)})$; if $\Psi \in H_E$, then such E is known for i.

We remark that the straightforward analog of the classical definition, see (11), would be based on condition $P_j^{(i)} \le E$ for

$$P_\Psi \le P_j^{(i)}, \tag{31}$$

instead of more general condition (30). However, it would trivialize the class of possible states of the world, because condition (31) is very restrictive.

We shall use the standard definition of common knowledge, see **ACN**, but based on the quantum representation of knowing an event, see Definition 1. As in the classical case, we have that "Where something is common knowledge, everybody knows it."

We recall that in the classical case, for each event E, there is considered the set of all states of the world for which E is common knowledge. It is denoted by the symbol κE.

---

[5] For example, the state space H is four dimensional with the orthonormal basis (e_1, e_2, e_3, e_4), the projectors P_1 and P_2 project H onto the subspaces with the bases (e_1, e_2) and (e_3, e_4), respectively. Here (P_1, P_2) is information representation of an agent. Let E be the projector onto the subspace with the basis (e_1, e_4) and let $\Psi = (e_1 + e_4)/\sqrt{2}$. Then $Q_\Psi = I$, the unit operator. Hence, E is not known for this agent, although it belongs to H_E.

This definition is naturally generalized to the quantum case. Here κE is defined as the projector on the subspace consisting of all states of the world for which E is common knowledge.

Similar to the set-theoretic framework, we introduce the system of projectors

$$\tilde{P} = \cap_i \tilde{P}^{(i)}.$$

We remark that (by definition) a projector $P \in \tilde{P}$ if and only if, for each $i = 1, ..., N$, it can be represented in the form

$$P = \sum_m P_{j_m}^{(i)}. \tag{32}$$

Examples illustrating how this common knowledge structure works can be found in Haven et al. (2017).

Now, before to formulate the quantum version of Aumann's theorem, we recall that while in the classical Aumann scheme the update of the prior probability distribution on the basis of information representations of agents plays the crucial role. The quantum analog of the Aumann scheme is based on the quantum procedure of the state update as the result of measurement, in the simplest case this is the projection postulate based update.

10.4 Quantum Version of Aumann's Theorem

The quantum common prior assumption is formulated naturally: both agents assign to possible states of the world the same quantum probability distribution given by the density operator ρ, a priori state. The agents do not know exactly the real state of the world which is always a pure state and in general a possible state of the world appears for them as a mixed quantum state ρ.

Theorem 1 (Khrennikov & Basieva, 2014b) *Under assumptions of common prior and common knowledge, the following interference version of Aumann's theorem holds:*

$$q_i - q_s = \frac{1}{\text{Tr}\rho^{\vee} C_{q_1...q_N}} \left(\sum_{j \neq m} \text{Tr} P_{k_j}^{(i)} \rho P_{k_m}^{(i)} E - \sum_{j \neq m} \text{Tr} P_{k_j}^{(s)} \rho P_{k_m}^{(s)} E \right). \tag{33}$$

If the amplitude of right-hand side of (33), the interference term between updates of the probability of the event E by two different agents, is small, we can say that the agents named i and s practically agree. The interference term can be considered as a measure of "agreement on disagree" between the agents.

11 Concluding Discussion

As was confirmed by plenty of experimental statistical data from cognitive psychology, decision-making, social and political sciences, generally human agents make their decisions by violating CP-laws (see introduction). It is natural to look for a more general formalism to model cognition and decision-making. The QP-formalism serves as the basis for such extension of CP-based approach. In this chapter, we discussed the consequences of the quantum-like model for rationality. Quantum rationality, i.e., based on quantum probability inference, differs crucially from classical Bayesian rationality. It has its advantages and disadvantages; some of them were discussed in this paper. The main reason for the use of quantum(-like) information processing is information overload, especially in the situation demanding quick decision-making under temporal and computation resource restrictions. Since the modern information society is characterized by intensification of information flows generated by mass-media, social networks, and working process, we can expect increasing use of quantum information processing by human agents and, hence, the phenomenon of quantum rationality will become more common. From the classical rationality viewpoint, this kind of rationality can be viewed as irrationality. However, it seems the only choice in the modern society.

The main danger of quantum rational behavior is that such agents become a very good medium for social engineering; in particular, a good active medium for social lasing. The latter can be used to generate instability throughout the world, in the form of mass protests and color revolutions.

The quantum-like approach to modeling of cognition, decision-making, and rationality will definitely find applications in artificial intelligence, for two types of AI-systems:

- Systems equipped with genuine quantum information processing devices, say quantum computers or simulators.
- Systems equipped with classical information processing devices, say classical digital or analog computers, realizing quantum(-like) information processing.

Personally I do not share the generally high expectation for successful realization of genuine quantum physical computing project, especially hopes that such quantum devices can be useful for AI-systems, say robots. I think that quantum information processing based on classical computational devices has better perspectives. But, since in science it is always difficult to make prognoses for future development, both types of AI-systems, genuine quantum and quantum-like, have to be studied. In future, the output of this paper may become useful for modeling rationality in collectives composed of quantum and quantum-like robots and other AI-systems.

References

Aerts, D., Khrennikov, A., Melucci, M., & Toni, B. (Eds.). (2019). Quantum-like models for information retrieval and decision-making. STEAM-H: Science, technology, engineering, agriculture, mathematics & health. Springer.

Arndt, M., Juffmann, T., & Vedral, V. (2009). Quantum physics meets biology. *HFSP Journal,3*(6), 386–400. https://doi.org/10.2976/1.3244985

Asano, M., Basieva, I., Khrennikov, A., & Yamato, I. (2017a). A model of differentiation in quantum bioinformatics. *Progress in Biophysics and Molecular Biology,130*, Part A, 88–98.

Asano, M., Basieva, I., Khrennikov, A., Ohya, M., Tanaka, Y., & Yamato, I. (2012). Quantum-like model for the adaptive dynamics of the genetic regulation of E. coli's metabolism of glucose/lactose. *Systems and Synthetic Biology, 6*, 1–7.

Asano, M., Basieva, I., Khrennikov, A., Ohya, M., Tanaka, Y., & Yamato, I. (2015b). Quantum information biology: From information interpretation of quantum mechanics to applications in molecular biology and cognitive psychology. *Foundations of Physics, 45*(10), 1362–1378.

Asano, M., Khrennikov, A., Ohya, M., Tanaka, Y., & Yamato, I. (2014). Violation of contextual generalization of the Leggett-Garg inequality for recognition of ambiguous figures. *Physica Scripta,T163*, 014006.

Asano, M., Khrennikov, A., Ohya, M., Tanaka, Y., & Yamato, I. (2015a). *Quantum adaptivity in biology: From genetics to cognition*. Springer.

Asano, M., Basieva, I., Khrennikov, A., Ohya, M., & Tanaka, Y. (2017b). A quantum-like model of selection behavior. *Journal of Mathematical Psychology, 78*, 2–12.

Asano, M., Ohya, M., Tanaka, Y., Basieva, I., & Khrennikov, A. (2011). Quantum-like model of brain's functioning: Decision making from decoherence. *Journal of Theoretical Biology, 281*(1), 56–64.

Aumann, R. J. (1976). Agreeing on disagree. *The Annals of Statistics,4*, 1236–1239.

Bagarello, F. (2019). *Quantum concepts in the social, ecological and biological sciences*. Cambridge University Press.

Bagarello, F., Basieva, I., Pothos, E. M., & Khrennikov, A. (2018). Quantum like modeling of decision making: Quantifying uncertainty with the aid of Heisenberg-Robertson inequality. *Journal of Mathematical Psychology, 84*, 49–56.

Bagarello, F., & Oliveri, F. (2013). A phenomenological operator description of interactions between populations with applications to migration. *Mathematical Models and Methods in Applied Sciences, 23*(03), 471–492.

Basieva, I., & Khrennikov, A. (2015). On the possibility to combine the order effect with sequential reproducibility for quantum measurements. *Foundations of Physics, 45*(10), 1379–1393.

Basieva, I., Pothos, E., Trueblood, J., Khrennikov, A., & Busemeyer, J. (2017). Quantum probability updating from zero prior (by-passing Cromwell's rule). *Journal of Mathematical Psychology, 77*, 58–69.

Bernroider, G. (2017). Neuroecology: Modeling neural systems and environments, from the quantum to the classical level and the question of consciousness. *Journal of Advanced Neuroscience Research, 4*, 1–9.

Bernroider, G., & Summhammer, J. (2012). Can quantum entanglement between ion transition states effect action potential initiation? *Cognitive Computation, 4*, 29–37.

Birkhoff, J., & von Neumann, J. (1936). The logic of quantum mechanics. *Annals of Mathematics, 37*(4), 823–843.

Boyer-Kassem, T., Duchene, S., & Guerci, E. (2015). Quantum-like models cannot account for the conjunction fallacy. *Theory and Decision, 10*, 1–32.

Busemeyer, J., & Bruza, P. (2012). *Quantum models of cognition and decision*. Cambridge University Press.

Busemeyer, J. R., Wang, Z., Khrennikov, A., & Basieva, I. (2014). Applying quantum principles to psychology. *Physica Scripta,T163*, 014007.

Busemeyer, J. R., Wang, Z., & Townsend, J. T. (2006). Quantum dynamics of human decision making. *Journal of Mathematical Psychology, 50,* 220–241.

Davies, E. B. (1976). *Quantum theory of open systems.* Academic Press.

Davies, E. B., & Lewis, J. T. (1970). An operational approach to quantum probability. *Communications in Mathematical Physics, 17,* 239–260.

Dzhafarov, E. N., & Kujala, J. V. (2012). Selectivity in probabilistic causality: Where psychology runs into quantum physics. *Journal of Mathematical Psychology, 56,* 54–63.

Dzhafarov, E. N., Zhang, R., & Kujala, J. V. (2015). Is there contextuality in behavioral and social systems? *Philosophical Transactions of the Royal Society A, 374,* 20150099.

Friedell, M. (1969). On the structure of shared awareness. *Behavioral Science, 14,* 28–39.

Hameroff, S. (1994). Quantum coherence in microtubules. A neural basis for emergent consciousness? *Journal of Consciousness Studies, 1,* 91–118.

Haven, E., & Khrennikov, A. (2013). *Quantum social science.* Cambridge University Press.

Haven, E., Khrennikov, A., & Robinson, T. R. (2017). *Quantum Methods in Social Science: A First Course.* WSP.

Haven, E. (2005). Pilot-wave theory and financial option pricing. *International Journal of Theoretical Physics, 44*(11), 1957–1962.

Kahneman, D., & Thaler., R. (2006). Utility maximization and experienced utility. *Journal of Economic Perspectives,20,* 221–232.

Kahneman, D. (2003). Maps of bounded rationality: Psychology for behavioral economics. *American Economic Review, 93*(5), 1449–1475.

Kahneman, D., & Tversky, A. (1972). Subjective probability: A judgment of representativeness. *Cognitive Psychology, 3*(3), 430–454.

Kahneman, D., & Tversky, A. (1979). Prospect theory: An analysis of decision under risk. *Econometrica, 47,* 263–291.

Kahneman, D., & Tversky, A. (1984). Choices, values and frames. *American Psychologist, 39*(4), 341–350.

Khrennikov, A. (2004b). *Information dynamics in cognitive, psychological, social, and anomalous phenomena.* Ser.: Fundamental Theories of Physics.

Khrennikov, A. (2010). *Ubiquitous quantum structure: From psychology to finances.* Springer.

Khrennikov, A. (2015a). Quantum-like model of unconscious-conscious dynamics. *Frontiers in Psychology,6,* Art. N 997.

Khrennikov, A. (2016). *Probability and randomness: Quantum versus classical.* Imperial College Press.

Khrennikov, A. (2020). *Social laser.* Jenny Stanford Publ.

Khrennikov, A., & Basieva, I. (2014a). Quantum model for psychological measurements: From the projection postulate to interference of mental observables represented as positive operator valued measures. *NeuroQuantology,12,* 324–336.

Khrennikov, A., Alodjants, A. Trofimova., A., & Tsarev, D. (2018). On interpretational questions for quantum-like modeling of social lasing. *Entropy,20*(12), 921.

Khrennikov, A., Basieva, I., Dzhafarov, E.N., Busemeyer, J. R. (2014). Quantum models for psychological measurements: An unsolved problem. *PLOS ONE, 9,* Art. e110909.

Khrennikov, A., Basieva, I., Pothos, E. M., & Yamato, I. (2018). Quantum probability in decision making from quantum information representation of neuronal states. *Scientific Reports, 8.* Article number: 16225.

Khrennikov, A. (1999). Classical and quantum mechanics on information spaces with applications to cognitive, psychological, social and anomalous phenomena. *Foundations of Physics, 29,* 1065–1098.

Khrennikov, A. (2003). Quantum-like formalism for cognitive measurements. *Biosystems, 70,* 211–233.

Khrennikov, A. (2004a). On quantum-like probabilistic structure of mental information. *Open Systems and Information Dynamics, 11*(3), 267–275.

Khrennikov, A. (2006). Quantum-like brain: Interference of minds. *BioSystems, 84,* 225–241.

Khrennikov, A. (2015a). Towards information lasers. *Entropy, 17*(10), 6969–6994.

Khrennikov, A. (2015b). Quantum version of Aumann's approach to common knowledge: Sufficient conditions of impossibility to agree on disagree. *Journal of Mathematical Economics, 60,* 89–104.

Khrennikov, A. (2016). Quantum Bayesianism as the basis of general theory of decision-making. *Philosophical Transactions of the Royal Society A, 374,* 20150245.

Khrennikov, A. (2016). Social laser: Action amplification by stimulated emission of social energy. *Philosophical Transactions of the Royal Society, 374*(2054), 20150094.

Khrennikov, A. (2018). Social laser model: From color revolutions to Brexit and election of Donald Trump. *Kybernetes, 47*(2), 273–278.

Khrennikova, P. (2014). A quantum framework for 'Sour Grapes' in cognitive dissonance. In H. Atmanspacher, E. Haven, K. Kitto, & D. Raine (Eds.), *Quantum interaction. QI 2013. Lecture Notes in Computer Science* (Vol. 8369). Springer.

Khrennikova, P. (2016). Quantum dynamical modeling of competition and cooperation between political parties: The coalition and non-coalition equilibrium model. *Journal of Mathematical Psychology,* to be published.

Khrennikova, P. (2017). Modeling behavior of decision makers with the aid of algebra of qubit creation-annihilation operators. *Journal of Mathematical Psychology, 78,* 76–85.

Khrennikov, A., & Basieva, I. (2014b). Possibility to agree on disagree from quantum information and decision making. *Journal of Mathematical Psychology, 62*(3), 1–5.

Khrennikov, A., Toffano, Z., & Dubois, F. (2019). Concept of information laser: From quantum theory to behavioural dynamics. *The European Physical Journal Special Topics, 227*(15–16), 2133–2153.

Kolmogorov, A. N. (1933). *Grundbegriffe der Wahrscheinlichkeitsrechnung.* Springer.

Melucci, M. (2015). *Introduction to information retrieval and quantum mechanics.* Springer.

Okamura, K., & Ozawa, M. (2016). Measurement theory in local quantum physics. *Journal of Mathematical Physics,57,* 015209.

Ozawa, M. & Khrennikov, A. (2020a). Application of theory of quantum instruments to psychology: Combination of question order effect with response replicability effect. *Entropy,22*(1), 37. 1-9436.

Ozawa, M. & Khrennikov, A. (2020b). Modeling combination of question order effect, response replicability effect, and QQ-equality with quantum instruments. arXiv:2010.10444

Ozawa, M. (1984). Quantum measuring processes for continuous observables. *Journal of Mathematical Physics, 25,* 79–87.

Ozawa, M. (1997). An operational approach to quantum state reduction. *Annals of Physics (N.Y.), 259,* 121–137.

Ozawa, M. (2004). Uncertainty relations for noise and disturbance in generalized quantum measurements. *Annals of Physics (N.Y.), 311,* 350–416.

Penrose, R. (1989). *The Emperor's new mind.* Oxford University Press.

Pothos, E., & Busemeyer, J. R. (2009). A quantum probability explanation for violations of 'rational' decision theory. *Proceedings of Royal Society B, 276,* 2171–2178.

Savage, L. J. (1954). *The foundations of statistics.* Wiley.

Surov, I. A., Pilkevich, S. V., Alodjants, A. P., & Khmelevsky, S. V. (2019). Quantum phase stability in human cognition. *Frontiers in Psychology, 10,* 929.

Tsarev, D., Trofimova, A., Alodjants, A., et al. (2019). Phase transitions, collective emotions and decision-making problem in heterogeneous social systems. *Scientific Reports, 9,* 18039.

Umezawa, H. (1993). *Advanced field theory: Micro, macro and thermal concepts.* AIP.

van Rijsbergen, C. J. (2004). *The geometry of information retrieval.* Cambridge University Press.

Vitiello, G. (2001). *My double unveiled: The dissipative quantum model of brain.* Advances in Consciousness Research, John Benjamins Publishing Company.

Vitiello, G. (1995). Dissipation and memory capacity in the quantum brain model. *International Journal of Modern Physics B, 9,* 973.

Von Neumann, J. (1955). *Mathematical foundations of quantum mechanics.* Princeton University Press.

Wang, Z., & Busemeyer, J. R. (2013). A quantum question order model supported by empirical tests of an a priori and precise prediction. *Topics in Cognitive Science, 5,* 689–710.

Wang, Z., Solloway, T., Shiffrin, R. M., & Busemeyer, J. R. (2014). Context effects produced by question orders reveal quantum nature of human judgments. *PNAS, 111,* 9431–9436.

White, L. C., Pothos, E. M., & Busemeyer, J. R. (2014). Sometimes it does hurt to ask: The constructive role of articulating impressions. *Cognition, 133*(1), 48–64.

Yuen, H. P. (1987). Characterization and realization of general quantum measurements. In M. Namiki et al. (Ed.), *Proceedings of the 2nd International Symposium of Foundations of Quantum Mechanics* (pp. 360–363).

Quantum Music, Quantum Arts and Their Perception

Volkmar Putz and Karl Svozil

Abstract The expression of human art, and supposedly sentient art in general, is modulated by the available rendition, receiving and communication techniques. The components or instruments of these techniques ultimately exhibit a physical, in particular, quantum layer, which in turn translates into physical and technological capacities to comprehend and utilize what is possible in our universe. In this sense, we can apply a sort of Church-Turing thesis to art, or at least to its rendition.

Keywords Quantum computer music · Quantum music theory · Music perception · Church–Turing thesis · Bose and Fermi modelling

1 Realm of Quantum Expressibility

A short glance at historic practices of music and artistic expression, in general, suggests that there has been, and still is, a fruitful exchange of ideas between craftsmanship, technology, and (material) sciences on the one hand, and entertainment, artistry, and creativity on the other hand. Impulses and ideas flow back and forth, very much like in the accompanying fields of mathematics and natural sciences. This is even true, in particular, for culinary subjects such as molecular gastronomy, where it has been argued that "food processing dominates cuisines": because even if all of the French recipes would have been erased from people and other memories, most if not all of these revered dishes could be "recovered" by merely following "reasonable" rules of food processing This (2005)—which strongly are linked to technology, such as the "domestication" of fire.

V. Putz (✉)
Pädagogische Hochschule Wien, Grenzackerstraße 18, 1100 Vienna, Austria
e-mail: volkmar.putz@phwien.ac.at

K. Svozil
Institute for Theoretical Physics, Vienna University of Technology, Wiedner Hauptstraße 8-10/136, 1040 Vienna, Austria
e-mail: svozil@tuwien.ac.at

© Springer Nature Switzerland AG 2022
E. R. Miranda (ed.), *Quantum Computing in the Arts and Humanities*,
https://doi.org/10.1007/978-3-030-95538-0_5

It thus comes of no surprise that the evolution of quantum physics brought about the quest for the quantum arts; and in particular, for quantum music Putz and Svozil (2017) and quantum fine arts, especially quantum visual art. Indeed, every aspect of human life can be re-evaluated and reframed in terms of the quantum paradigm.

In our (not so humble) opinion there are two immediate issues: One issue is the tendency to re-introduce irrational "magic", a sort of "quantum hocus pocus" Svozil (2016) that brings it close to the esoteric, and fosters a kind of pseudo-religion allegedly justified by the most advanced contemporary physics.

Another, converse, issue is the temptation to argue that, just like in quantum computing (Mermin, 2007, Section 1.1), "any art is quantum" as the "underlying physical layer" of any (classical) artistic expression is governed by the laws of quantum mechanics. However, we emphasize upfront that we have to resist this temptation towards a premature flattening and folding of the quantum phenomena into classical molds. Rather we consider quantum arts, and, in particular, quantum music, as operations exploiting certain very special transformations of physical internal states, subject to very carefully controlled conditions.

So what exactly are these very special transformations that characterize quantum art? In this regard, we can proceed almost in parallel to the development of quantum computation Fortnow (2003); Nielsen and Chuang (2010); Mermin (2007), and point out some central assets or capacities:

(i) parallelization through coherent superposition (aka simultaneous linear combination) of classically mutually exclusive tones or signals that are acoustic, optic, touch, taste, or otherwise sensory;

(ii) entanglement not merely by classical correlation Peres (1978) but by relational encoding Schrödinger (1935); Brukner and Zeilinger (1999); Zeilinger (1999); Brukner et al. (2002) of multi-partite states such that any classical information is "scrambled" into relational, joint multi-partite properties while at the same time losing value definiteness about the single constituents of such multi-partite states —this can be seen as a sort of zero-sum game, a tradeoff between individual and collective properties;

(iii) complementarity associated with value (in)definiteness of certain tones or signals that is acoustic, optic, touch, taste, or otherwise: if one such observable is definite, another is not, and *vice versa*;

(iv) contextuality is an "enhanced" form of complementarity and value indefiniteness that can be defined in various ways Dzhafarov et al. (2017); Abramsky (2018); Grangier (2002); Aufféves and Grangier (2018, 2020); Grangier (2020); Budroni et al. (2021), in particular, emphasizing homomorphic, structure-preserving nonembeddability into classical schemes Specker (1960); Kochen and Specker (1967); Svozil (2021)

Those criteria or observables constitute significant signatures of quantum behavior. The transformations and processing of classical-to-quantum states or quantum states exhibiting these features can be considered musical, optical, or other instruments or "transmitters" for the creation of quantum art. Similarly, assorted trans-

Fig. 1 (Color online) Temporal succession of quantum tones $|\Psi_c\rangle$, $|\Psi_d\rangle$, ..., $|\Psi_b\rangle$ in the C major scale forming the octave basis $\mathfrak{B}$

formations process quantum art. Finally, the process of information transmission requires instruments of perception or "receivers" (Shannon, 1948, Fig. 1).

Let us mention typical components and theoretical entities as example transformations. For instance, Hadamard transformations produce perfect "mixtures" of classically mutually exclusive signals. Quantum Fourier transforms produce generalized mixtures. All of them have to be uniformly unitary—that is, in terms of the various equivalent formal definitions, they have to transform orthonormal basis into orthonormal ones, they have to preserve scalar products or norms, and their inverse is the adjoint. One of the physical realizations is in terms of generalized beam splitters Reck et al. (1994); Zukowski et al. (1997).

Depending on whether we are willing to contemplate genuine quantum receivers or merely classical ones we end up with either a quantum cognition or with merely a classical cognition of this quantum art; and, in particular, of quantum music. In the first, radical deviation from classical music, we would have to accept the possibility of human or sentient consciousness or audience to perceive quantum impressions.

This is ultimately a neurophysiologic question. It might well be that the processing of signals exterior to the receiving and perceiving "somewhere along those channels" requires a breakdown to classicality; most likely through the introduction of stochasticity Glauber (1986). This is very much in the spirit of Schrödinger's cat Schrödinger (1935) and (later) quantum jellyfish Schrödinger (1995) metaphors based on the assumption that, ultimately, even if decoherence by environmental intake can be controlled, there cannot be any simultaneous co-experience of being both dead and alive, just as there might not be any co-experience of passing into a room by two separate doors simultaneously.

On the other hand, nesting of the Wigner's friend type von Neumann (1996); Everett (1957); Wigner (1995); Everett (2012), suggests that there might be substance to a sort of mindful co-experience of two classical distinct experiences. Whether such experiences remain on the subconscious primordial level of perception, or whether this can be levied to a full cognitive level is a fascinating question on its own that exceeds the limited scope of this article.

2   Quantum Musical Tones

In what follows we closely follow our nomenclature and presentation of quantum music Putz and Svozil (2017). Those formal choices are neither unique nor comprehensive. Alternatives are mentioned.

We consider a quantum octave in the C major scale, which classically consists of the tones c, d, e, f, g, a, and b, represented by eight consecutive white keys on a piano. (Other scales are straightforward.) At least three ways to quantize this situation can be given:

(i) bundling octaves by coherent their superposition (aka simultaneous linear combination), as well as
(ii) considering pseudo-field theoretic models treating notes as field modes that are either bosonic or fermionic.

The seven tones c, d, e, f, g, a, and b of the octave can be considered as belonging to disjoint events (maybe together with the null event 0) whose probabilities should add up to unity. This essentially suggests a formalization by a seven (or eight) dimensional Hilbert space $\mathbb{C}^7$ or $\mathbb{C}^8$) with the standard Euclidean scalar product. The respective Hilbert space represents a full octave.

We shall study the seven-dimensional case $\mathbb{C}^7$. The seven tones forming one octave can then be represented as an orthonormal basis $\mathfrak{B}$ of $\mathbb{C}^7$ by forming the set theoretical union of the mutually orthogonal unit vectors; that is, $\mathfrak{B} = \{|\Psi_c\rangle, |\Psi_d\rangle, \ldots |\Psi_b\rangle\}$, where the basis elements are the Cartesian basis tuples

$$|\Psi_c\rangle = (0, 0, 0, 0, 0, 0, 1),$$
$$|\Psi_d\rangle = (0, 0, 0, 0, 0, 1, 0),$$
$$\ldots$$
$$|\Psi_b\rangle = (1, 0, 0, 0, 0, 0, 0)$$

of $\mathbb{C}^7$. Figure 1 depicts the basis $\mathfrak{B}$ by its elements, drawn in different colors.

2.1   Bundling Octaves into Single Tones

Pure quantum musical states could be represented as unit vectors $|\psi\rangle \in \mathbb{C}^7$ which are linear combinations of the basis $\mathfrak{B}$; that is,

$$|\psi\rangle = \alpha_c|\Psi_c\rangle + \alpha_d|\Psi_d\rangle + \cdots + \alpha_b|\Psi_b\rangle, \tag{1}$$

with coefficients α_i satisfying

$$|\alpha_c|^2 + |\alpha_d|^2 + \cdots + |\alpha_b|^2 = 1. \tag{2}$$

Equivalent representations of $|\psi\rangle$ are in terms of the one-dimensional subspace $\{|\phi\rangle \mid |\phi\rangle = \alpha|\psi\rangle, \ \alpha \in \mathbb{C}\}$ spanned by $|\psi\rangle$, or by the projector $\mathbf{E}_\psi = |\psi\rangle\langle\psi|$.

A musical "composition"—indeed, and any succession of quantized tones forming a "melody"—would be obtained by successive unitary permutations of the state $|\psi\rangle$. The realm of such compositions would be spanned by the succession of all unitary transformations $\mathbf{U} : \mathfrak{B} \mapsto \mathfrak{B}'$ mapping some orthonormal basis $\mathfrak{B}$ into another orthonormal basis $\mathfrak{B}'$; that is Schwinger (1960), $\mathbf{U} = \sum_i |\Psi'_i\rangle\langle\Psi_i|$.

2.2 Coherent Superposition of Tones as a New Form of Musical Parallelism

One of the mind-boggling quantum field theoretic features of a "bundling" within single modes is the possibility of the simultaneous "co-existence" of classically excluding musical states, such as a 50:50 quantum g in the C major scale obtained by sending $|0_g\rangle$ through the Hadamard gate $\mathbf{H} = \frac{1}{\sqrt{2}} \begin{pmatrix} 1 & 1 \\ 1 & -1 \end{pmatrix}$, resulting in $\frac{1}{\sqrt{2}}\left(|0_g\rangle - |1_g\rangle\right)$, and depicted in Fig. 2 by a 50 white 50 black; that is, gray, tone (though without the relative "$-$" phase).

This novel form of musical expression might contribute to novel musical experiences; in particular, if any such coherent superposition can be perceived by the audience in full quantum uniformity. This would require the cognition of the recipient to experience quantum coherent superpositions—a capacity that is highly speculative. It has been mentioned earlier that any such capacity is related to Schrödinger's cat Schrödinger (1935) and quantum jellyfish Schrödinger (1995) metaphors, as well as to nestings of the Wigner's friend type von Neumann (1996); Everett (1957); Wigner (1995); Everett (2012).

2.3 Classical Perception of Quantum Musical Parallelism

In the following, we shall assume that quantum music is "reduced" to the continuous infinity of its classical forms. Then, if a classical auditorium listens to the quantum musical state $|\psi\rangle$ in Eq. 1, the individual classical listeners may perceive $|\psi\rangle$

$|\Phi_g\rangle$

Fig. 2 Representation of a 50:50 quantum tone $|\Phi_g\rangle = \frac{1}{\sqrt{2}}\left(|0_g\rangle - |1_g\rangle\right)$ in gray (without indicating phase factors)

very differently; that is, they will hear only a *single one* of the different tones with probabilities of $|\alpha_c|^2$, $|\alpha_d|^2$, ..., and $|\alpha_b|^2$, respectively.

Indeed, suppose that classical recipients (aka "listeners") modeled by classical measurement devices acting as information-theoretic receivers are assumed. Then any perception (aka "listening" or reception) of a quantum musical state that is in a coherent superposition—with some coefficients $0 < |\alpha_i| < 1$—because of the supposedly irreducably stochastic Zeilinger (2005) quantum-to-classical translation Svozil (2004) represents an "irreducible" Peres (1980); Scully and Drühl (1982); Greenberger and YaSin (1989); Scully et al. (1991); Zajonc et al. (1991); Kwiat et al. (1992); Pfau et al. (1994); Chapman et al. (1995); Herzog et al. (1995) stochastic measurement. This can never render a unique classical listening experience, as the probability to hear the tone i is $|\alpha_i|^2$. Therefore, partitions of the audience will hear different manifestations of the quantum musical composition made up of all varieties of successions of tones. These experiences multiply and diverge as more tones are perceived.

For the sake of a demonstration, let us try a two-note quantum composition. We start with a pure quantum mechanical state in the two-dimensional subspace spanned by $|\Psi_c\rangle$ and $|\Psi_g\rangle$, specified by

$$|\psi_1\rangle = \frac{4}{5}|\Psi_c\rangle + \frac{3}{5}|\Psi_g\rangle = \frac{1}{5}\begin{pmatrix} 4 \\ 3 \end{pmatrix}. \tag{3}$$

$|\psi_1\rangle$ would be detected by the listener as c in 64% of all measurements (listenings), and as g in 36% of all listenings. Using the unitary transformation $\mathbf{X} = \begin{pmatrix} 0 & 1 \\ 1 & 0 \end{pmatrix}$, the next quantum tone would be

$$|\psi_2\rangle = \mathbf{X}|\psi_1\rangle = \frac{3}{5}|\Psi_c\rangle + \frac{4}{5}|\Psi_g\rangle = \frac{1}{5}\begin{pmatrix} 3 \\ 4 \end{pmatrix}. \tag{4}$$

This means for the quantum melody of both quantum tones $|\psi_1\rangle$ and $|\psi_2\rangle$ in succession—for the score, see Fig. 3—that in repeated measurements, in $0.64^2 = 40.96\%$ of all cases $c - g$ is heard, in $0.36^2 = 12.96\%$ of all cases $g - c$, in $0.64 \cdot 0.36 = 23.04\%$ of all cases $c - c$ or $g - g$, respectively.

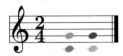

Fig. 3 (Color online) A two-note quantum musical composition—a natural fifth

3 Quantum Musical Entanglement

Quantum entanglement Schrödinger (1935) is the property of multipartite quantum systems to code information "across quanta" in such a way that the state of any individual quantum remains irreducibly indeterminate; that is, not determined by the entangled multipartite state Schrödinger (1935); Brukner and Zeilinger (1999); Zeilinger (1999); Brukner et al. (2002). Thus the entangled whole should not be thought of as composed of its proper parts. Formally, the composite state cannot be expressed as a product of separate states of the individual quanta.

A typical example of an entangled state is the *Bell state*, $|\Psi^-\rangle$ or, more generally, states in the Bell basis spanned by the quantized notes e and a; that is

$$|\Psi^\pm\rangle = \frac{1}{\sqrt{2}}\left(|0_e\rangle|1_a\rangle \pm |1_e\rangle|0_a\rangle\right),$$

$$|\Phi^\pm\rangle = \frac{1}{\sqrt{2}}\left(|0_e\rangle|0_a\rangle \pm |1_e\rangle|1_a\rangle\right),$$

(5)

A necessary and sufficient condition for entanglement among the quantized notes e and a is that the coefficients α_1, α_2, α_3, α_4 of their general composite state $|\Psi_{ga}\rangle = \alpha_1|0_e\rangle|0_a\rangle + \alpha_2|0_e\rangle|1_a\rangle + \alpha_3|1_e\rangle|0_a\rangle + \alpha_4|1_e\rangle|1_a\rangle$ obey $\alpha_1\alpha_4 \neq \alpha_2\alpha_3$ (Mermin, 2007, Sec. 1.5). This is clearly satisfied by Eqs. (5). Figure 4 depicts the entangled musical Bell states.

Fig. 4 Quantum musical entangled states $|\Psi_{ea}^-\rangle$ and $|\Psi_{ea}^+\rangle$ in the first bar, and $|\Phi_{ea}^-\rangle$ and $|\Phi_{ea}^+\rangle$ in the second bar (without relative phases)

Fig. 5 (Color online) Quantum musical entangled states for bundled octaves $|\Psi_{ea'}^-\rangle$ and $|\Psi_{ea'}^+\rangle$ in the first bar, and $|\Phi_{ea'}^-\rangle$ and $|\Phi_{ea'}^+\rangle$ in the second bar (without relative phases)

Entanglement between different octaves can be constructed similarly. Figure 5 depicts this configuration for an entanglement between e and a'.

4 Quantum Musical Complementarity and Contextuality

Although complementarity Pauli (1933) is mainly discussed in the context of observables, we can present it in the state formalism by observing that, as mentioned earlier, any pure state $|\psi\rangle$ corresponds to the projector $\mathbf{E}_\psi = |\psi\rangle\langle\psi|$. In this way, any two nonvanishing nonorthogonal and noncollinear states $|\psi\rangle$ and $|\phi\rangle$ with $0 < |\langle\phi|\psi\rangle| < 1$ are complementary. For the dichotomic field approach, Fig. 6 represents a configuration of mutually complementary quantum tones for the note a in the C major scale (a), and mutually complementary linear combinations as introduced in Sect. 2 (b).

Complementarity can be extended to more advanced configurations of contexts. These quantum configurations and their associated quantum probability distributions, if interpreted classically, either exhibit violations of classical probability theory, classical predictions, or nonisomorphic embeddability of observables into classical propositional structures Dzhafarov et al. (2017); Abramsky (2018); Grangier (2002); Aufféves and Grangier (2018, 2020); Grangier (2020); Budroni et al. (2021); Specker (1960); Kochen and Specker (1967); Svozil (2021).

5 Bose and Fermi Model of Tones

An alternative quantization to the one discussed earlier is in analogy to some fermionic or bosonic—such as the electromagnetic—field. Just as the latter one in quantum optics Glauber (1970, 2007) and quantum field theory Weinberg (1977) is quantized by interpreting every single mode (determined, for the electromagnetic

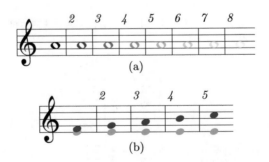

Fig. 6 Temporal succession of complementary tones (a) for binary occupancy $|\phi_a\rangle = \alpha_a|0_a\rangle + \beta_a|1_a\rangle$, with $|\alpha_a|^2 + |\beta_a|^2 = 1$ with increasing $|\alpha_a|$ (decreasing occupancy), (b) in the bundled octave model, separated by bars

Fig. 7 Temporal succession of tones $|\Psi_c\rangle$, $|\Psi_d\rangle$, ..., $|\Psi_b\rangle$ in an octave in the C major scale with dicreasing mean occupancy

field for instance by a particular frequency and polarization) as a sort of "container"—that is, by allowing the occupancy of that mode to be either empty or any positive integer (and a coherent superposition thereof)—we obtain a vast realm of new musical expressions which cannot be understood in classical terms.

Whereas in a "bosonic field model" occupancy of field modes is easy to be correlated with the classical volume of the corresponding tone, in what follows we shall restrict ourselves to a sort of "fermionic field model" of music which is characterized by a binary, dichotomic situation, in which every tone has either null or one occupancy, represented by $|0\rangle = (0, 1)$ or $|1\rangle = (1, 0)$, respectively. Thus every state of such a tone can thus be formally represented by entities of a two-dimensional Hilbert space, $\mathbb{C}^2$, with the Cartesian standard basis $\mathfrak{B} = \{|0\rangle, |1\rangle\}$.

Any note $|\Psi_i\rangle$ of the octave consisting of $|\Psi_c\rangle$, $|\Psi_d\rangle$, ..., $|\Psi_b\rangle$, in the C major scale can be represented by the coherent superposition of its null and one occupancies; that is,

$$|\Psi_i\rangle = \alpha_i|0_i\rangle + \beta_i|1_i\rangle, \tag{6}$$

with $|\alpha_i|^2 + |\beta_i|^2 = 1$, $\alpha_i.\beta_i \in \mathbb{C}$.

Every tone is characterized by the two coefficients α and β, which in turn can be represented (like all quantized two-dimensional systems) by a Bloch sphere, with two angular parameters. If we restrict our attention (somewhat superficially) to real Hilbert space $\mathbb{R}^2$, then the unit circle, and thus a single angle φ, suffices for a characterization of the coefficients α and β. Furthermore, we may very compactly notate the mean occupancy of the notes by gray levels. Now, in this "fermionic setting", with the mean occupation number of any tone between 0 and 1 the gray level does not indicate the volume of the corresponding tone but the mere chance of it being present or not, see also Sect. 2. Figure 7 depicts a sequence of tones in an octave in the C major scale with decreasing occupancy, indicated as gray levels.

In this case, any nonmonotonous unitary quantum musical evolution would have to involve the interaction of different tones; that is, in a piano setting, across several keys of the keyboard.

6 Quantum Visual Arts

Just as for the performing arts such as music one could contemplate the quantum options and varieties for the visual arts. Suffice it to say that the notion of "color" experience can be extended to the full quantum optical varieties that result from the electromagnetic field quantization, as already mentioned earlier. Incidentally, Schrödinger published a series of papers on classical color perception Schrödinger (1924); Schrödinger and Niall (2017) until around 1925. Yet to our best knowledge he never considered the particular quantum aspects of human color and light perception.

Human rod cells respond to individual photons Hecht et al. (1942); Westheimer (2016). Moreover, recent reports suggest that humans might be capable of "being aware" of the detection of a single-photon incident on the cornea with a probability significantly above chance Tinsley et al. (2016). It thus may be suspected that this area of perception presents the most promising pathway into truly quantum perception. Speculations how this issue may be transferred to the perception of sound are compelling.

Let us state up front that quantum visual art, and, in particular, quantum parallelism, is not about additive color mixing, but it is about the simultaneous existence of different, classically mutually exclusive "colors", or visual impressions in general. Quantum visual arts use the same central assets or capacities (i)–(iv) mentioned earlier in Sect. 1. It can be developed very much in parallel to quantum music but requires the creation of an entirely new nomenclature. The perception of quantum visual art is subject to the same assumptions about the cognitive capacities to comprehend these artifacts fully quantum mechanically or classically. This will be shortly discussed in the following section.

7 Can Quantum Art Render Cognitions and Perceptions Beyond Classical Art?

Suppose for a moment that humans are capable to sense, receive and perceive quantum signals not only classically but in a fully quantum mechanical way. Thereby, they would, for instance, be capable of simultaneously "holding" different classically distinct tones at once—not just by interference but by parallel co-existence. This would result in a transgression of classical art forms, and in entirely novel forms of art.

The existence of such possibilities depends on the neurophysiology of the human, or, more generally, sentient, perception apparatus. Presently the question as to whether or not this is feasible is open; the answer to it is unknown.

In the case that merely classical perceptions are feasible, we would end up with a sort of Church-Turing thesis for music. In particular, quantum music would not be able to "go beyond" classical music for a single observer, as only classical renditions could be perceived. Of course, as we mentioned earlier, quantum music might "sound differently for different observers". To this end, we might conceptualize a kind of universal musical instrument that is capable of rendering all possible classical notes. Pianos and organs might be "for all practical purposes good" approximations to such a universal device.

Quantum music and quantum arts, just like quantum computing Deutsch (1985), or computations starting and ending in real numbers but using imaginary numbers as intermediaries Musil (1906), might be a sort of bridge crossing elegantly a gap between two classical domains of perception. And yet they could be so much more if only the quantum could be "heard" or "sensed".

8 Summary

We have contemplated the various extensions of music, and arts in general, to the quantum domain. Thereby we have located particular capacities which are genuine properties. These involve parallelization through coherent superposition (aka simultaneous linear combination), entanglement, complementarity and contextuality. We have reviewed the nomenclature introduced previously Putz and Svozil (2017) and considered particular instances of quantum music. Then we have briefly discussed quantum visual arts.

The perception of quantum arts depends on the capacity of the audience to either perceive quantum physical states as such, or reduce them to classical signals. In the first case, this might give rise to entirely novel artistic experiences. We believe that these are important issues that deserve further attention, also for sentient perception in general and human neurophysiology, in particular.

Acknowledgements This research was funded in whole, or in part, by the Austrian Science Fund (FWF), Project No. I 4579-N. For the purpose of open access, the author has applied a CC BY public copyright licence to any Author Accepted Manuscript version arising from this submission.

References

This, H. (2005). Modelling dishes and exploring culinary 'precisions': the two issues of molecular gastronomy. *British Journal of Nutrition, 93*, S139.

Putz, V., & Svozil, K. (2017). Quantum music. *Soft Computing, 21*, 1467. arXiv:1503.09045.

Svozil, K. (2016). Quantum hocus-pocus, Ethics in Science and Environmental Politics (ESEP). *16*, 25. arXiv:1605.08569.

Mermin, D. N. (2007). *Quantum Computer Science*. Cambridge: Cambridge University Press.

Fortnow, L. (2003). One complexity theorist's view of quantum computing. *Theoretical Computer Science, 292*, 597.

Nielsen, M. A., & Chuang, I. L. (2010). *Quantum Computation and Quantum Information*. Cambridge: Cambridge University Press. 10th Anniversary Edition.

Peres, A. (1978). Unperformed experiments have no results. *American Journal of Physics, 46*, 745.

Schrödinger, E. (1935). Die gegenwärtige Situation in der Quantenmechanik. *Naturwissenschaften, 23*, 807.

Brukner, Č, & Zeilinger, A. (1999). Operationally invariant information in quantum measurements. *Physical Review Letters, 83*, 3354. quant-ph/0005084.

Zeilinger, A. (1999). A foundational principle for quantum mechanics. *Foundations of Physics, 29*, 631.

Brukner, Č., Zukowski, M., & Zeilinger, A. (2002). The essence of entanglement. arXiv:quant-ph/0106119, translated to Chinese by Qiang Zhang and Yond-de Zhang, New Advances in Physics (Journal of the Chinese Physical Society).

Dzhafarov, E. N., Cervantes, V. H., & Kujala, J. V. (2017). Contextuality in canonical systems of random variables. *Philosophical Transactions of the Royal Society A: Mathematical, Physical and Engineering Sciences, 375*, 20160389. arXiv:1703.01252 .

Abramsky, S. (2018). Contextuality: At the borders of paradox. In E. Landry (Ed.) *Categories for the Working Philosopher* (pp. 262–285). Oxford University Press, Oxford, UK. arXiv:2011.04899.

Grangier, P. (2002). Contextual objectivity: a realistic interpretation of quantum mechanics. *European Journal of Physics, 23*, 331. arXiv:quant-ph/0012122.

Aufféves, A., Grangier, P. (2018). Extracontextuality and extravalence in quantum mechanics. *Philosophical Transactions of the Royal Society A: Mathematical, Physical and Engineering Sciences, 376*, 20170311. arXiv:1801.01398.

Auffèves, A., & Grangier, P. (2020). Deriving born's rule from an inference to the best explanation. *Foundations of Physics, 50*, 1781. arXiv:1910.13738.

Grangier, P. (2020). Completing the quantum formalism in a contextually objective framework. arXiv:2003.03121.

Budroni, C., Cabello, A., Gühne, O., & Kleinmann, M. J. (2021). Quantum contextuality: Åke Larsson. arXiv:2102.13036 [quant-ph].

Specker, E. (1960). Die Logik nicht gleichzeitig entscheidbarer Aussagen. *Dialectica, 14*, 239. arXiv:1103.4537 .

Kochen, S., Specker, E. P. (1967). The problem of hidden variables in quantum mechanics. *Journal of Mathematics and Mechanics (now Indiana University Mathematics Journal), 17*, 59.

Svozil, K. (2021). Varieties of contextuality emphasizing (non)embeddability. arXiv:2103.06110.

Shannon, C. E. (1949). Bell System Technical Journal 27, 379 (1948), reprinted in C. E. Shannon and W. Weaver: The Mathematical Theory of Communication, University of Illinois Press, Urbana, Illinois.

Reck, M., Zeilinger, A., Bernstein, H. J., & Bertani, P. (1994). Experimental realization of any discrete unitary operator. *Physical Review Letters, 73*, 58.

Zukowski, M., Zeilinger, A., & Horne, M. A. (1997). Realizable higher-dimensional two-particle entanglements via multiport beam splitters. *Physical Review A, 55*, 2564.

Glauber, R. J. (1986). Amplifiers, attenuators, and schrödinger's cat. *Annals of the New York Academy of Sciences, 480*, 336.

Schrödinger, E. (1995). *The Interpretation of Quantum Mechanics. Dublin Seminars (1949-1955) and Other Unpublished Essays*. Woodbridge, Connecticut: Ox Bow Press.

von Neumann, J. (1932, 1996). *Mathematische Grundlagen der Quantenmechanik*, (2nd ed.). Berlin, Heidelberg: Springer, English translation in [54].

Everett, H., III. (1957). Relative State formulation of quantum mechanics. *Reviews of Modern Physics, 29*, 454.

Wigner, E. P. (1961, 1962, 1995) Remarks on the mind-body question. In I. J. Good (Ed.), *The Scientist Speculates* (pp. 284–302). London, New York, and Berlin: Heinemann, Basic Books, and Springer-Verlag.

Everett, H., III. (2012). *The Everett Interpretation of Quantum Mechanics: Collected Works 1955–1980 with Commentary*, edited by J. A. Barrett & P. Byrne. Princeton, NJ: Princeton University Press.

Schwinger, J. (1960). Unitary operators bases. *Proceedings of the National Academy of Sciences (PNAS), 46*, 570.

Zeilinger, A. (2005). The message of the quantum. *Nature, 438*, 743.

Svozil, K. (2004). Quantum information via state partitions and the context translation principle. *Journal of Modern Optics, 51*, 811. arXiv:quant-ph/0308110.

Peres, A. (1980). Can we undo quantum measurements? *Physical Review D, 22*, 879.

Scully, M. O., & Drühl, K. (1982). Quantum eraser: A proposed photon correlation experiment concerning observation and "delayed choice" in quantum mechanics. *Physical Review A, 25*, 2208.

Greenberger, D. M., & YaSin, A. (1989). "Haunted" measurements in quantum theory. *Foundation of Physics, 19*, 679.

Scully, M. O., Englert, B.-G., & Walther, H. (1991). Quantum optical tests of complementarity. *Nature, 351*, 111.

Zajonc, A. G., Wang, L. J., Zou, X. Y., & Mandel, L. (1991). Quantum eraser. *Nature, 353*, 507.

Kwiat, P. G., Steinberg, A. M., & Chiao, R. Y. (1992). Observation of a "quantum eraser": a revival of coherence in a two-photon interference experiment. *Physical Review A, 45*, 7729.

Pfau, T., Spälter, S., Kurtsiefer, C., Ekstrom, C. R., & Mlynek, J. (1994). Loss of spatial coherence by a single spontaneous emission. *Physical Review Letters, 73*, 1223.

Chapman, M. S., Hammond, T. D., Lenef, A., Schmiedmayer, J., Rubenstein, R. A., Smith, E., & Pritchard, D. E. (1995). Photon scattering from atoms in an atom interferometer: Coherence lost and regained. *Physical Review Letters, 75*, 3783.

Herzog, T. J., Kwiat, P. G., Weinfurter, H., & Zeilinger, A. (1995). Complementarity and the quantum eraser. *Physical Review Letters, 75*, 3034.

Pauli, W. (1933). Die allgemeinen Prinzipien der Wellenmechanik. In H. Geiger & K. Scheel (Ed.), *Handbuch der Physik* (Vol. 24, p. 126). Berlin: Springer.

Glauber, R. J. (1969). Quantum theory of coherence. In S. M. Kay & A. Maitland (Ed.), *Quantum Optics: Proceedings of the Scottish Universities' Summer School in Physics 1969*. London: Academic Press.

Glauber, R. J. (2007). Amplifiers, attenuators and Schrödingers cat. In *Quantum Theory of Optical Coherence* (pp. 537–576). Wiley-VCH Verlag GmbH & Co. KGaA.

Weinberg, S. (1977). The search for unity: Notes for a history of quantum field theory. *Daedalus, 106*, 17.

Schrödinger, E. (1924). Über den Ursprung der Empfindlichkeitskurven des Auges. *Die Naturwissenschaften, 12*, 925.

Schrödinger, E., Niall, K. K. (2017). *Erwin Schrödinger's Color Theory*. Springer International Publishing.

Hecht, S., Shlaer, S., & Pirenne, M. H. (1942). Energy, quanta, and vision. *Journal of General Physiology, 25*, 819.

Westheimer, G. (2016). *History of physiological optics in the twentieth century, in Handbook of Visual Optics: Fundamentals and Eye Optics, Volume One* (Chap. 1, pp. 1–10). CRC Press, Taylor & Francis Group.

Tinsley, J. N., Molodtsov, M. I., Prevedel, R., Wartmann, D., Espigulé-Pons, J., Lauwers, M., & Vaziri, A. (2016). Direct detection of a single photon by humans. *Nature Communications, 7*. https://doi.org/10.1038/ncomms12172.

Deutsch, D. (1985). Quantum theory, the Church-Turing principle and the universal quantum computer. *Proceedings of the Royal Society of London. Series A, Mathematical and Physical Sciences (1934–1990), 400*, 97.

Musil, R. (1906). *Die Verwirrungen des Zöglings Törleß*. Wien und Leipzig: Wiener Verlag, project Gutenberg ebook # 3471.

von Neumann, J. (1955). *Mathematical Foundations of Quantum Mechanics*. Princeton, NJ: Princeton University Press, German original in [27].

Quanta in Sound, the Sound of Quanta: A Voice-Informed Quantum Theoretical Perspective on Sound

Maria Mannone and Davide Rocchesso

Abstract Humans have a privileged, embodied way to explore the world of sounds, through vocal imitation. The Quantum Vocal Theory of Sounds (QVTS) starts from the assumption that any sound can be expressed and described as the evolution of a superposition of vocal states, i.e., phonation, turbulence, and supraglottal myoelastic vibrations. The postulates of quantum mechanics, with the notions of observable, measurement, and time evolution of state, provide a model that can be used for sound processing, in both directions of analysis and synthesis. QVTS can give a quantum-theoretic explanation to some auditory streaming phenomena, eventually leading to practical solutions of relevant sound-processing problems, or it can be creatively exploited to manipulate superpositions of sonic elements. Perhaps more importantly, QVTS may be a fertile ground to host a dialogue between physicists, computer scientists, musicians, and sound designers, possibly giving us unheard manifestations of human creativity.

1 Sound $\overset{voice}{\leftrightarrow}$ Quanta

Sometimes, when kids imitate sounds around them, they are blamed for producing weird noises. However, they are unknowingly using their own voice as a probe to investigate the world of sounds, and thus they are probably performing some experiments. Some of the these kids will become sound designers, other ones composers, other sound engineers and physicists; some other ones, will blame future kids, and the cycle repeats.

M. Mannone
European Centre for Living Technology (ECLT), Ca' Foscari University of Venice, Venice, Italy

M. Mannone (✉) · D. Rocchesso
Department of Mathematics and Computer Sciences, University of Palermo, Palermo, Italy
e-mail: maria.mannone@unive.it; mariacaterina.mannone@unipa.it

D. Rocchesso
e-mail: davide.rocchesso@unipa.it

© Springer Nature Switzerland AG 2022
E. R. Miranda (ed.), *Quantum Computing in the Arts and Humanities*,
https://doi.org/10.1007/978-3-030-95538-0_6

What is a vocal imitation? It is the attempt to reproduce some essential features of a sound, thought of or actually heard, with the human voice. The imitation can refer to characteristics of the sound, or to its hypothetical sources (Gaver 1993).

The human brain catches some salient sound features, and the voice attempts to reproduce them. Sometimes, even poets (some of the kids above became poets as well) coined new words as to include auditory dimensions in their poetry, producing examples of onomatopoeia. This happened for example at the beginning of 20th Century, with the poems by the futurist Filippo Tommaso Marinetti, where the words *Zang Tumb Tumb* imitate motor and war noises (Marinetti 1914), and with *La fontana malata* (The sick fountain) by Aldo Palazzeschi, where the words *Clof, clop, cloch* (Palazzeschi 1909) mimic the intermittent flow of water and the noise of falling drops.

Onomatopoeia gives poetical dignity to vocal imitations. Vocal imitations raised the interest of science as well. In the framework of a recent European project, the voice has been shown to be a powerful means to produce sound sketches, which can be transformed into refined sound designs through interactive voice-driven sound manipulations. In this sense, the machine extracts sound from the embodied imagination of the sound designer (Delle Monache et al. 2018). Tools of this kind, taking the form of an augmented microphone, have been prototyped (Rocchesso et al. 2016) and, with the purpose of turning the microphone into a music controller, even developed into products (Vochlea microphone 2022).

A few scholars may argue that there are precise tools to investigate sounds, such as the Fourier formalism, which uses decompositions based on sinusoidal functions, and all formalisms inspired by Fourier's approach (von Helmholtz 1870; Koenig and Delwin 2015). However, this formalism is not immediately understandable in everyday communications, and it is less directly manipulable than vocal imitations. While the Fourier formalism is powerful at the level of persons with some education in sound and music, it is not the way laypersons communicate and reason about sonic realities.

A powerful support to both qualitative and quantitative communication and reasoning on sound is given by sound visualizations. Spectrograms display the spectrum of frequencies through time of a sound. With spectrograms, we can easily compare sounds and investigate how their properties change through time. However, the vocal imitations of a natural or an artificial sound, which appears as completely intuitive to humans (again, think about the kid giving voice to a toy car), might be hard to find by comparison of the spectrogram of the vocal imitation with the spectrogram of the original sound. It can be possible to investigate some emerging properties of both sounds, but it can be really hard while dealing, for example, with a vocal imitation of a motor, or some other mechanical noise, that has a really different spectral profile than human voice. Thus, Fourier-driven sound visualization has some limitations in revealing the embodied perceptual features of sounds.

Which are the characteristics of human voice? The utterances of humans and many mammals can be decomposed into overlapping chunks that fall within three primitive classes: phonation, turbulence, and supraglottal myoelastic vibrations (Friberg et al. 2018). In phonation, the source is in the vocal folds. In turbulence, the source is in chaotic motion of inhaled or exhaled air. Supraglottal myoelastic vibrations include several kinds of low-frequency oscillations or pulse trains generated with different parts of the vocal apparatus, such as lips or tongue. We can build up a new formalism to describe the sound based on these components.

And what are the characteristics of sound as it is produced out of our body? Sound is made of waves of rarefaction and compression, produced by vibrating strings, air-filled pipes, vibrating membranes or plates, and so on. Consider the simplest of these systems, which is probably the flexible string fastened to a rigid support at both ends. This is one of the most important models in physics, which has been used to demonstrate fundamental phenomena, in acoustics as well as in other areas of physical sciences. In fact, whilst vibrating strings have often been used as a paradigm for quantum mechanics, the vice versa, that is, using quantum mechanics as a paradigm to understand sound, was proposed in the nineteen-forties by Gabor (1947), who imagined how sound analysis and synthesis could be based on acoustical quanta, or wavelets.[1] His seminal work has been extensively carried on and expanded both by scientists and musicians, and is certainly at the root of granular approaches to sound and music (Roads 2001).

A variety of ideas and methods of quantum mechanics have been applied to describe forms and phenomena pertaining to that form of art whose medium is sound, that is music. For example, tonal attractions have been modeled as metaphorical forces (Blutner and beim Graben 2020), quantum parallelism has been proposed to model music cognition (Dalla Chiara et al. 2015), the quantum formalism has been proposed as a notational tool for music-visual "atomic" elements (Mannone et al. 2020), the non-Markovianity of open quantum systems has been proposed as a measure of musical memory within a score (Mannone and Compagno 2014). Quantum computing, that is computation based on actual physical quantum processes (Nielsen 2010), starts being used to control sound synthesizers and computer-generated music (Miranda 2021a, b). The opposite practice, that is using sonification as a means to make the actions of quantum algorithms perceivable as musical lines and recognizable as patterns, has appeared with the flourishing of quantum computing as an area of theoretical computer science (Weimer 2010).

This chapter is part of a book on quantum-theoretical and -computational approaches to art and humanites, and its first chapter provides an excellent introduction to quantum theory and quantum computing. Nevertheless, we give a few basic notions, essentially the postulates of quantum mechanics, in Sect. 1.1. For now, we can say that quantum mechanics is a branch of physics, where, in a nutshell:

- Matter and energy, seen at the level of subatomic particles, are described as discrete;
- We describe particles as points or as probability waves to find them in some places;
- If we know the momentum of a particle, we don't know its position, and vice versa;
- The measurement influences the state of what is measured: the observer (subject) influences the observed (object).

---

[1] A wavelet is a wave-like oscillation under a finite temporal envelope.

In 1935, Albert Einstein tried to resist to quantum mechanics, postulating hidden variables to justify such a bizarre behavior (Einstein et al. 1935). Some further studies showed that, if Einstein was right, some inequalities should be satisfied (Bell 1964), but quantum-mechanical systems can be conceived and implemented that actually violate such inequalities (Aspect et al. 1982). This means that a local realistic view of the world does not apply to quantum phenomena. According to another Nobel prize, Richard Feynman, nobody really understood quantum mechanics (Feynman 1995).

One might ask: if quantum mechanics is so difficult to be interpreted and understood, why is it so often invoked to explain mundane affairs that have nothing to do with particle physics? As a possible explanation, the formalism is based on a few postulates, it assumes linearity and unitary (energy-preserving) time evolution, and it gives a probabilistic framework capable to explain concurrent and interfering phenomena.

Let us go back to sound and voice. If quantum mechanics can be joined with the sound, and the sound with the voice, thus quantum mechanics can be joined with the voice, and this is our idea: a Quantum Vocal Theory of Sound (Rocchesso and Mannone 2020).

This approach is not opposed to the richness and complexity of Fourier formalism, spectrograms, and so on. It presents a different paradigm, a different starting point, using the primitive components of human voice. The novelty is that these components appear as useful not only to investigate the voice itself, but also to face the complexity of the world of general sound. It is a strong statement, but it actually follows the intuition: each kid knows well how to imitate the *vroom vroom* of a car, a long time before learning how to read an equation and how to interpret a graph.

The QVTS approach can be exploited to investigate sound, decomposing it into its essential features through the analysis step. But QVTS can also help do the opposite, that is, create new sounds, in the synthesis step. Sound synthesis can lead to creative applications; some possible applications are described later on in this chapter.

The structure of the chapter is the following. In Sect. 1.1, we remind of some basics of quantum mechanics. In Sect. 2, we present the fundamental ideas of QVTS. In Sect. 3, examples of sound processing based on the QVTS, with audible and interpretable outcomes, are given. In Sect. 4, we describe our vision on the future of QVTS keeping an eye on interdisciplinary collaborations and creative applications. As an example of possible creative applications, we sketch the structure of a piece based on vocal states.

1.1 Some Postulates to Live by

An observable is a physical quantity, that can be described as a mathematical (Hermitian, linear) operator. Each operator acts on a complex vector space, the state space. The space where quantum observables live is the separable Hilbert space. It's separable, because we can distinguish the components along different axes. In the case of QVTS, the space is related with the vocal primitives, and it is separable as well, as we will see in the next section.

A quantum state, that is a unit-length vector in the state space, can be seen as a superposition of values with some probabilities. An eigeinstate is a characteristic state of some operator. After the measurement process, the probability wave collapses to a certain value: it is the eigenvalue of a certain operator, and the system is in an eigenstate of that operator.

Probability is a key concept in quantum mechanics. According to the principle of uncertainty, we cannot know, let's say, the position and the momentum of a particle with the same precision. The more is the information we have on position, the less we know about momentum, and vice versa.

Let us say more on the idea of quantum measurement. Consider a Cartesian framework with axes x, y, and z—a tridimensional space with three mutually orthogonal axes. We can perform measurements along each of the axes. Quantum measure implies a change in the measured entity. If the measurement along the direction x can give a positive result, in all subsequent measurements along the same direction we will have a positive value. A measure along x would zero out the y and z components, while leaving only the x component with value 1. If, before measurement, there is a given probability to get a specific outcome, after the effective measurement of that outcome, the probability to get the same value in each subsequent measurement along the same direction is 100%. In fact, in quantum mechanics, the measurement of a state implies the destruction of part of the initial information, and thus the process is called destructive measure. A quantum state is a superposition of eigenstates, which are reduced to a single state after the measurement. Such state collapse happens in the context including both the system and the measuring entity, through the interaction of the two (Rovelli 2021). Intuitively, it's like observing and taking a picture of a person, and blocking him or her as the represented image along that specific shooting direction. (Be careful the next time you'll take pictures).

Dennis Gabor first exploited the paradigm of quantum theory to investigate sound (Gabor 1947), instead of doing the usual vice versa, with sound and strings used as metaphors to understand quantum waves. Gabor proposed the concept of quantum of sound, as a unit-area cell in the time-frequency plane, which could be called *phon*, from the Greek φωνή. On the other hand, we start from a vocal description of sound, to define the *phon* as the set of vocal primitive operators.

2   The Quantum Vocal Theory of Sound

In a recent article (Rocchesso and Mannone 2020), we have proposed the basics for a Quantum Vocal Theory of Sound (QVTS). Here, we summarize its main ideas, and then we propose some hints for future developments.

First of all, let us define the phon formalism, where the word *phon* indicates the quantum of sound, expressed in the state space of vocal primitives. With the phon

Fig. 1 The Bloch sphere
adapted to QVTS, to
represent the phon space.
Hand-drawing by
D. Rocchesso

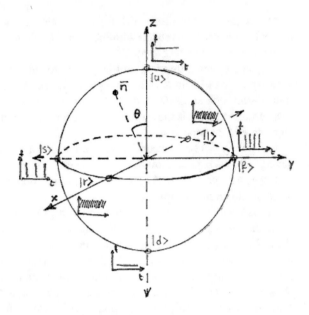

formalism, we can define vocal states, and extend the quantum bit (qubit) language
to the human voice. Some quantum-mechanical concepts, such as state preparation
and measurement, can be extended to the domain of voice as well.

Consider a space with three independent directions: x, y, and z. In the QVTS, the
three axes of this "phonetic space" have a vocal meaning:

- z is the phonation, giving waveforms of different pitches;
- x is the turbulence, giving noises of different brightnesses;
- y is the myoelasticity, giving pulsations at different tempos (thought of as slow
  pulse trains).

Such three-dimensional space is sketched in Fig. 1 where, at the intersections between
each of the axes and the unit (called Bloch) sphere, we can find two mutually orthogonal vectors, each depicted as a tiny sketchy spectrogram.

Given a sound recording of human voice, if we measure phonation using a specific
computational tool (such as SMS-tools Bonada et al. 2011), it is possible to separate
such component from the rest, and all subsequent measurements of phonation will be
giving the same result. If we measure a primitive component first, and then another
one, the result is generally dependent on the order of the two operations: A fact that
is known as non-commutativity. Figure 5 shows a couple of example spectrograms
illustrating the difference.

A vocal state can be described as a superposition of phonation, turbulence, and
myoelasticity with certain probabilities. We can thus define a phon operator $\bar{\sigma}$ as a
3-vector operator, providing information on the x, y, and z components through its
specific directions in the 3d phonetic space. Each component of $\bar{\sigma}$ is represented by
a linear operator, so we have σ_x, σ_y, and σ_z.

2.1 Preparation and Measurement Along Axes

According to the postulates of quantum mechanics, it is possible to perform measurements along one arbitrary axis of the 3d phonetic space and, as a result, we will have prepared the phon along that specific axis.

A quantum measurement is represented by an operator, called a projector, acting on the state, and provoking its collapse onto one of its eigenvectors. If the system is in a state $|\psi\rangle$ and then we make a measurement, the probability to get the result j is given by:

$$p_j := p_m(j|\psi) = \langle \psi | M_j | \psi \rangle = \langle \psi | M_j^\dagger M_j | \psi \rangle, \tag{1}$$

where the set $\{M_j\}$ is a projector system in the Hilbert space of states. $\{M_j\}$ is a complete set of Hermitian and idempotent matrices. An Hermitian matrix is a complex matrix, that is equal to its transposed conjugate (indicated by the † symbol in Eq. 1). It has real eigenvalues. Idempotent means that, if we apply multiple times an operator, the result is the same as if we applied the operator just once. With an orthonormal basis of measurement vectors $|a_j\rangle$, the elementary projectors are $M_j = |a_j\rangle\langle a_j|$, and the system collapses into $|a_j\rangle$.

2.1.1 Measurement Along Z

A measurement along the z axis is performed through the operator σ_z. The eigenvectors (or eigenstates) of σ_z are $|u\rangle$ and $|d\rangle$, corresponding to pitch-up phonation and pitch-down phonation, with eigenvalues $\lambda_u = +1$ and $\lambda_d = -1$, respectively:

$$\sigma_z |u\rangle = |u\rangle, \quad \sigma_z |d\rangle = -|d\rangle.$$

The eigenstates $|u\rangle$ and $|d\rangle$ are orthogonal, i.e., $\langle u|d\rangle = 0$, and they can be represented as column vectors

$$|u\rangle = \begin{bmatrix} 1 \\ 0 \end{bmatrix}, \quad |d\rangle = \begin{bmatrix} 0 \\ 1 \end{bmatrix}. \tag{2}$$

The operator σ_z can also be represented in matrix form as

$$\sigma_z = \begin{bmatrix} 1 & 0 \\ 0 & -1 \end{bmatrix}. \tag{3}$$

Applying a measurement along the z direction to a generic phon state $|\psi\rangle$ corresponds to pre-multiply it by one of the measurement operators (or projectors)

$$M_u = |u\rangle \langle u| = \begin{bmatrix} 1 & 0 \\ 0 & 0 \end{bmatrix}$$

or

$$M_d = |d\rangle \langle d| = \begin{bmatrix} 0 & 0 \\ 0 & 1 \end{bmatrix},$$

and to normalize the resulting vector to have length one. Such operators satisfy the completeness relation $M_u + M_d = I$, summing up to the unit operator.

A generic phon state $|\psi\rangle$ can be expressed as

$$|\psi\rangle = \alpha_u |u\rangle + \alpha_d |d\rangle, \tag{4}$$

where the coefficients are complex numbers, $\alpha_u = \langle u|\psi\rangle$, and $\alpha_d = \langle d|\psi\rangle$. Being the system in state $|\psi\rangle$, the probability to measure pitch-up is

$$p_u = \langle \psi| M_u^\dagger M_u |\psi\rangle = \langle \psi|u\rangle \langle u|u\rangle \langle u|\psi\rangle = \langle \psi|u\rangle \langle u|\psi\rangle = \alpha_u^* \alpha_u \tag{5}$$

and, similarly, the probability to measure pitch-down is $p_d = \langle \psi|d\rangle \langle d|\psi\rangle = \alpha_d^* \alpha_d$, where $*$ denotes complex conjugation. The completeness relation ensures that p_u and p_d sum up to one.

If we repeatedly prepare a state ψ and measure it along the z direction, we get the average value

$$\langle \sigma_z \rangle := \sum_{m=\{u,d\}} \lambda_m p_m = \langle \psi| \left(\sum_{m=\{u,d\}} \lambda_m M_m^\dagger M_m \right) |\psi\rangle = \langle \psi|\sigma_z|\psi\rangle, \tag{6}$$

where the sum within brackets is called the observable of the measurement.

In quantum computing terminology, the vectors[2] (2) give the computational basis of a qubit vector space. The operator (3) corresponds to a Z gate, which acts as a phase flip on the second state of the computational basis.

2.1.2   Measurement Along X

The eigenstates of the operator σ_x are $|r\rangle$ and $|l\rangle$, corresponding to turbulent primitive sounds having different spectral distributions, one with the rightmost (or highest-frequency) centroid and the other with the lowest-frequency centroid. Their respective eigenvalues are $\lambda_r = +1$ and $\lambda_l = -1$, such that

$$\sigma_x |r\rangle = |r\rangle, \quad \sigma_x |l\rangle = -|l\rangle.$$

---

[2] In quantum computing, the vectors of the computational basis are normally called $|0\rangle$ and $|1\rangle$.

If the phon is prepared $|r\rangle$ (turbulent) and then the measurement apparatus is set to measure σ_z, there will be equal probabilities of getting pitch-up or pitch-down phonation as an outcome. This measurement property is satisfied if $|r\rangle$ is defined as

$$|r\rangle = \frac{1}{\sqrt{2}}|u\rangle + \frac{1}{\sqrt{2}}|d\rangle . \tag{7}$$

A similar definition is given for $|l\rangle$, such that the two eigenstates of turbulence are orthogonal ($\langle r|l\rangle = 0$):

$$|l\rangle = \frac{1}{\sqrt{2}}|u\rangle - \frac{1}{\sqrt{2}}|d\rangle . \tag{8}$$

In matrix form, the turbulence operator is expressed as

$$\sigma_x = \begin{bmatrix} 0 & 1 \\ 1 & 0 \end{bmatrix}, \tag{9}$$

and its eigenvectors are

$$|r\rangle = \begin{bmatrix} \frac{1}{\sqrt{2}} \\ \frac{1}{\sqrt{2}} \end{bmatrix}, \quad |l\rangle = \begin{bmatrix} \frac{1}{\sqrt{2}} \\ -\frac{1}{\sqrt{2}} \end{bmatrix}. \tag{10}$$

Applying a measurement along the x direction to a generic phon state $|\psi\rangle$ corresponds to pre-multiply it by one of the measurement operators

$$M_r = |r\rangle \langle r| = \frac{1}{2} \begin{bmatrix} 1 & 1 \\ 1 & 1 \end{bmatrix}$$

or

$$M_l = |l\rangle \langle l| = \frac{1}{2} \begin{bmatrix} 1 & -1 \\ -1 & 1 \end{bmatrix},$$

and to normalize the resulting vector to have length one. Such operators satisfy the completeness relation $M_r + M_l = I$.

In quantum computing, the operator (9) corresponds to a X gate, which is the equivalent of the NOT gate in classical logic circuits, as it flips the states of the computational basis. The vectors (7) and (8) form the Hadamard basis, often denoted with the symbols $\{|+\rangle, |-\rangle\}$.

Preparation in one of the states of the Hadamard basis $\{|r\rangle, |l\rangle\}$, followed by measurement along the z axis, results in an operation that is equivalent to coin flipping, $+1$ or -1 being obtained with equal probability.

2.1.3 Measurement Along Y

The eigenstates of the operator σ_y are $|f\rangle$ and $|s\rangle$, corresponding to slow myoelastic pulsations, one faster and one slower,[3] with eigenvalues $\lambda_u = +1$ and $\lambda_d = -1$, such that

$$\sigma_y |f\rangle = |f\rangle \quad \sigma_y |s\rangle = -|s\rangle.$$

If the phon is prepared $|f\rangle$ (pulsating) and then the measurement apparatus is set to measure σ_z, there will be equal probabilities for $|u\rangle$ or $|d\rangle$ phonation as an outcome. This measurement property is satisfied if

$$|f\rangle = \frac{1}{\sqrt{2}} |u\rangle + \frac{i}{\sqrt{2}} |d\rangle, \tag{11}$$

where i is the imaginary unit.

Likewise, the $|s\rangle$ state can be defined in such a way that the two eigenstates of pulsation are orthogonal ($\langle f|s\rangle = 0$):

$$|s\rangle = \frac{1}{\sqrt{2}} |u\rangle - \frac{i}{\sqrt{2}} |d\rangle. \tag{12}$$

In matrix form, the pulsation operator is expressed as

$$\sigma_y = \begin{bmatrix} 0 & -i \\ i & 0 \end{bmatrix}, \tag{13}$$

and its eigenvectors are

$$|f\rangle = \begin{bmatrix} \frac{1}{\sqrt{2}} \\ \frac{i}{\sqrt{2}} \end{bmatrix}, \quad |s\rangle = \begin{bmatrix} \frac{1}{\sqrt{2}} \\ -\frac{i}{\sqrt{2}} \end{bmatrix}.$$

Applying a measurement along the y direction to a generic phon state $|\psi\rangle$ corresponds to pre-multiply it by one of the measurement operators

$$M_f = |f\rangle \langle f| = \frac{1}{2} \begin{bmatrix} 1 & -i \\ i & 1 \end{bmatrix}$$

or

$$M_s = |s\rangle \langle s| = \frac{1}{2} \begin{bmatrix} 1 & i \\ -i & 1 \end{bmatrix},$$

and to normalize the resulting vector to have length one. Such operators satisfy the completeness relation $M_f + M_s = I$.

---

[3] In describing the spin eigenstates, the symbols $|i\rangle$ and $|o\rangle$ are often used, to denote the in–out direction.

The matrices (3), (9), and (13) are called the Pauli matrices. In quantum computing, these are all useful one-qubit gates.

2.2 Measurement Along an Arbitrary Direction

Orienting the measurement apparatus in the phonetic space along an arbitrary direction $\bar{n} = [n_x, n_y, n_z]'$ means taking a weighted mixture of Pauli matrices:

$$\sigma_n = \bar{\sigma} \cdot \bar{n} = \sigma_x n_x + \sigma_y n_y + \sigma_z n_z = \begin{bmatrix} n_z & n_x - i n_y \\ n_x + i n_y & -n_z \end{bmatrix}. \tag{14}$$

2.2.1 Sines + Models and the Phon Space

The Harmonic plus Noise model (Bonada et al. 2011) is well suited to describe measurement and preparation in the phonation-turbulance planar section of the 3d phonetic space. An arbitrary direction in such plane is described by the operator

$$\sigma_n = \begin{bmatrix} \cos\theta & \sin\theta \\ \sin\theta & -\cos\theta \end{bmatrix}, \tag{15}$$

where θ is the angular direction, pointing to a superposition of phonation and turbulence (see Fig. 1). The eigenstate for eigenvalue $+1$ is

$$|\lambda_1\rangle = [\cos\theta/2, \sin\theta/2]', \tag{16}$$

the eigenstate for eigenvalue -1 is

$$|\lambda_{-1}\rangle = [-\sin\theta/2, \cos\theta/2]', \tag{17}$$

and the two are orthogonal. Suppose we prepare the phon to pitch-up $|u\rangle$. If we rotate the measurement system along $\bar{n}$, the probability to measure $+1$ is

$$p(+1) = \langle u|\lambda_1\rangle\langle\lambda_1|u\rangle = |\langle u|\lambda_1\rangle|^2 = \cos^2\theta/2, \tag{18}$$

and the probability to measure -1 is

$$p(-1) = |\langle u|\lambda_{-1}\rangle|^2 = \sin^2\theta/2. \tag{19}$$

The expectation (or average) value of measurement is therefore

$$\langle\sigma_n\rangle = \sum_j \lambda_j p(\lambda_j) = (+1)\cos^2\theta/2 + (-1)\sin^2\theta/2 = \cos\theta. \tag{20}$$

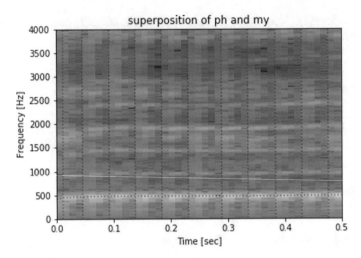

Fig. 2 Spectrogram of a vocal sound which is a superposition of phonation and supgraglottal myoelastic vibration. A salient pitch (horizontal dotted line) as well as quasi-regular train of pulses (vertical dotted lines) are automatically extracted

Symmetrically, if we prepare the phon in state $|\lambda_1\rangle$ and we measure along the z axis, we get a pitch-up with probability $\cos^2 \theta/2$ and a pitch-down with probability $\sin^2 \theta/2$.

More generally, the Sines plus Noise plus Transients model (Verma et al. 1997) may be suitable to describe measurement and preparation in the whole 3d phonetic space, where supraglottal myoelastic vibrations are made to correspond to transient pulse trains. For example, consider the vocal fragment[4] whose spectrogram is represented in Fig. 2. An extractor of pitch salience and an extractor of onsets[5] have been applied to highlight respectively the phonation (horizontal dotted line) and myoelastic (vertical dotted lines) components in the spectrogram. In the $z - y$ plane, there would be a measurement orientation and a measurement operator that admit such sound as an eigenstate.

2.3 Purity and Mixing

In quantum mechanics, the density operator is a mathematical object that describes the statistical (pure or mixed) state of a quantum system, and it is usually represented as a matrix. A pure state is not referred to a moral condition, but to a separability of states. A mixed state indicates an inseparability of states from the viewpoint of

---

[4] It is one of the example vocal sounds considered in Rocchesso et al. (2016), and taken from Newman (2004).

[5] The feature extractors are found in the Essentia library (Bogdanov et al. 2013).

the observer, who has some degree of epistemic uncertainty. Thus, the concept of density matrix generalizes the concept of state superposition. The density operator is defined as

$$\rho = \sum_j p_j \, |\psi_j\rangle \, \langle\psi_j| \, , \tag{21}$$

where p_j indicates the probability for the j-state. The density operator for a pure state is $\rho = |\psi\rangle\langle\psi|$, and the trace of the associated density matrix is $\mathrm{tr}[\rho^2] = 1$. For a mixed state, $\mathrm{tr}[\rho^2] < 1$. It can be shown that the density matrix (21) can be expressed as a composition of Pauli matrices as in (14), with the addition of the identity matrix. From such representation, pure states can be shown to lay on the surface of the Bloch sphere, while mixed states stay inside the sphere, with the completely chaotic state being found at the origin (Cariolaro 2015). A pure state can contain a superposition, but such a composition is defined with certainty. A mixed state is a probabilistic mixing. The mixed state is inseparable. The generalization introduced by mixed states can represent the audio concept of mixing, thus coming useful in composition of auditory scenes.

2.4 Not Too Sure? Uncertainty Can Be Measured

In the wonderland of quantum mechanics, it can happen that, the better we know something, the lesser we know something else. In QVTS, the more precise our knowledge of phonation, the less precise our measurement of turbulence. In quantum mechanics, if we measure two observables $\mathbf{L}$ and $\mathbf{M}$ simultaneously in a single experiment, the system is left in a simultaneous eigenvector of the observables only if $\mathbf{L}$ and $\mathbf{M}$ commute, i.e., if their commutator $[\mathbf{L}, \mathbf{M}] = \mathbf{LM} - \mathbf{ML}$ vanishes. When the commutator is different from zero, we say that the two operators do not commute. This happens with measurement operators along different axes. It is the case of $[\sigma_x, \sigma_y] = 2i\sigma_z$. As a consequence for QVTS, phonation and turbulence cannot be simultaneously measured with certainty.

The uncertainty principle is one of the key ideas of quantum mechanics. It is based on Cauchy–Schwarz inequality in complex vector spaces. According to the uncertainty principle, the product of the two uncertainties is at least as large as half the magnitude of the commutator:

$$\Delta\mathbf{L}\Delta\mathbf{M} \geq \frac{1}{2} \, |\langle\psi| \, [\mathbf{L}, \mathbf{M}] \, |\psi\rangle| \tag{22}$$

Let $\mathbf{L} = \mathscr{T} = t$ be the time operator, and $\mathbf{M} = \mathscr{W} = -i\frac{d}{dt}$ be the frequency operator. Applying them to the complex oscillator $Ae^{i\omega t}$, we get a time-frequency uncertainty, where the uncertainty is minimized by the Gabor function, a sinusoid windowed by a Gaussian (Irino and Patterson 1997).

2.4.1 The Order Matters

Kids learn that multiplying a times b, with a, b natural numbers, is the same as multiplying b times a—and, since early age, they think that commutativity is always verified. Reading a book and then going for a walk might be the same stuff as going for a walk and then reading a book (maybe). Quantum mechanics does not work that way, and the same for QVTS. If we record a singer, we take away vowels, and then we take again away vowels, the result is the same—the recording is in an autostate of no-vowels. If we take away vowels, and then we take away the noise, the result is different from what we could hear if we do the opposite, that is, taking away the noise and then the vowels. More precisely, the measurement operators oriented along different axes do not commute. For example, let A be an audio segment. The measurement (by extraction) of turbulence by the measurement operator turbulence-right $M_r = |r\rangle\langle r|$ leads to $M_r(A) = A'$. A successive measurement of phonation by the measurement operator pitch-up $M_u = |u\rangle\langle u|$ gives $M_u(A') = A''$, thus $M_u(A') = M_u M_r(A) = A''$. If we perform the measurements in the opposite order, with phonation first and turbulence later, we obtain $M_r M_u(A) = M_r(A^*) = A^{**}$. We expect that $[M_r, M_u] \neq 0$, and thus, that $A^{**} \neq A''$. The diagram in Fig. 3 shows non-commutativity in the style of category theory.

In bra-ket notation, this fact can be expressed as

$$M_r M_u |A\rangle = |r\rangle \langle r|u\rangle \langle u|A\rangle = \langle r|u\rangle |r\rangle \langle u|A\rangle \neq$$
$$M_u M_r |A\rangle = |u\rangle \langle u|r\rangle \langle r|A\rangle = \langle u|r\rangle |u\rangle \langle r|A\rangle . \tag{23}$$

Given that $\langle r|u\rangle$ is a scalar and $\langle u|r\rangle$ is its complex conjugate, and that $|u\rangle\langle r|$ is generally non-Hermitian, we get

$$[M_r, M_u] = |r\rangle \langle r|u\rangle \langle u| - |u\rangle \langle u|r\rangle \langle r| =$$
$$= \langle r|u\rangle |r\rangle \langle u| - \langle u|r\rangle |u\rangle \langle r| \neq 0, \tag{24}$$

or, in terms of matrices

Fig. 3 A non-commutative diagram representing the non-commutativity of measurements of phonation (M_u) and turbulence (M_r) on audio A

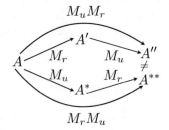

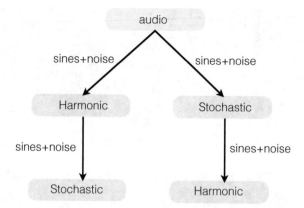

Fig. 4 On the left, an audio segment is analyzed via the sines+noise model. Then, the noise part is submitted to a new analysis. In this way, a measurement of phonation follows a measurement of turbulence. On the right, the measurement of turbulence follows a measurement of phonation. This can be described via projectors through equation (23), and diagrammatically in Fig. 3

$$[M_r, M_u] = \frac{1}{2} \begin{pmatrix} 1 & 1 \\ 1 & 1 \end{pmatrix} \begin{pmatrix} 1 & 0 \\ 0 & 0 \end{pmatrix} - \frac{1}{2} \begin{pmatrix} 1 & 0 \\ 0 & 0 \end{pmatrix} \begin{pmatrix} 1 & 1 \\ 1 & 1 \end{pmatrix} =$$
$$= \frac{1}{2} \begin{pmatrix} 1 & 0 \\ 1 & 0 \end{pmatrix} - \frac{1}{2} \begin{pmatrix} 1 & 1 \\ 0 & 0 \end{pmatrix} = \frac{1}{2} \begin{pmatrix} 0 & -1 \\ 1 & 0 \end{pmatrix} = \frac{i}{2} \sigma_y \neq 0. \tag{25}$$

On audio signals, measurements of phonation and turbulence can be performed using the sines + noise model (Bonada et al. 2011), as described in Fig. 4. The measurement of phonation is performed through the extraction of the sinusoidal component, while the measurement of turbulence is performed through the extraction of the noise component with the same model. The spectrograms for A'' and A^{**} in Fig. 5 show the results of such two sequences of analyses on a segment of female speech, confirming that the commutator $[M_r, M_u]$ is non-zero.

Consider again Fig. 1, which shows a representation of the phon space using the Bloch sphere. There are small spectrograms at the extremities, in correspondence of $|s\rangle$, $|f\rangle$, $|u\rangle$, $|d\rangle$, $|r\rangle$, and $|l\rangle$. Applying $\sigma_z\sigma_x$ to a state, we get the flipped state we would obtain if we had applied $\sigma_x\sigma_z$. If we apply a pitch operator and then a turbulence operator (or vice versa) to a slow impulse train ($|s\rangle$ or $|f\rangle$), we get another impulse train.

2.5 Time Flies

The variation of quantum states in time can be obtained through the application of time evolution operators on them. Similarly, suitable time operators can make the density matrix vary in time as well. Given a density operator $\rho(t_0)$ at time t_0, its time variation is obtained applying a unitary operator $U(t_0, t)$:

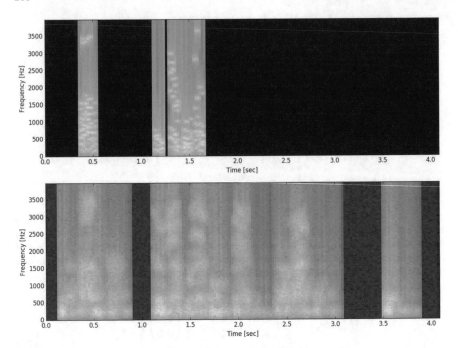

Fig. 5 Top: spectrogram corresponding to a measurement of phonation M_u following a measurement of turbulence M_r, leading to $M_u M_r(A) = A''$; bottom: spectrogram corresponding to a measurement of turbulence M_r following a measurement of phonation M_u, leading to $M_r M_u(A) = A^{**}$

$$\rho(t) = \mathbf{U}^{\dagger}(t_0, t)\rho(t_0)\mathbf{U}(t_0, t).$$ (26)

This is the most general definition: There are no assumptions on states (mixed or pure), and the only assumptions on the operator $\mathbf{U}$ are that it is unitary, i.e., $\mathbf{U}^{\dagger}\mathbf{U} = \mathbf{I}$, with $\mathbf{I}$ the identity matrix, and that it depends only on t and t_0.

But actually there is more. The unitary operator $\mathbf{U}$, evaluated at a tiny time increment ϵ, is related to the Hamiltonian $\mathbf{H}$, describing the energy of the system:

$$\mathbf{U}(\epsilon) = \mathbf{I} - i\epsilon\mathbf{H}.$$ (27)

For a closed and isolated system, $\mathbf{H}$ is time-independent, and the unitary operator becomes $\mathbf{U}(t) = e^{i\mathbf{H}(t-t_0)}$. However, nature is more complex, things are not isolated, and usually $\mathbf{H}$ is time-dependent, and the time evolution is given by an integral. To complicate things even more, with a non-commutative Hamiltonian, an explicit solution cannot be found. The problem can be circumvented by considering local time segments where the Hamiltonian is locally commutative.

An evolving state can, at a certain time, be subject to measurement. The quantum measurement operator (or projector) acts on the state and make it collapse onto one of its eigenvectors. If we have a mixed state, the system collapses into an ensemble of states.

In the QVTS, the phon state evolution is subject to restoring forces, and the Hamiltonian depends on the state orientation in the phon space. Such evolution is alike that of a spin in a magnetic field. The Hamiltonian can be expressed as

$$\mathbf{H} = \frac{\omega}{2}\vec{\sigma}\cdot\vec{n} = \frac{\omega}{2}\begin{bmatrix} n_z & n_x - in_y \\ n_x + in_y & -n_z \end{bmatrix}, \tag{28}$$

whose energy eigenvalues are $E_j = \pm\frac{\omega}{2}$, with energy eigenvectors $|E_j\rangle$. An initial phon $|\psi(0)\rangle$ can be expanded in the energy eigenvectors as

$$|\psi(0)\rangle = \sum_j \alpha_j(0)|E_j\rangle, \tag{29}$$

where $\alpha_j(0) = \langle E_j|\psi(0)\rangle$, and the time evolution of the state is

$$|\psi(t)\rangle = \sum_j \alpha_j(t)|E_j\rangle = \sum_j \alpha_j(0)e^{-iE_jt}|E_j\rangle. \tag{30}$$

Where do the restoring forces come from, in the sound domain? Broadly speaking, they come from the local temporal sound production context. Similarly to the concept of coarticulation in phonetics, the locally-defined Hamiltonian is determined by neighboring sounds, extending their effects in the short-term future or past (Daniloff and Hammarberg 1973). In practice, we can rely on an audio analysis system, such as the Short-Time Fourier Transform (STFT), to extract and manipulate slowly-varying features such as pitch salience or spectral energy to determine the components of the Hamiltonian (28). Considered a slice of time and an audio signal, the initial phon state can be made to evolve subject to a time-dependent yet commutative Hamiltonian expressed as

$$H(t) = e^{-kt}\mathbf{S}, \tag{31}$$

where $\mathbf{S}$ is a time-independent Hermitian matrix and k governs the spreading of coarticulating features. Such Hamiltonian evolution has been inspired by a quantum approach to image segmentation (Youssry et al. 2015), or figure-ground segregation. For evolution in the phon space, the matrix $\mathbf{S}$ can be set to assume the structure (28), where the components of potential energy can be extracted as audio features through time-frequency analysis. For example, the n_z component can be made to correspond to the extracted pitch salience, and the n_x component can be made to correspond to the extracted noisiness. In the time slice under examination, an initial $|u\rangle$ state will evolve to a final state

$$|\psi(t)\rangle = e^{-i\int_0^t H(\tau)d\tau}|u\rangle = \mathbf{U}(0,t)|u\rangle, \tag{32}$$

which in general will be a superposition (4) in the phon space. A measurement in the computational (phonation) basis will make it collapse to $|u\rangle$ or $|d\rangle$ according to the probabilities $\alpha_u^* \alpha_u$ or $\alpha_d^* \alpha_d$, respectively. If there are two competing and concurrent pitch lines, the Hamiltonian evolution followed by measurement may thus make a pitch following process stay on one line or jump to the other one. In this way, auditory streaming processes (Bregman 1994) and figure-ground segregation can be mimicked.

3   Quantum Vocal Sound Processing

In this section, we present some examples that show how the quantum formalism, as assimilated by the QVTS, can be used together with classical signal processing, for creative yet controllable analysis/synthesis tasks.   Given the time-frequency representation of an audio signal, as provided by the STFT, the elements of the **S** matrix of the Hamiltonian (31) can be computed from decimated audio features. For example, pitch salience can be extracted from time-frequency analysis (Salamon and Gomez 2012), and used as the n_z component. The exponential factor can be set to $g(m) = e^{-km}$, where m is the frame number within a segment of M frames. The time evolution (32) can be computed by approximating the integral with a cumulative sum. Starting from an initial state (e.g., $|u\rangle$), the phon goes through repeated cycles of Hamiltonian evolution, measurement, and collapse. The decision to measure phonation or turbulence can be based on the degree of pitchiness that the evolution within a certain audio segment has reached. Since the observable σ_z has eigenvalues ± 1 for eigenvectors $|u\rangle$ and $|d\rangle$, a measure of the degree of pitchiness can be given by the distance of $\|\sigma_z |\psi\rangle\|$ from $\||\psi\rangle\|$. The degrees of noisiness and transientness can be similarly determined using the observables σ_x and σ_y, respectively.

When doing actual audio signal processing based on the QVTS, several degrees of freedom are available to experiment with: The decimation factor or number M of frames in a segment; The exponential damping factor k; The thresholds for switching to a certain measurement direction in phon space; The decision to collapse or not—this is a freedom we have if we are using a classical computer!

3.1   Playing with Pure States

3.1.1   Fugue Following

Consider the beginning of the Fugue from the *Toccata and Fugue* in D Minor, BWV 565, by Johann Sebastian Bach (Fig. 6). In this fragment, there is only one voice, played by the left hand. However, the design of this sequence actually creates the illusion of two voices: an upper line with an *ostinato* A, and a lower line with notes G, F, E, D, C♯, D, E, F, and so on.

Fig. 6 *Toccata and Fugue* in D minor BWV 565 by J. S. Bach: beginning of the Fugue

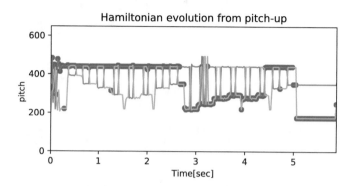

Fig. 7 The synthetic recording of the excerpt in Fig. 6 through Hamiltonian evolution. The upper line with the repeated A is evident in the first part of the graph, while the second part contains a fragment of the melody of the lower line

The score fragment of Fig. 6 was automatically rendered with piano samples at 100 bpm and analyzed via the STFT,[6] with pitch salience and noise energy extracted via the SMS-tools (Bonada et al. 2011). Setting the parameters frame decimation $M = 10$, exponential damping $k = 0.1$, threshold of pitchiness 0.9, collapse decimation 5, we obtain a phon evolution from pitch-up phonation represented by the green dots of Fig. 7. The red and yellow lines represent the two most salient pitches, as features extracted from the STFT. In the first part of the evolution, the green dots mainly correspond to the *ostinato* of note A, while, in the second part, they mainly follow the ascending scale fragment (A B C♯ D...). Even without any noise added, transient and noise components are inherent in piano samples (e.g., key noise) and, therefore, the phon is subject to non-negligible forces and actually moves from the initial $|u\rangle$ state. Different runs will give different evolutions, as the collapse is governed by probability amplitudes, but in general we observe that the Hamiltonian listener follows the upper or the lower melodic line for some fractions of time. Interestingly, the melodic line following is stable, or even more stable, if we add a strong white noise to the signal, with an amplitude that is about one tenth of the signal. An example evolution is depicted in Fig. 8, where the effect of the added noise is only apparent after second 5, when effectively there is no signal. Figure 9 shows the phon evolution when the fugue is drowned into noise. In this case the melodic contour following is more easily disrupted. The zero-pitch green dots represent points where measurement and collapse have been oriented to the x direction, for the effect of thresholding in

---

[6] Sample rate 44100Hz, window size 2048, transform size 4096, hop size 1024.

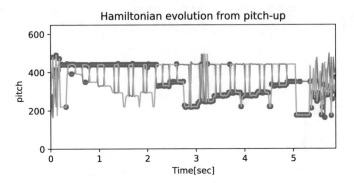

Fig. 8 The synthetic recording of the excerpt in Fig. 6 through Hamiltonian evolution, with the added noise (amplitude 0.1). The upper line with the repeated A is evident in the first part of the graph, while the second part contains a fragment of the melody of the lower line

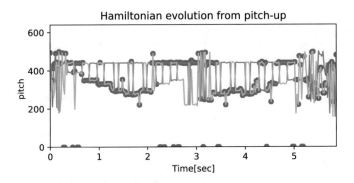

Fig. 9 The same fragment of Fig. 6 completely drowned into noise (amplitude 1.0)

pitchiness. In a resynthesis, these dots correspond to noise bursts, while the other dots come from z-oriented measurements and produce pitched notes.

In repeated runs of Hamiltonian fugue following, we can see multiple melodic lines emerging as the time evolution of some initial sound/vocal state. The collapse of the phon to a state or another can be interpreted as the attention shifts from figure to ground, or vice versa (Bregman 1994; Bigand et al. 2000).

The proposed example can be the starting point for a wider investigation in the field of auditory bistability. Bistability is an intriguing topic in cognition. As a reference for quantum effects in cognition, especially regarding superposition and non-classical probability, we may refer to Yearsley and Busemeyer (2016) for a theoretical quantum-based approach to explain cognitive acts. The idea of bistability is also faced in an article on mathematics and image/music (Mannone et al. 2020). It exploits the Dirac formalism used in quantum mechanics to represent images as superpositions of essential visual forms. There is a minimum number of forms which allows the recognizability of the form. With a little abuse of terminology, we can consider this limiting, minimum value of simple forms as the limit of a Gestalt neighborhood, as the discrete version of a topological neighborhood, having in the center the initial, complete, not-approximated visual form. When an image is bistable, we can imagine to have two neighborhoods as the two faces of a thin cylinder. We can see one face or the other one; but we cannot see the two faces together. This is the core idea of bistability. While classic examples of bistability are visual, also auditory illusions can be constructed, e.g., with different auditory streamings (Byrne et al. 2019). These cases might be analyzed with the help of QVTS, as we did with the beginning of the Fugue from the *Toccata and Fugue* BWV 565.

3.1.2   Glides Tunneling

Continuity effects have been very important to derive a perceptual organization of sound for auditory scene analysis (Bregman 1994). Gestalt principles such as proximity or good continuation are often used to describe how listeners follow concurrent pitch lines and extract temporal patterns from a scene. A simple yet significant case is that of two gliding and crossing tones interrupted, at the crossing point, by a short burst of noise (Ciocca and Bregman 1987).

Figure 10 (top) shows the spectrogram of two gliding and crossing tones, interrupted by a 200ms-band of white noise, intervening at time 1.5s. The red and yellow lines are the traces of the two most salient pitches, as extracted using the Essentia library (Bogdanov et al. 2013). With stimuli such as this, listeners most often report perceiving a single frequency-varying auditory object tunneling (Vicario 1960) the interruption. Depending on the temporal extension and intensity of the noise burst, a perceived V-shaped trajectory may be predominant over a rectilinear continuation, thus making proximity prevail over good continuation.

It is interesting to use the case of crossing glides interrupted by noise as a test for Hamiltonian evolution, with the matrix **S** of the Hamiltonian (31) computed from decimated audio features such as pitch salience and noise energy. As a result of such feature extraction from a time-frequency representation, we obtain two potentials, for phonation and turbulence, which drive the Hamiltonian evolution. Figure 10 also displays (middle) the computed salience for the two most salient pitches and (bottom) the energy traces for two bands of noise (1kHz–2kHz and 2kHz–6kHz). It is clear how the pitch extractor becomes uncertain when the two tones get close in frequency and start beating, and it wiggles around during the noisy interruption.

Figure 11 shows an example evolution of the phon state, starting from $|u\rangle$. In this specific run of the evolution, the phon sticks to phonation (one of the two pitches) until well inside the noise band, even though pitch has very little salience within the interruption (see Fig. 10, middle), with only occasional switches to turbulence (the zero-pitch green dots in Fig. 11). Right after the noise interruption, the phon evolution is still uncertain, until it steadily takes a $|u\rangle$ state, thus giving an overall V-shaped bouncing trajectory. In this instance, proximity is shown to prevail over good continuation. Due to the statistical nature of quantum measurement, another run

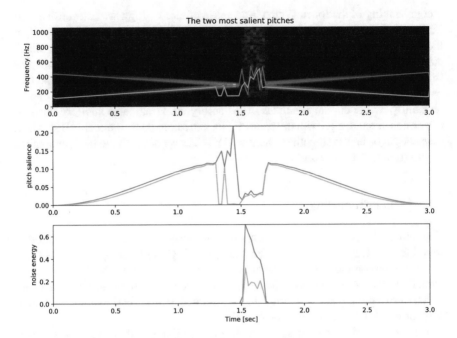

Fig. 10 Tracing the two most salient pitches and noise energy for two crossing glides interrupted by noise

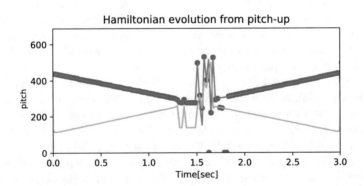

Fig. 11 Tracking the phon state under Hamiltonian evolution from pitch-up

of the evolution may well produce a downward-crossing trajectory, and turbulence bursts may be found at different times. Such uncertainty on noise location is consistent with the known perceptual fact that bursts of noise overlapped to a noise transition are not precisely located, with errors that can be up a few hundred milliseconds (Vicario 1963).

With this example, we have given a demonstration of how quantum evolution of the phon state can be set to reproduce relevant phenomena in auditory perception, with possible applications in computational auditory scene analysis.

3.2 Playing with Mixed States

The quantum concept of mixing, briefly described in Sect. 2.3, can be related to the familiar audio concept of mixing. At the start of a Hamiltonian evolution, the initial state may be mixed, i.e., known only as a probabilistic mixture. For example, at time zero we may start from a mixture having $\frac{1}{3}$ probability of $|u\rangle$ and $\frac{2}{3}$ probability of $|d\rangle$. The density matrix would evolve in time according to equation (26).

When a pitch measurement is taken, the outcome is up or down according to

$$P[m = i|\rho] = Tr[\rho M_i], \qquad (33)$$

and the density matrix that results from collapsing upon measurement is given by

$$\rho^{(i)} = \frac{M_i \rho M_i}{Tr[\rho M_i]}. \qquad (34)$$

The density matrix can be made audible in various ways, thus sonifying the Hamiltonian evolution. For example, the completely chaotic mixed state, corresponding to the half-identity matrix $\rho = \frac{1}{2}\mathbf{I}$, can be made to sound as noise, and the pure states can be made to represent distinct components of an audio mix.

3.2.1 Glides Tunneling

Given the same audio scene of the two crossing glides interrupted by noise (Fig. 10), we may follow the Hamiltonian evolution from an initial mixed state. We can choose to make the pure states to sound like the upper or the lower of the most salient pitches, and the completely mixed state to sound like noise. These three components can be mixed for states with intermediate degrees of purity. If p_u and p_d are the respective probabilities of $|u\rangle$ and $|d\rangle$ as encoded in the mixed state, the resulting mixed sound can be composed by a noise having amplitude $\min(p_u, p_d)$, by the upper pitch weighted by $p_u - \min(p_u, p_d)$, and by the lower pitch weighted by $p_d - \min(p_u, p_d)$. Figure 12 shows an example of evolution from the mixed state having probabilities $\frac{1}{3}$ and $\frac{2}{3}$, with periodic measurements and collapses ruled by equation (34).

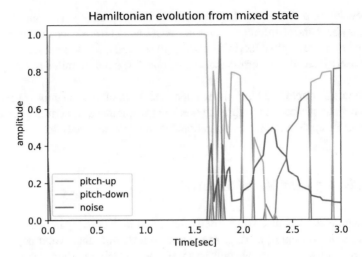

Fig. 12 Amplitudes of components $|u\rangle$, $|d\rangle$, and noise resulting from a Hamiltonian evolution from a mixed state

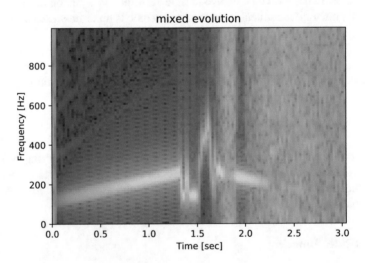

Fig. 13 Sound synthesis obtained from the density matrix evolution from a mixed state, using the component amplitudes depicted in Fig. 12

The analyzed audio scene and the model parameters, including the computed Hamiltonian, are the same as used in the evolution of pure states described in Sect. 3.1.2. The amplitudes of the three components can be used as automated knobs to control two oscillators and a noise generator, producing the sound of spectrogram Fig. 13, characterized by a prevailing upward tone with a downward bifurcation and a noisy tail.

3.2.2 Vocal Superposition

As another example of mixed state evolution, we consider again the vocal sound whose spectrogram is depicted in Fig. 2. It was chosen as an example of actual superposition of phonation and slow myoelastic vibration. Despite the presence of only one definite pitch, we can prepare the phon in an initial mixed state, having $\frac{1}{3}$ probability of $|u\rangle$ and $\frac{2}{3}$ probability of $|d\rangle$, and compute a Hamiltonian evolution based on potentials deduced from the time-frequency analysis, namely pitch salience, noise component, and detected onsets. As in the example of Sect. 3.2.1, we chose to assign phonation amplitudes equal to $p_u - \min(p_u, p_d)$ and $p_d - \min(p_u, p_d)$ to the components $|u\rangle$ and $|d\rangle$, respectively, and turbulence amplitude $\min(p_u, p_d)$ to the noise component. In addition, here we extract a pulsating component as well, corresponding to slow myoelastic vibration, whose amplitude is derived from the probabilities p_f and p_s of fast or slow pulsation. For example, p_f is derived from $Tr[\rho\,|f\rangle\,\langle f|]$, which is similar to equation (33).

Figure 14 shows the amplitude profiles that are extracted from the Hamiltonian evolution, where we chose to measure phonation when $\min(p_u, p_d) > 0.5$, otherwise measuring along the slow myoelastic vibration axis. With non-physical freedom, we collapsed the mixed state, along $|u\rangle$, $|d\rangle$, $|f\rangle$, or $|s\rangle$, using an equation similar to (34), once every five measurements. The resulting sound, which can be considered as a quantum-inspired audio effect, has the spectrogram depicted in Fig. 15, where the most salient pitch and the onsets have been extracted again and superimposed.

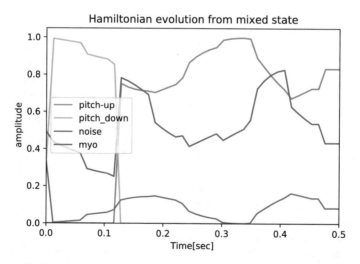

Fig. 14 Amplitudes of components $|u\rangle$, $|d\rangle$, turbulence, and slow myoelastic vibration, resulting from a Hamiltonian evolution from a mixed state, run on the vocalization of Fig. 2

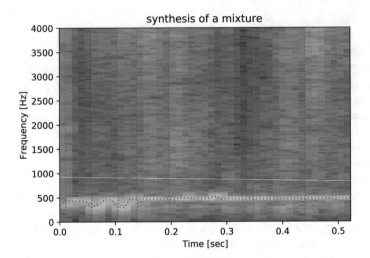

Fig. 15 Sound synthesis obtained from the density matrix evolution from a mixed state, using the component amplitudes depicted in Fig. 14, and all three components of phonation, turbulence, and slow myoelastic vibration. The most salient pitch and onsets, as extracted from the synthetic sound, are displayed as red dashed lines

3.3 From Signal Processing to Quantum Computing, and Back

In digital audio, talking about quantization does not mean referring to quantum theory. Instead, quantization is meant to be the reduction of a continuous range of signal amplitude values to a finite set of discrete values, with a cardinality that depends on the number of bits dedicated to represent each discrete-time sample. Signal quantization introduces a kind of noise, which tends to have a spectrotemporal structure that somehow follows the signal, thus becoming audible as a distortion for low-amplitude signals. A cure for quantization noise is dithering, i.e., adding some tiny broadband noise to the audio signal itself, before quantization, thus making quantization noise more spectrally uniform and perceptually tolerable (Pohlmann 1995). That injecting dither noise to signals and systems can make human and machine processing more robust is a fact that has been known for a long time, and widely applied in a variety of fields, including audio and image processing. In quantum-inspired sound processing, as illustrated in the example of Sect. 3.1, dithering can be used to control how erratic leading-pitch attribution can be, in auditory scenes of competing sources.

As opposed to low-amplitude noise, that may actually make the pitch evolution of a phon more stable, when a high-amplitude noise burst is encountered, it actually acts as a bounce on the phon state, making it rotate by an angle θ. A sequence of bursts, such as that of the example in Sect. 3.2, is much like a sequence of bounces between billiard balls of highly different weights. Recently, the classically mechanic behavior of balls, whose weights are in ratios of powers of 100, has been shown to be perfectly analogous to the kernel of the Grover algorithm for quantum search (Brown 2020), which is based on unitary reflections in the state space.

In the examples of Sects. 3.1 and 3.2, the quantum evolution is driven by potentials that are derived from the same audio that is being processed. To turn these evolutions into quantum algorithms we should freeze a reference audio segment, extract the feature-based potentials from the time-frequency representation, and convert the elementary unitary transformations into quantum gates, arranged along left-to-right wires. Each stage of the quantum algorithm would represent a bounce or a measurement in the phon space, as long as the operators are consistent with the postulates of quantum mechanics. It should be noted that we have only been considering single-qubit (or single-phon) operators. The universe of multiple and entangled phons remains to be explored.

In both quantum mechanics and sound signal processing, unitary operators and unitary transformations have a central role. In fact, in physically-inspired sound synthesis and digital audio effects, unitary matrix transformations are often found, as scattering elements in feedback delay networks for artificial reverberation (Pulkki et al. 2011). In these structures, if the feedback matrix $\mathbf{A}$ is chosen to be unitary, an initial pulse bounces indefinitely, at each bounce scattering into a multiplicity of other pulses. In the ball-within-the-box (BaBo) model (Rocchesso 1995), the matrix $\mathbf{A}$ can be interpreted as a scattering ball that redistributes the energy from incoming wavefronts into different directions, each corresponding to a planar wave loop, which produces a harmonic series. Indeed, the matrix $\mathbf{A}$ does not have to be unitary for the feedback structure to be lossless (Schlecht and Habets 2017). However, even staying within the class of unitary matrices, $\mathbf{A}$ can be chosen for its scattering properties, ranging from the identity matrix (no scattering at all) to maximally-diffusive structures (Rocchesso 1997; Schlecht 2020). A promising perspective for future quantum sound processing, is to find realizable quantum operators for such matrices. In particular, the Hadamard operator and the Householder reflection are extensively used in quantum algorithms, and these were proposed as reference matrices for feedback delay networks with maximally-diffusive properties (Jot 1997). In the context of QVTS, a Hadamard gate $\mathbf{H}$ converts a phon from $|r\rangle$ to $|u\rangle$, and from $|l\rangle$ to $|d\rangle$. If followed by a measurement in the computational basis, it can be used to discriminate between the two turbulent states. If inserted in a phon manipulation sequence, it determines a switch in the vocal state of sound. Loops are not allowed in quantum computing (Nielsen 2010), but by spatially unfolding feedback, a reverberator based on feedback delay networks may be converted to a quantum algorithm with several stages of unitary operators, acting as scattering elements on a multiplicity of phons. As a non-negligible detail, banks of delay lines of different lengths, in the order of tens of milliseconds, should be interposed between consecutive scattering sections.

Shor's algorithm for factorization of a large integer N relies on an efficient way of finding the periodicity (modulo N) of the function a^x, constructed from a randomly chosen integer a that is smaller than N and coprime with it. To compute the periodicity of a function, the Quantum Fourier Transform (QFT) operator is used, which transforms a superposition of $N = 2^n$ computational basis states on n qubits, with coefficients $\mathbf{x} = [x_0, x_1, \ldots, x_{N-1}]$, into another superposition with coefficients $\mathbf{X} = [X_0, X_1, \ldots, X_{N-1}] = DFT(\mathbf{x})$, that is the Discrete Fourier Transform of $\mathbf{x}$. Using quantum parallelism, such DFT is implemented with $O(n^2)$ quantum gates, while classically that would take $O(n2^n)$ steps. Recently, a direct transposition of the Fast Fourier Transform (FFT) into the form of a quantum circuit has been proposed and called the Quantum FFT (QFFT) (Asaka et al. 2020). Instead of the amplitude encoding used for the QFT, a basis encoding is used, where a data sequence $\mathbf{x}$ is expressed as a tensor product of vector spaces $\bigotimes_{j=0}^{N-1} |x_j\rangle$. A potential impact on audio signal processing would be that quantum parallelism would allow to perform all frames of a N-bins STFT simultaneously, with $O(N \log_2 N)$ gates. The aggregate features of an audio segment would then be encoded in the resulting vector of qubits.

4   Quantum Evolution of the State (of the Art)

A respectable scientific theory helps find new results, confirms expectations, extends the validity of known laws bringing them toward the realm of the unknown and (formerly) inexplicable, and so on.

An exciting scientific theory leaves room for imagination and artistic creativity. New ideas can arise from the interdisciplinary dialogue between people of different fields. QVTS is intrinsically interdisciplinary, and we think it can enhance the dialogue between worlds.

Interchanges between music and quantum mechanics constitute a relatively new and flourishing research area. Our contribution to this field is the addition of the human voice, and the use of vocal primitives as a probe to more generally investigate the world of sounds. In Sect. 3 we have proposed some examples of a creative use of QVTS, where the Hamiltonian evolution is the starting point for sound synthesis. In this section, we suggest some further creative applications.

The density matrix can be exploited to improve source separation techniques. In fact, the operation of partial trace on density matrix allows us to separate a system from the environment, the *reservoir*. Choosing on which part of the whole system we are making the operation of partial trace, we can interchangeably choose which part we are neglecting. For example, given a polyphonic vocal recording, we can establish that singer 1 is the system and singers 2, 3, and 4 are the environment (thus, we can perform the partial trace on singers 2-3-4), or that singer 2 is the system and singers 1, 3, and 4 are the environment, and so on. In fact, as a practical interest in the domain of QVTS, we can think of a general recording, with multiple voices, and interpret it as a statistic mixture of states. Voices might be organized as a solo singing voice against a a background of several other voices of a choir—a quantum choir. Therefore, QVTS may help analyze choral music. In addition, it can give us hints also on how to create music. Creativity can precisely take off from mixtures of states and vocal polyphony.

Because QVTS constitutes a bridge between sounds and quantum formalism, we can play with the symmetries of particle processes and transform them to musical symmetries, thus giving voice to quantum processes. We can create correspondences between certain quantum properties of particles and the sounds, their transformations, and musical transformations. For example, an inversion of the spin could be musically rendered with an inversion of the pitch interval; a quantum superposition can be rendered with the simultaneous playing of different orchestral sections to create a "cloud of sound." A quantum measurement, with the subsequent collapse of the sound wave, could be rendered with the sudden silence of other orchestral sections, and with the remaining sound of a section, or even only one instrument sound. Musical structures can be thought of as transformations over time of "states" (short musical sequences or essential musical ideas for example). According to this metaphor, we might describe the time evolution of quantum states, including density matrices describing inseparable state superpositions through generated musical structures. These hints should be compared with perceptual criteria, to create an idea of the processes in the mind of the listener.

Finally, we may imagine an interface where the user can modify states on the Bloch sphere, modifying the synthesis in real time. Such an interface might allow a "Quantum Synthesis," maybe the Gabor's dream. A quantum synthesizer with potential for development has indeed been recently proposed by Costa Hamido et al. (2020), where the quantum circuits such as the one for Grover's search can be run on a simulator or a quantum computer, and probability distributions and computation steps can be heard, with auditory exploitation of quantum noise.

We end this section with a fun, original musical fragment, or, better, a set of instructions to make music directly out of the QVTS-Bloch's sphere. A suitable synthesizer as the one hypothesized above could make this attempt a concrete tool for creative purposes. Let us imagine a short musical composition with two vocal (not instrumental) lines, created out of moving states on the Bloch's sphere. As another homage to Bach, we can be inspired by the structure of the *Two-Part Inventions*, with the parts imitating each other, as in a simple counterpoint. Thus, we can provocatively call our attempt *Two-Part Quantum Invention* No. 1. Figure 16 shows a tentative notation, with a schematic sphere derived from Fig. 1, and the sequence of state variations. Generalization of the proposed idea to more voices and intricate counterpoints is up the reader. This structure could be used as a set of instructions for vocal improvisation, similarly to the "quantum improvisations" Pauline Oliveros used to conduct (Oliveros 2002). Conceptually, the two voices can be instances of the same evolving phon, from which we can, in principle, extract infinite counterpoint lines. If the parallel motion of parts causes troubles in classical counterpoint because of the feeling of sameness, intrinsic parallelism is the real advantage of quantum computation, eventually leading to quantum supremacy for some computational problems. Music counterpoint may actually give voice to quantum parallel computations.

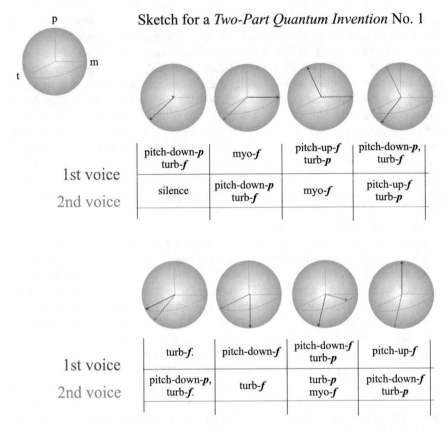

Sketch for a *Two-Part Quantum Invention* No. 1

	pitch-down-*p* turb-*f*	myo-*f*	pitch-up-*f* turb-*p*	pitch-down-*p*, turb-*f*
1st voice				
2nd voice	silence	pitch-down-*p* turb-*f*	myo-*f*	pitch-up-*f* turb-*p*

	turb-*f*.	pitch-down-*f*	pitch-down-*f* turb-*p*	pitch-up-*f*
1st voice				
2nd voice	pitch-down-*p*, turb-*f*.	turb-*f*	turb-*p* myo-*f*	pitch-down-*f* turb-*p*

Fig. 16 Sketch for a music composition based on Bloch's sphere for QVTS (by M.M.).

4.1 Concluding Remarks

Starting from kids' playing and moving toward futuristic scenarios of quantum choirs and Quantum Inventions, in this chapter we presented the fundamental ideas of the Quantum Vocal Theory of Sound (QVTS), along with some proposal of future developments.

We aimed to discuss a supplemental formalism to describe sound (with the vocal probe), rather than proving any "wrongness" or obsolescence of the classical formalism, such as Fourier analysis, for sound. Our supplemental formalism is an alternative one, it gives a new perspective, and it has the advantage of providing more information, especially regarding Gestalt-related phenomena, as in the case of bistable states.

QVTS is interdisciplinary in nature, as it provides a bridge between sound sources, sound production, human perception, and the intuitive identification of sounds and sound sources. In addition, the uncertainty (or fuzziness) that is proper to quantum thinking might be compared with the approximation of human intuitive assessments about sounds, sound sources, and sound identification.

Beyond the theoretical foundation, our aim is to foster the creation of machines that measure sounds in terms of vocal primitives and express sonic processes as evolving superpositions of vocal primitives. We hope that our presentation of QVTS may lead to further questions, research, developments, as well as to artistic contributions.

May it be a simple metaphor, or a quantitative tool, the core of quantum mechanics is more and more inspiring for musicians, performers, and scientists, shedding light on new and unexplored collaborations and insights.

All of that, can lead to *the sound of quanta*.

References

Asaka, R., Sakai, K., & Yahagi, R. (2020). Quantum circuit for the fast Fourier transform. *Quantum Information Processing, 19,* 277.

Aspect, A., Dalibard, J., & Roger, G. (1982). Experimental test of Bell's inequalities using time-varying analyzers. *Physical Review Letters, 49*(25), 1804.

Bell, J. S. (1964). On the Einstein Podolsky Rosen Paradox. *Physics Physique Fizika, 1*(3), 195–200. https://doi.org/10.1103/PhysicsPhysiqueFizika.1.195.

Bigand, E., McAdams, S., & Forêt, S. (2000). Divided attention in music. *International Journal of Psychology, 35*(6), 270–278.

Blutner, R., & Beim Graben, P. (2020). Gauge models of musical forces. *Journal of Mathematics and Music*. https://doi.org/10.1080/17459737.2020.1716404.

Bogdanov, D., Wack, N., Gómez Gutiérrez, E., Gulati, S., Herrera Boyer, P., Mayor, O., Roma Trepat, G., Salamon, J., Zapata González, J. R., & Serra, X. (2013). Essentia: An audio analysis library for music information retrieval. In *Proceedings of the 14th Conference of the International Society for Music Information Retrieval (ISMIR)* (pp. 493–498), Curitiba, Brazil

Bonada, J., Serra, X., Amatriain, X., & Loscos, A. (2011). Spectral processing. In U. Zölzer (Ed.), *DAFX: Digital Audio Effects* (pp. 393–445). John Wiley & Sons Ltd.

Bregman, A. S. (1994). *Auditory scene analysis: The perceptual organization of sound*. Cambridge, MA: MIT Press.

Brown, A. R. (2020) Playing Pool with $|\psi\rangle$: from Bouncing Billiards to Quantum Search. Quantum 4, 357, Verein zur Förderung des Open Access Publizierens in den Quantenwissenschaften.

Byrne, Á., Rinzel, J., & Rankin, J. (2019). Auditory streaming and bistability paradigm extended to a dynamic environment. *Hearing Research,383*, 107807.

Cádiz, R., & Ramos, J. (2014). Sound synthesis of a gaussian quantum particle in an infinite square well. *Computer Music Journal, 38*(4), 53–67.

Cariolaro, G. (2015). *Quantum Communications*. Cham: Springer.

Ciocca, V., & Bregman, A. S. (1987). Perceived continuity of gliding and steady-state tones through interrupting noise. *Perception & Psychophysics,* **42**(5), 476–484, Springer .

Costa Hamido, O., Cirillo, G. A., & Giusto, E. (2020). Quantum synth: a quantum-computing-based synthesizer. In *Proceedings of the 15th International Conference on Audio Mostly* (pp. 265–268) (2020)

Dalla Chiara, M. L., Giuntini, R., Leporini, R., Negri, E., & Sergioli, G. (2015). Quantum information, cognition, and music. *Frontiers in Psychology, 6,* 1583.

Daniloff, R. G., Hammarberg, R. E. (1973). On defining coarticulation. *Journal of Phonetics, 1*(3), 239–248. http://www.sciencedirect.com/science/article/pii/S0095447019313889

Delle Monache, S., Rocchesso, R., Bevilacqua, F., Lemaitre, G., Baldan, S., & Cera, A. (2018). Embodied sound design. *International Journal of Human-Computer Studies, 118,* 47–59.

Einstein, A., Podolsky, B., & Rosen, N. (1935). Can quantum-mechanical description of physical reality be considered complete? *Physical Review, 41,* 777. http://www.drchinese.com/David/EPR.pdf

Feynman, R. (1995). *The Character of Physical Law.* Cambridge (MA): MIT Press.

Friberg, A., Lindeberg, T., Hellwagner, M., Helgason, P., Salomão, G. L., Elowsson, A., Lemaitre, G., & Ternström, S. (2018). Prediction of three articulatory categories in vocal sound imitations using models for auditory receptive fields. *The Journal of the Acoustical Society of America, 144*(3), 1467–1483

Gabor, D.: Acoustical quanta and the theory of hearing. *Nature, 159,* 591–594 (1947)

Gaver, W. W. (1993). What in the world do we hear? An ecological approach o auditory event perception. *Ecological Psychology, 5*(4), 285.

Ghirardi, G. (1999). Quantum superpositions and definite perceptions: envisaging new feasible experimental tests. *Physics Letters A, 262*(1), 1–14.

Irino, T., & Patterson, R. D. (1997). A time-domain, level-dependent auditory filter: The gammachirp. *The Journal of the Acoustical Society of America, 101*(1), 412–419.

Jot, J. -M. (1997) Efficient models for reverberation and distance rendering in computer music and virtual audio reality. In *Proceedings: International Computer Music Conference 1997*, Thessaloniki, Hellas, 25–30 September 1997, 236–243. The International Computer Music Association (1997)

Juett, J. C. (2010). Pauline Oliveros and quantum sound. *Liminalities: A Journal of Performance Studies, 6*(2), 1–10. http://liminalities.net/6-2/oliveros.pdf

Koenig, D. M., & Delwin, D. F. (2015). *Spectral Analysis of Musical Sounds with Emphasis on the Piano.* New York: Oxford University Press.

Kulpa, J. (2019). *QuBits, an interactive virtual reality project and compositional space for sound and image.* Advisor: E. Campion. Ph.D. dissertation, Berkeley, California. https://escholarship.org/uc/item/5tq9x2hb

Kulpa, J., Campion, E., & Cella, C. (2020) QuBits, a system for interactive sonic virtual reality. In *Proceedings of ICMC 2020, Santiago, Chile.* http://www.carminecella.com/research/2020_ICMC_QuBits.pdf

Lindeberg, T., & Friberg, A. Idealized computational models for auditory receptive fields. *PLoS One, 10*(3), e0119032. https://doi.org/10.1371/journal.pone.0119032

Mannone, M. (2018). Introduction to gestural similarity. An application of category theory to the orchestra. *Journal of Mathematics and Music, 12*(2), 63–87.

Mannone, M., & Compagno, G. (2014). Characterization of the degree of Musical non-markovianity. https://arxiv.org/abs/1306.0229

Mannone, M., Favali, F., Di Donato, B., & Turchet, L. (2020). Quantum GestART: identifying and applying correlations between mathematics, art, and perceptual organization. *Journal of Mathematics and Music.* https://doi.org/10.1080/17459737.2020.1726691.

Marinetti, F. T. (1914) Zang tumb tumb. Milano, Edizioni futuriste di poesia. http://parliamoitaliano.altervista.org/zang-tumb-tumb/.

Miranda, E. R. (2021) Creative quantum computing: Inverse FFT sound synthesis, adaptive sequencing and musical composition. In: A. Adamatzky (Ed.), *Alternative Computing, World Scientific* (2021)

Miranda, E. R. (2021). Quantum computer: hello, music! In E. Miranda (Ed.), *Handbook of Artificial Intelligence for Music: Foundations, Advanced Approaches, and Developments for Creativity.* Springer Nature

Newman, F. (2004). *MouthSounds: How to Whistle Pop: Boing, and Honk... for All Occasions and Then Some.* Workman Publishing.

Nielsen, M. A. & Chuang. I. L. (2010). *Quantum Computation and Quantum Information*. Cambridge University Press.

Oliveros, P. (2002) Quantum listening: From practice to theory (to practise practice). In K. Siu Tong & C. Sin-wai (Eds.), *Culture and Humanity in the New Millennium: The Future of Human Values* (pp. 27–44). The Chinese University Press.

Palazzeschi, A. (1909). La fontana malata, from Poemi. Firenze, Cesare Blanc. https://www. pensieriparole.it/poesie/poesie-d-autore/poesia-38384

Pohlmann, K. C. (1995). *Principles of Digital Audio* (3rd ed.). USA: McGraw-Hill Inc.

Pulkki, V., Lokki, T., & Rocchesso, D. (2011) *Spatial Effects. DAFX: Digital Audio Effects* (pp. 139–183). Wiley.

Roads, C. (2001). *Microsound*. Cambridge, MA: MIT Press.

Rocchesso, D. (1995). The ball within the box: A sound-processing metaphor. *Computer Music Journal, 4*(19), 47–57. The MIT Press. http://www.jstor.org/stable/3680990

Rocchesso, D., Mauro, D. A., & Delle Monache, S. (2016). MiMic: The Microphone as a Pencil. In *Proceedings of the TEI '16: Tenth International Conference on Tangible, Embedded, and Embodied Interaction, in Eindhoven, Netherlands* (pp. 357–364). New York, NY, USA: Association for Computing Machinery, TEI '16. https://doi.org/10.1145/2839462.2839467

Rocchesso, D., Mauro, D., & Drioli, C. (2016). Organizing a sonic space through vocal imitations. *Journal of the Audio Engineering Society, 64*(7/8), 474–483. http://www.aes.org/e-lib/browse. cfm?elib=18333

Rocchesso, D. (1997). Maximally diffusive yet efficient feedback delay networks for artificial reverberation. *IEEE Signal Processing Letters, 4*(9), 252–255. https://doi.org/10.1109/97.623041.

Rocchesso, D., & Mannone, M. (2020). A quantum vocal theory of sound. *Quantum Information Processing, 19*, 292. https://doi.org/10.1007/s11128-020-02772-9.

Rovelli, C. (2021). Helgoland. Penguin Books.

Salamon, J., & Gomez, E. (2012). Melody extraction from polyphonic music signals using pitch contour characteristics. *IEEE Transactions on Audio, Speech, and Language Processing, 20*(6), 1759–1770.

Schlecht, S. J. (2020). FDNTB: The feedback delay network toolbox. In *Proceedings of the 23rd International Conference on Digital Audio Effects DAFx-20*.

Schlecht, S. J., & Habets, E. A. P. (2017). On lossless feedback delay networks. *IEEE Transactions on Signal Processing, 65*(6), 1554–1564. https://doi.org/10.1109/TSP.2016.2637323.

Shamma S. (2013). Spectro-temporal receptive fields. In D. Jaeger , R. Jung (Eds.), *Encyclopedia of Computational Neuroscience*. New York, NY: Springer. https://doi.org/10.1007/978-1-4614-7320-6_437-1.

Spence, C. (2011) Crossmodal correspondences: A tutorial review. *Attention, Perception, & Psychophysics, 73*, 971–995. https://doi.org/10.3758/s13414-010-0073-7

Spence, C., Senkowski, D., & Röder, B. (2009). Crossmodal processing. *Experimental Brain Research, 198*(107), 107–111 (2009). https://link.springer.com/article/10.1007/s00221-009-1973-4.

Verma, T. S., Levine, S. N., & Meng, T. H. (1997). Transient modeling synthesis: A flexible analysis/synthesis tool for transient signals. In *Proceedings of the International Computer Music Conference* (pp. 48–51)

Vicario, G. B. (1963). La "dislocazione temporale" nella percezione di successioni di stimoli discreti (The "time displacement" in the perception of sequences of discrete stimuli). *Rivista di Psicologia,57*(1), 17–87.

Vicario, G. (1960). L'effetto tunnel acustico (The acoustic tunnel effect). *Rivista di Psicologia, 54*, 41–52.

Vochlea microphone. Retrieved from https://vochlea.com/.

von Helmholtz, H. (1870). *Die Lehre von den Tonempfindungen als physiologische Grundlage für die Theorie der Musik*. Braunschweig: F. Vieweg und sohn.

Weimer, H. (2010). Listen to quantum computer music. Retrieved Jan. 20, 2021, from http://www. quantenblog.net/physics/quantum-computer-music.

Yearsley, J. M., & Busemeyer, J. R. (2016). Foundations of probability theory in psychology and beyond (special issue). *Journal of Mathematical Psychology, 74,* 99.

Youssry, A., El-Rafei, A., & Elramly, S. (2015). A quantum mechanics-based framework for image processing and its application to image segmentation. *Quantum Information Processing, 14*(10), 3613–3638.

Quantum Cinema and Quantum Computing

Renate C.-Z. Quehenberger

It cannot help striking me that this extension of Algebra ought to lead to a further extension similar in nature to the Geometry of three dimensions; & that again perhaps to a further extension into some unknown region & so on ad infinitum possibly! (Ada Lovelace in a letter to her tutor Augustus De Morgan, 1832, quoted in Padua, 2015, 60).

Abstract The author discusses philosophical, ontological and phenomenological questions of quantum phenomena and their possible visualization by means of quantum geometry as well as links to the complex number space, prime numbers and lattices used for quantum computation. The chapter engages with intuitively developed images realized as digital dynamic 3D geometry. Some of these images, which could be relevant in the context of Quantum Computing, are introduced. The author discusses discoveries that became the basics of Quantum Computing through a journey into the history of Quantum Physics, which lead to the notion of quantum information and the conception of quantum computers.

Keywords Quantum humanities · Higherdimensional geometry of Quantum states · 3D Penrose patterns · Heisenberg's uncertainty principle · Poincaré conjecture

1 Introduction

This chapter on Quantum Cinema and Quantum computing engages with intuitively developed images realized as digital dynamic 3D geometry. Some of these images, which could be relevant in the context of quantum computing, are presented here.

R. C.-Z. Quehenberger (✉)
Independent SciArt Researcher, QC:L, Vienna, Austria

Luxor, Egypt
e-mail: epitaten@pm.me; office@qc-l.eu
URL: https://quantumcinema.uni-ak.ac.at/

The exciting development of quantum computation is rooted in the fundamentally new interpretation of quantum physics which became the "it from bit" doctrine by John Archibald Wheeler, who linked information, physics and quantum phenomena by suggesting all things physical are information-theoretic in origin (Wheeler, 1989). According to Claude Shannon's communication protocol, information is neg-entropy (Shannon, 1948). His model of the communication process defines Wheeler's information paradigm. Shannon's original communication model limits the interest in signalling systems that include transmitter, receiver, channel and noise (Abel and Trevor, 2005). It gave rise to the new discipline of quantum information and quantum information science, the study of how a quantum computer works.

Quantum computers rely on qubits as their basic unit of information. Whereas traditional computers use bits, which is a unit of information based on binary and can either be in an 'on' state or an 'off' state, quantum computers use qubits instead. A qubit introduces a new state: both 'off' and 'on', at the same time. This creates three states instead of two, allowing for much higher quantities of data to be transferred. The current issue (which has to be overcome) is that quantum machines can generate control signals for multiple qubits at cryogenic temperatures—100 millikelvin, or—273.05 °C, which is almost absolute zero (Pauka et al., 2021). The world's biggest quantum computers currently operate with just 50 or so qubits, whereas the control platform with the cryogenic chip to control thousands of qubits is the next big step to scale up quantum computing. While Google is predicting the first error-free quantum computer by the end of the decade.

Here we will discuss some philosophical ontological and phenomenological questions of quantum phenomena and their possible visualization by means of quantum geometry as well as links to the complex number space, prime numbers and lattices used for quantum computation.

We will tackle discoveries that became the basics of quantum computing in a parcour through the history of quantum physics, which lead to quantum information and the conception of quantum computers.

2   Quantum Cinema—A Digital Vision

Linear algebra is the mathematical language of quantum computation. It is represented by high-dimensional vector spaces as developed by Hermann Grassmann, who was not only inspired by his father, a crystallographer, but also by ancient Indian philosophy that speaks of the infinite dimensional Net of Indra.— How can we imagine the complex network of 1-dimensional vectors embedded in a higher-dimensional spatial continuum?

This question about visualization of higher mathematics and the imagination of quantum phenomena was the task of the interdisciplinary SciArt research project

Quantum Cinema—a digital vision (2010–2013) [1] mentored by the then most eminent Austrian physicist Helmut Rauch, who first proved the 4π-value of the spin of a neutron (Rauch et al., 1975). As a pioneer in the field of neutron interferometry, he said,

> There are strong indications, that we may understand the physical world as interference pattern (Rauch, in a private conversation, 2012).

The Quantum Cinema project aimed to develop visual tools for the perception of quantum phenomena. After all, computer animation is the perfect medium for overcoming projective geometry, which is foundational for many scientific concepts. We hold that only the invention of digital media makes the appropriate outlook of higher mathematics possible, and animated 3D graphics are the new medium for a holographic depiction model which was realized with the help of a group of artists with trained imagination skills.

Quantum Cinema is based on the idea that geometry must not remain restricted to the plane, 2-dimensional page—merely a shadow of rigid shapes projected onto a plane as it is the case with projective geometry—but may create an abstract movie of previously unimaginable quantum phenomena in order to gain cognitive and visual access to the quantum realms.—Hence, the title Quantum Cinema.

The term Quantum Cinema was originally coined by the Austrian media theorist Peter Weibel—who was so kind as to head this project—for his (then) futuristic vision of a cinema that can be perceived without further external device directly in the brain (Weibel, 2005). Nowadays, human enhancement projects like Neuralink, brain–machine interfaces (BMIs) for full-bandwidth data streaming from a device are under development (Musk, 2019).

Questions of imagination and visualization of quantum phenomena relevant in the context of quantum computation shall be discussed here:

– How can we understand quantum phenomena such as the wave-particle duality and the superposition of particles and the entanglement of two or more particles?

These are important features for the application of quantum algorithms since they are produced by a discrete Fourier transform over the amplitudes of a particle's wave function. Fourier transform allows for any wave to be uniquely decomposable into an infinite superposition of trigonometric waves of different frequencies (Fourier, 1822).

– How can we imagine phenomena on the foundations of nature which serve as the most effective ingredients for quantum computing technology?
– How can we imagine the tiniest processes in nature, known as wave-particle duality, that serve as bits in quantum computers?
– How can we imagine quantum fluctuations, the wavelike state of a particle that produces a numeric output by means of a quantum computer?

---

[1] I am grateful to Quantum Cinema artists Nikola Tasic, Christian Magnes, Rudi Friemel, Kathrin Stumreich and the mathematician and civil engineer Dr. Hans Katzgraber.

We hope that 3D animated geometry as performed in the Quantum Cinema project could be interesting in regard to topological quantum systems, especially our new visualization of a hypersphere, a geometrical object that is used in Quantum geometry even though only insufficient construction methods exist.

Finally, features used in quantum computation such as Fourier transform, Bernhard Riemann's zeta function and Fuchsian functions will be discussed in relation to the famous Poincaré's conjecture about the fundamental polyhedron and the homology sphere which turns out to be homeomorphic to a 3-sphere, as we will show below.—Is this an alternative answer to the Clay math millennium's problem?

Different lines of thought from brilliant minds have been creating challenges as well as obstacles that have nevertheless led to the current state of development of quantum computers.

3 Lady Lovelace—Poetical Science and the Role of Imagination

Lady Lovelace is meanwhile widely recognized for the conception of the first computer algorithm, published as a footnote to her translation of Italian military engineer Luigi Menabrea article into English (Lovelace, 1842).

Although the further development of algebras remained widely abstract, in the 1830s, Augusta Ada King, the Countess of Lovelace, developed her imagination of algebraic operations in higher dimensions.

Her interest in mathematics was fostered by her mother, Lady Annabella Byron, dubbed "the Princess of Parallelograms" for her own fascination with mathematics. She tried to keep her, as Lord Byron's daughter, away from poetry to prevent her from following in his footsteps. So Lady Lovelace was tutored by the mathematician Augustus De Morgan, who was, with George Boole, one of the founders of formal logic and developers of modern algebra. In Boolean algebra, True and False are replaced by 0 and 1, which became the cornerstones of a binary decision-making system to which the third indeterminate value, the state of superposition, was added for quantum computation. In quantum logic, De Morgan's Laws play an important role in the projection-valued truth value assignment of observational propositions in quantum mechanics.

By 1841, Ada Lovelace was developing a concept of "Poetical Science" in which scientific logic would be driven by imagination. "Poetical Science" can also be understood as a precursor of SciArt, a collaboration between the arts and sciences as it is accomplished partly nowadays. For Lovelace, imagination is "the discovering faculty, pre-eminently—science and religion were the same and imagination itself was divine.

Imagination penetrates into the unseen worlds around us, the worlds of Science. She saw mathematics metaphysically, as "the language of the unseen relations between things" (Holmes, 2015).

The development of an appropriate geometrical imagination for these (still) unknown regions of most abstract higher mathematics was also the topic of the SciArt project called Quantum Cinema—a digital vision.

It is a gift to the art world that Ada Lovelace's prediction, that one day we will be able to make music with the mathematical engine, was realized in a wide range of electronic devices and music computer programs (Lovelace in Toole, 2010, 366–67). While the vision of her comrade, the engineer of the analytical engine, Charles Babbage, was about a machine with divine abilities which apparently became reality. He spoke of an engine that like god—whom he called the 1st programmer—will detect and record each whisper and every movement of a leaf (Babbage, 1838). Nowadays, we may well recognize in his words a metaphor for surveillance programs currently harnessed by quantum computing technologies.

3.1 Art and Science

The physicist David Bohm spoke of an art form of science. The scientist is an artisan who has to create sensitive instruments in order to discover oneness and totality—like quantum computation technology, one might add—which aid perception and so make possible both the testing of new ideas for their truth or falsity, and the disclosure of new and unexpected kinds of facts (Bohm, 2004, 105).

On the other hand, the artist may create visual access to subscendent structures, to what Plato called the world of ideas, that express the harmony and beauty in nature—categories which remained more valid in the arts than in science.—The beautiful formula as a guarantee for truth became obsolete.

Computer-generated art is based on mathematical algorithms which were previously considered as an art form.

Algorism is the art by which at present we use those Indian figures, which number two times five. (Alexandre de Villedieu, poem Carmen de Algorismo, 1240)

This phrase alludes to the Indian origin of the numerals (Indorum ars numerandi) or Hindu numerals. The term *algorithm* goes back to the name of Persian mathematician Muhammad ibn Musa al-Khwārizmī (780–850), who is arguably the most important mathematician of the Middle Ages. Besides calculating the most exact calendar of his time and his contributions to architecture and poetry, he developed two distinct branches of mathematics, both of which owe their name to him: algebra and algorithms (Brezina, 2006).

For the French, thirteenth century poet Alexander de Villa Dei mathematics was part of the Liberal Arts and algorithms an art form. A century later, Leonardo da Vinci pursued painting as a scientific method. For example, his advice on the mode of studying anatomy was,

Study the science first, and then follow the practice which results from that science. (Leonardo da Vinci [1490] 1877, 13)

The Austrian quantum physicist Wolfgang Pauli confirmed that at the time of the "re-opening" of the Platonic Academy in Florence (1438–39) sponsored by Cosimo de' Medici, Leonardo was the scientist while Marsilio Ficino, the "first academic", and Pico della Mirandola founder of esotericism, did not practice sober science, but religious mysticism, which originated in a new cosmic feeling and expressed itself in particular in the deification of space (Pauli, 1955, 8). In the fifteenth century, the German philosopher and theologian Nicolas of Cusa proclaimed that space is infinite and boundless, like the n-dimensional space in mathematics later on.

The invention of perspective and projective geometry centuries before by Renaissance artists, such as Paolo Uccello and Albrecht Dürer, only gained importance in mathematics in the nineteenth century. Projective geometry was the basis for the invention of cinema in the nineteenth century. The projection of an image onto a screen by a beam of light led to new worlds' arising at the point at infinity. Whereas according to Gilles Deleuze narrative cinema produces "time crystals" by freezing space and time into moving images taken from the real world (Deleuze, 1991, 132). 3D digital quantum cinema renders higher-dimensional virtual space configurations composed of 3-dimensional moving spaces in time.

It is less known that Dürer also explored shapes of higher-dimensional spaces. He designed Penrose type irregular five-fold patterns, one of them exactly matching the Ammann-Beenker Tiling (Dürer, folio 142, in Ruppich, 1969, 73). The astronomer Johannes Kepler also designed 4-dimensional polyhedra during his search for the shape of the space with the potential for the formation of matter and the emergence of motion (Lück, 2000). A closer look tells us that Kepler's Pythagorean approach towards the discovery of the three laws of planetary motion lead him in fact into higher-dimensional space configurations (Quehenberger, 2022).

The history of arts and literature is full of comments on scientific shortcomings. For instance, in the second half of the nineteenth century, during the hype about the 4th dimension, the poet Edwin Abbott Abbott made fun of limited mathematical imaginations in his satirical novel about "Flatlanders" who cannot imagine a sphere (Abbott, 1884). H. G. Wells invented a "Time Machine" (1895) for travelling on the time axis that inspired relativity theory significantly, while during the cubist art movement in the 1910s, Marcel Duchamp and Pablo Picasso interpreted aspects of the 4th dimension by depicting the simultaneity of different perspectives in their paintings. Inspired by conversations with Henri Poincaré, Duchamp also criticized the "free choice of metrics" as applied in metric tensors with wooden rulers carved in the arbitrary shape of dropped threads in his artwork (*3 Standard Stoppages*, 1913–14).

The famous "melting clocks" (*The Persistence of Memory*, 1931) paintings by the surrealist Salvatore Dalí are a reaction to General relativity, wherein "time and space lose the last remnant of physical reality." (Einstein, 1916, 117). It appears that the pseudo 4-dimensional Minkowski space led to a mental loss of a proper imagination of 4-dimensional systems. Quantum Cinema models are trying to overcome this deficit. In 2012, we produced the Quantum Fluxus Dada Duett as *Art's Birthday* present, a satirical dadaist sound poem on 100s of different currently used space concepts with Erwin Schrödinger's sentence, "this is absurd" as refrain (Quehenberger & Weibel, 2012).

We like to recall *pataphysique* as proclaimed by the French artist and writer Alfred Jarry at the beginning of the twentieth century. He called for an ideology-free space in science and announced a science beyond physics and metaphysics, a science of imagination that underlies everything (Jarry, [1911] 1996). Only recently, SciArt has emerged as the 3rd culture, opening up new collaborations between art and science.

3.2 The Role of Imagination

According to the nineteenth century German philosopher, Arthur Schopenhauer, we perceive the world and its representations, we create it as a whole using the imagination (1819). Schopenhauer, as well as a century later the quantum physicist Erwin Schrödinger, was inspired by the ancient Indian philosophy of the Vedanta School with its most important underlying proposition that matter has no essence independent of mental perception; that existence and perceptibility are convertible terms (Jones, 1794).

Schopenhauer criticized the abolition of Euclidean geometry by comparing it to somebody's cutting off their legs in order to walk with crutches (Schopenhauer, 1892, vol. I, 17). Since the emergence of many different non-Euclidean geometries during the nineteenth century, geometry completely lost its authority for proofing algebraic theories as it? had been the case in antiquity with Euclidean geometry. The result is that sophisticated algebraic geometry concepts remain abstract. From the artist's perspective, it is astonishing that mathematics and physics has no reliable image of the 4th dimension, not to speak of higher dimensions.

By developing a higher-dimensional Euclidean geometry by means of 3D animation we have tried to overcome this deficit, which was addressed by the German mathematician Felix Klein, famous for his Erlangen program, which aimed to unite all geometries (Klein, 1872)—150 years ago:

> The developments of recent decades have meant that in Germany in many cases the focus has been on abstract, logical inquiries of geometry, whereas the training of the appropriate outlook has been neglected. (Klein, 1892, 32)

It seems that the training of the imagination for higher-dimensional mathematical objects has still not been accomplished. Therefore, we claim that it is a task for artists to employ their visual expertise to engage with this problem of the visualization of higher mathematics, and computer animation is the appropriate medium here.

The most eminent mathematician, engineer, and physicist around 1900, the polymath Henri Poincaré spoke about the usefulness of geometric figures in the theory of imaginary functions and integrals caught between imaginary limits. He strived for visual access to higher mathematics that has not even been achieved today.

> N-dimensional geometry has a real object; nobody doubts it today. The beings of hyperspace are susceptible to definitions precisely like those in ordinary space, and if we cannot represent

them, we can design them and study them. If therefore, by example, Mechanics with more than three dimensions must be condemned as devoid of any object, it is not the same with the hypergeometry (Poincaré, 1895, 1).

What Poincaré called "hypergeometry" in his "Analysis situs" articles later became the field of topology. The most well-known figures in topology are tori and bezels, many holed tori, trousers, among other figures obviously unable to provoke mental pictures of a dynamic discrete higher-dimensional space-time continuum which we need in order to understand what is going on the Planck scale.

That's why we engaged in inventing a hypergeometry based on Poincaré's original ideas of interconnected spaces with digital artistic tools that allowed the building of a dynamic spatial continuum by means of a newly (re-)discovered heptahedron, the 3-dimensional representation of the Penrose kites and darts tiling named epitahedron (E±) (Quehenberger, 2013; 2014a, 2014b, 2014c; 2016, 2019). Penrose patterns are irregular patterns of the golden ratio filling the plane with arbitrarily large regions, with fivefold and tenfold rotational symmetry first designed by Roger Penrose (Penrose, 1974).

We've investigated quantum phenomena by means of geometrical experiments using these unit cells of 5-dimensional space along with the topology of the (hyper-) complex number space (Weibel-Katzgraber-Quehenberger, 2012).

Quantum Cinema visualizations are inspired by David Bohm so as to visually and geometrically test the assumption of the Penrose pattern as an answer to ontological questions, concerning the *implicate order* as David Bohm called it:

> In short, the implicate order, as I would state it, is the unitary deep structure (holo-archy) of level-1, which subscends or underlies the explicate surface structures of elementary particles and waves. (Bohm, 1982, 168)

3.3   How Can We Imagine a Simple Quantum System?

Quantum Cinema research worked on the assumption that the aperiodic Penrose tiling—a 2-dimensional slice of the 5- or 10-dimensional space (depending on its fivefold or tenfold symmetries) which is used as models for quasicrystals (Shechtman et al., 1984) can be adopted to quantum physics (Fig. 1).

The dual of the Penrose patterns is a cubic grid formed by parallels into five distinct directions of space, named Ammann-bars after their explorer, Robert Ammann (Grünbaum & Shephard, 1987, 571ff.). This grid is associated with the icosahedral group, isomorphous to the group A5 (de Brujin, 1981). Penrose tilings are defined in the 6-dimensional space through a strip and projection method. This gives rise to the idea that is appropriate for the visualization for the quantum space which is differing from the usual position and momentum-space. Accordingly, our first intuitively designed quantum space looks like as shown in Fig. 2.

In quaternionic quantum mechanics, the spin–orbit interaction demands that the (Euclidean) space–time dimension be 6-dimensional. Pauli correspondence is characterized by the fact that the state space HP^1 n S^4 of a spin 1/2 particle system admits

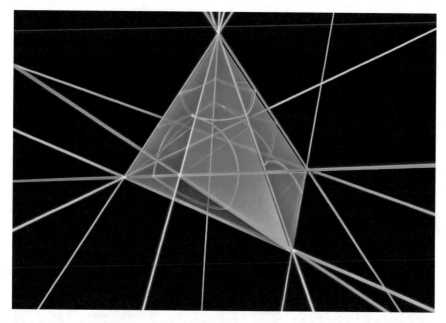

Fig. 1 The convex epitahedron E+ whose edges are partitions of the pentagrid; a polyhedron in 4-dimensional space (image: Quantum Cinema/Renate Quehenberger 2012© Bildrecht, Vienna 2021)

a natural embedding in R^5 (Brody and Gräfe, 2011). Also, Louis de Broglie was convinced of a 5-dimensional wave mechanics (de Broglie, 1927).

These facts seem to support our visualization method based on the Penrose tiling, the 2-dimensional slice of the 5-dimensional space.

The British physicist Alain Mackay showed that a set of six vectors lies in parallel to the fivefold rotation axis of the icosahedron (Mackay, 1981). This could possibly be useful for the hexagonal grids used in quantum computation.

4 The Heisenberg Uncertainty Principle and the Copenhagen Interpretation

Since the beginnings of modern quantum theory (around 1926) there has been a dispute about whether or which "images" of quantum events are, in principle, possible.

We are tempted to regard it as the biggest irony in the history of quantum mechanics, that the aim for "Anschaulichkeit" (German for "visual accessibility") as expressed by Werner Heisenberg in his first published article on the uncertainty principle (Heisenberg, 1927), turned into the very opposite when the ramifications of the principle were codified in the Copenhagen Interpretation, which states that the

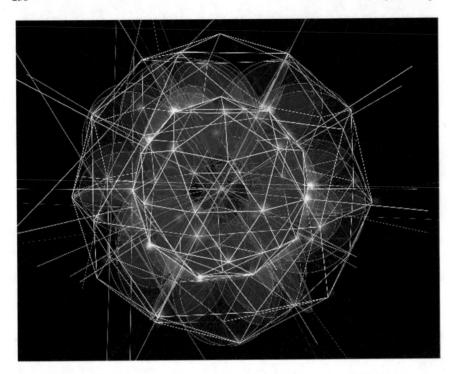

Fig. 2 The complex space model of a particle with straight lines, parallel into 5 directions constructed from two interlocked icosahedra (image: Quantum Cinema/Renate Quehenberger, 2011© Bildrecht, Vienna 2021)

wave equation is not a real phenomenon but allows only statistical predictions.—How would a quantum computer work then?

The Copenhagen Interpretation, which appears to be "orthodox" quantum theory today (cf. Bohr, Heisenberg, and others), called for a ban on images: In principle, the Heisenberg Uncertainty Principle prohibits thinking about processes and quantum mechanical formalism in general. This was compared with the Byzantine iconoclasm of the eight and the ninth centuries, when the "magical power" of images was opposed and the divine principle was promoted as an obscured power of a non-imaginative entity. In orthodox quantum theory, formalism itself has ascribed a "magic" (Feynman et al.), whereby "images" generated in computer models (of causal, so-called "hidden" parameters) the result of their materialistic approach, threatening that magic by demystification (Grössing, 2003).

Various "no-go theorems" were cited, which supposedly proved that such "images" (which could, via "hidden parameters", to a certain extent constitute a "deeper" explanatory authority "behind" the observed quantum mechanical processes) are impossible, although the "causal" interpretation after Louis de Broglie—who first proposed matter waves (de Broglie, 1925), David Bohm and others—refuted exactly this as viable counterexample.

Since then a "realism" on the bottom of reality is disputed although proved in a gedankenexperiment by Schrödinger's grandson and friends, known as the Barrett-Rudolph (PBR) theorem (Pusey et al., 2012). Matthew F. Pusey, Jonathan Barrett and Terry Rudolph have shown that the distributions for $|0\rangle$ and $|+\rangle$ cannot overlap. If the same can be shown for any pair of quantum states $|\psi_0\rangle$ and $|\psi_1\rangle$ then the quantum state can be inferred uniquely from λ. In this case, the quantum state is a physical property of the system. This results in the introduction of a new no-go theorem for hidden variables models of quantum theory. They find it surprising that although quantum states are the key mathematical objects in quantum theory, physicists have been unable to agree on what a quantum state represents.

Terry Rudolph is currently working on the application of his grandfather's entanglement phenomenon to enable photonic quantum computing. Instead of entangling individual photons to perform calculations, this technology involves the entanglement of entire light beams to create a "continuous-variable cluster state" over which computations can take place (Pool, 2020).

4.1 Heisenberg's Movie Frame Analogy

Perhaps one has to conceive the transition from one stationary state to the next in a similar way to the transition from one image to another in films. The transition does not take place suddenly. Rather, the one image becomes weaker and the other slowly emerges and becomes stronger. For a while, both images permeate each other and one does not know which is actually meant. Perhaps there is thus a state in which one does not know whether an atom is in the upper or the lower state. (Heisenberg, 1989, 31)

The young Heisenberg illustrated his ideas about quantum indeterminacy to Albert Einstein with a movie frame analogy: Maybe we have to imagine the transition from one stationary state to the next in a similar way like the transition from one picture to the next in some films. There are not enough discrete frames available in order to achieve a clear image and the same happens with measurement. Only an ensemble of particles allows the appropriate measurement outcome, just as you need a series of images to make pin-sharp movie pictures (Heisenberg, 2011 [1979], 39).

Heisenberg's analogy argument that we will never be able to see the still picture between one and the next film frame in terms of mathematical "Anschaulichkeit", was the topic of our geometrical experiment applied to a geometrical model.

In the Quantum Cinema project, Heisenberg's movie frame analogy was applied to moving straight lines, partitions of the 6-dimensional space grid, which must be considered to be in constant movement. The still picture shows all possible positions of the points on the two circles merged in the white lines. Our geometrical digital experiment comprised of possible positions of two points connected to two circles in blue and red. The straight lines rotating at different speeds generated traces of white lines that can be interpreted as probability distribution for the localization of a particle, as shown in the still images (Fig. 3).

(a)   **(b)**

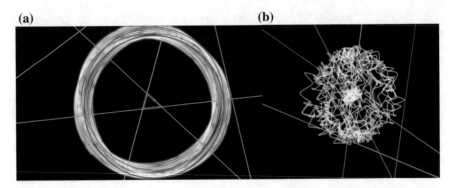

Fig. 3 3D animated digital experiment: slowly moving centre points of circles create distorted white lines (b) the traces of fast movement points creating a distorted bundle of white lines (image: Quantum Cinema/R. Quehenberger 2012© Bildrecht, Vienna 2021)

We take two circles (a, b) with centre points A and B on a triangle. The two circles perform rotations over these 3 points. The distance AB can be regarded as minimum distance h. Thereby the points on the circle outline run into turbulence and the line is no longer a straight line but generates different forms of loops on the circle line. Any point and position of the point on this loop will be in a fast wavelike motion. One cannot "measure" both the position and momentum of the point if the simulation is played at full speed.

Any point and position of the point on this loop will be in fast motion so that one cannot measure the position and momentum simultaneously. But Heisenberg's idea that the still image between one and the next picture cannot be seen in terms of mathematical visual access turns into something different: in 3D digital geometry, one can see all images in-between if the movie is run in very slow motion. The still image shows all possible positions of the points on the two circles merged in the white lines.

The Quantum Cinema 3D digital geometry examples on the uncertainty principle were actually inspired by Johannes Kepler's concept of quantum and fluxus with the idea of "exire"—points in motion that create lines. He was relying on Proclus' dynamic interpretation of Euclid's "Elements" which was also the formation principle of the hyper-Euclidean geometry we have supposedly only (re-)invented.

Alan Mackay also mentioned that unfortunately, Kepler lived too early to learn that the five regular polyhedra are in fact (as we now believe!) eigenfunctions of Schrödinger's equation and thus do describe the symmetries of the planetary electrons surrounding an atomic nucleus (Mackay, 1990, 3).

Based on our higher-dimensional geometry, we reached the conclusion that the cause for the Heisenberg indeterminacy lies in the fact that higher-dimensional space itself has to be regarded as being in constant motion. On a moving background of a 6-dimensional space grid, it is impossible to fix a moving particle and measure the position and momentum of an object.

5 Up, Up to Higher Dimensions

The discussion on higher dimensions goes back to Ancient Greece, while evidence for higher-dimensional space models can be found in ancient Egypt. Meanwhile, Plato's 5th element could be identified as unit cells of the 5-dimensional space (Quehenberger, 2016).

Plato's description of the composition of triangles that form the well-known Platonic solids starts with the tetrahedron, the octahedron and the icosahedron, and ends with the cube. He does not mention the dodecahedron, which does not fit into his concept because it does not consist of triangles and was ascribed only to his scholar Theaetetus, who also proved that there are only five regular convex polyhedra, the Platonic solids, before Euclid. Plato himself added instead this obscure remark:

> [...] since there is another fifth configuration the god used it for the delineation of the universe." (Plato, Timaios, 55c)

Only the possible compositions of dodecahedra from the 3D representation of the Penrose Kites and Darts make it clear that he kept the secret, which he and Pythagoras probably found in ancient Egyptian mystery schools. Only by not telling the name of the 5th regular solid, he left it open that the shape of the 5th element leads to a range of dodecahedral spaces in three and more dimensions which may well be used as models for the Universe, as Poincaré suggested in 1904.

We should not forget that the modern Standard Theory describes particles in terms of Group theory, which relies on the symmetries of Platonic solids conforming Plato's idea that the formation of matter is caused by traces of shapes in motion.

We also have to consider that Group theory requires a 5-dimensional space, as Poincaré pointed out:

> An immediate induction shows us that the classification of algebraic surfaces and the theory of their birational transformations are intimately connected with the classification of real closed hypersurfaces in the space of five dimensions from the point of view of Analysis situs. (Poincaré, 1900, 19)

Echoing Pythagorean harmonies, Alan L. Mackay, who coined the term "generalized crystallography", anticipated the existence of aperiodic fivefold crystals before the discovery of quasicrystals, revealed:

> What the Pythagoreans really discovered was group theory and of course this can be applied to a vast range of circumstances. (Mackay, 1990, 3)

At this point, we have to consider that Plato was visualizing Pythagoras, and both are evidently relying on ancient Egyptian mystery schools and mythology that explains the origami of 5-dimensional space conveyed as sex and crime stories about gods and goddesses (Quehenberger, 2013, 2019). Pythagoras understood the significance of making the connection between the sensory aspects of our world and the realm of numbers with the goal of realizing the pre-established harmony of the cosmos (Miller, 1996, 181).

Pythagoras' theorem also plays a key role in quantum mechanics. It is used to find the length of the vector when complex numbers are used in quantum mechanics. It is fundamentally impossible to predict with certainty the outcome of a future measurement on a system prepared in a pure state, unless the state is an eigenstate of the observable to be measured. The best prediction is probabilistic, given by the Born rule, which states that the probability density of finding a particle at a given point, when measured, is proportional to the square of the magnitude of the particle's wave function at that point. Hence, the Born rule is a consequence of the complex Pythagorean theorem: the measure of a subset of the complex line is the sum of the measures of its projections on all eigenspaces (Mandolesi, 2019).

In Euclidean geometry, the Pythagorean theorem is represented among the three sides of a right triangle. It states that the area of the square whose side is the hypotenuse (the side opposite the right angle) is equal to the sum of the areas of the squares on the other two sides. The probability in quantum mechanics must then take the form $P_c = P_a + P_b$, since the two possibilities are mutually exclusive due to orthogonality of measuring either the value associated with the subspace A or the value associated with the subspace B, is given by: $|c|^2 = |a|^2 + |b|^2$. The quaternions, discovered by Hamilton, are a 4-dimensional algebra, $H = \{a + bi + cj + dk: a, b, c, d \in R\}$. It is a normed division algebra, meaning that the norm $|a + bi + cj + dk| = \sqrt{(a^2 + b^2 + c^2 + d^2)}$ obeys $|qq'| = |q||q'|$ for all $q, q' \in H$. The unit sphere in H, a 3-sphere, is, therefore, a group (Baez, 2017).

The dynamics of quaternions remind us of the Greek term *dynamis*. In his dialogue, *Theaitetos* Plato speaks of "dynamis", sometimes translated as square roots, to indicate the transformation from a quadrangle (*tetragonos*) to a square, of an imaginary meaning and also that these numbers possess a potential and dynamics for returning to the real numbers (Tydecks, 2020).

> Even though the particulars of Plato's 'theory of everything' as outlined in Timaeus, strike the modern sensibility as absurd (if not borderline psychotic), there are many parallels between his picture of the universe and that embodied in string theory. (Yau & Nadis, 2010, 326)

The mathematicians Shing-Tung Yau and Steve Nadis mention that geometrization, the idea that the physics we observe springs directly from geometry, stands at the core of both approaches. Would they be surprised to read that Plato's 5th element could be composed as 26-dimensional space, as required by String Theory for one particle to inhabit in order to achieve the unification?

On digesting this higher-dimensional information from Plato's world of ideas, we may well convince ourselves that the Pythagoreans also worked with this higher-dimensional space concept that was lost in religious doctrines in the ensuing centuries.

Higher dimensions were again considered at the turn of the seventeenth century by Renée Descartes, who was the first to introduce the notion of higher dimensions in relation to equations of higher order (Descartes [1637] 1902, 13). Descartes is well-known for the x, y coordinate system, however, as early as 1610, he also stated that a combination of systems of coordinate axes is a prerequisite for the depiction of higher-dimensional spaces. He worked with the German weaver Johann Faulhaber, who also produced calculations for Johannes Kepler—on higher order simplices. Faulhaber worked on cubic equations based on Girolamo Cardano, who had first published Tartaglia's solution a century earlier (1545) and defined formulas for sums up to the 17th power, which is known as the Faulhaber Formula (Faulhaber, 1631).

Faulhaber's work is still relevant for computation because his works influenced Jakob Bernoulli, and Bernoulli numbers are the core ingredients of the first so-called software program of Ada Lovelace that she developed for the Analytical Engine designed by Charles Babbage.

> The Analytical Engine weaves algebraic patterns like the Jaquard loom weaves flowers and leaves. (Lovelace, in Padua, 2015, 88)

It is notable that punched cards were invented for weaving on a loom in 1804 by Jean Marie Jacquard and were once commonly used in data storing and data processing applications or to directly control automated machinery even after the development of electronic computers in the 1950s and until recently.

5.1   Higher-Dimensional "Quantum Stuff" and the Curse of Dimensionality

Until now, dimensionality in computational software programs like mathematica is solved by assigning different degrees of freedom from 1-dimensional lines but not from a spatial compound. In contrast to this, we took inspiration from Henri Poincaré, who was primarily interested in the physical and philosophical implications of the meaning of the concept of dimension, states:

> The fundamental proposition of Analysis Situs is, that space is a continuum of three dimensions. (Poincaré, 1912)

He inspired the Dutch mathematician Luitzen E. J. Brouwer, who founded intuitionism to a dimension concept based on "lawless sequences" (German: *Wahlfolgen*). We assume that we can apply this concept to our hyper-Euclidean space compounds of interconnected spaces that can be ascribed to a certain dimensionality (Brouwer, 1913).

Alas, there is currently no appropriate concept of dimensionality that can be easily applied to digital computation. For this lack of functionality, the term "curse of dimensionality" was coined by Richard E. Bellman. The problems arose in dynamic programming when analyzing and organizing data in high-dimensional spaces was not possible, but do not occur in low-dimensional settings, such as the three-dimensional physical space of everyday experience (Bellman, 1957). Consequently, in computer science, higher dimensions are avoided by the process of dimension reduction. It is the transformation of data from a high-dimensional space into a low-dimensional space so that the low-dimensional representation retains some meaningful properties of the original data. While this method seems viable for signal processing, its application in natural sciences—e.g., in bioinformatics—seems rather problematic, because important information concerning complex structure components gets lost.

Although applying our continuous higher-dimensional space concept to computer programs remains a challenge.

The founding fathers of quantum physics, Erwin Schrödinger and Louis de Broglie, had good imaginations for quantum phenomena taking place in a higher-dimensional configuration space:

> One actually no longer has to assign a psi function to this totality of particles, which represents a wave propagating in ordinary 3-dimensional physical space but a psi function that represents a wave propagating in abstract space, the configuration space, which has three dimensions as many times as there are parts in the system. (de Broglie, 1943, 145f. Transl. RCZQ)

Erwin Schrödinger, who first proposed entanglement imagined the wave function—which he formulated for the first time—as the description of a phenomenon located in higher-dimensional space. He regarded the purely statistical Copenhagen interpretation, which takes the wave function merely as probability, introduced by Max Born after Heisenberg discovered quantum indeterminacy, as a terrible misinterpretation of his theory. This development of quantum theory went in a completely different direction than Schrödinger had wanted. Abstract statistics doesn't demand any kind of imagination. Even the Nobel laureate Gerard 't Hooft admitted, speaking about higher dimensions, "I leave them out whenever I can" ('t Hooft in QC-Interview, Quehenberger, 2011).

Contributing to its complexity, the American physicist Richard Feynman was the first to imagine a quantum computer that would be able to calculate the simulation of complex quantum systems:

> But the full description of quantum mechanics for a large system with R particles is given by a function $\psi(x_1, x_2, \ldots, x_R, t)$ which we call the amplitude to find the particles $x_1, \ldots, x_R$, and therefore, because it has too many variables, it cannot be simulated with a normal computer with a number of elements proportional to R or proportional to N. (Feynman, 1982)

Only recently, the discussion on higher dimensions gained new momentum. In the operational approach to general probabilistic theories, one distinguishes two spaces, the state space of the "elementary systems" in the quantum world and the physical space in which "laboratory devices" are embedded. Each of those spaces has its own dimensionality. This means there is a different minimal number of real parameters (coordinates) needed to specify the state of the system and for a point within the physical space (Dakic and Brukner, 2016).

6 Schrödinger's "Entanglement"

Entanglement is the most important quantum phenomenon applied to quantum encryption since entangled particle pairs serve as the primary technical component in quantum computers.

The rather romantic term entanglement was coined by the Austrian physicist— and notorious womanizer who lived with two wives—Erwin Schrödinger, whose name is tied to the famous Schrödinger wave equation.

Schrödinger tried to understand the possible unification of matrix mechanics by Heisenberg, Jordan algebra and his wave mechanics, since partial differential equations are equivalent to quadratic forms of infinitely many variables (Schrödinger in a letter to Born, April 14, 1926, in Meyenn, 2010, 201).

Finally, Paul Dirac's attempt to prove the equivalence of matrix mechanics and wave mechanics by essential use of the delta function and his bra-ket formalism based on hypercomplex numbers similar to William Rowan Hamilton's quaternions led to a deeper understanding of the foundations (Dirac, 1930). As artists, we are trying to understand mathematical abstract formalism in a way that can be visualized geometrically. We are less focussed on measurement outcomes than on ontological questions—what is really happening on the subscendent *level-1*?

In the information interpretation of quantum mechanics, information is the most fundamental basic entity (Svozil, 2000). Every quantized system associated with a definite discrete amount of information can be traced back to Erwin Schrödinger's "catalogue of expectation values" mentioned in his seminal papers "On the Current Situation of Quantum Mechanics" (Schrödinger, 1935). Here, he also introduced the famous simultaneously dead and alive "Schrödinger's cat" metaphor.

In this drastically cruel thought experiment about a quantum cat in a closed box with cyanide, while hidden from the observer, the cat is dead and alive at the same time. It is intended to serve a better understanding of the phenomenon of the superposition of a quantum state.

> Well, 'state' is a term that everybody uses today, even the holy PAM, but it doesn't make it more rich in content (Schrödinger in a letter to Pauli, Dublin, beginning of July 1935, in Meyenn, 1985, 406)

The "holy PAM" is Paul Adrien Maurice Dirac, the British physicist who first defined the spin of a particle. Wolfgang Pauli famously said of him: "There is no God and Dirac is his prophet." (Pauli, in Meyenn, 2010, 406).

Schrödinger's (1935) articles were immediate answers to Albert Einstein's question "Can quantum–mechanical description of physical reality be considered complete?" wherein he assumed a local deterministic realism of quantum states (Einstein et al., 1935).

Schrödinger also reacted to the forbidden determinism by introducing the term "entanglement" for the undetermined state depending on an earlier random event instead of Einstein's "spooky action at a distance" of which Isaac Newton had already stated:

> I believe no man who has in philosophical matters a competent faculty of thinking, can ever fall into it. (Newton in a letter to Richard Bentley in 1693, quoted by Price, 2013)

Entanglement is a property of correlations between two or more quantum systems that defy classical description.

Quantum non-locality does not prove that "signals" travel "faster than light". Rather, it shows that at a deep level of reality the speed of light as a limiting factor is irrelevant because phenomena are instantaneously connected regardless of distance. Because the spin of a particle does not exist until a measurement is made, the act of making the measurement and determining the axis of spin of particle 1 will also determine the spin of particle 2 no matter how far apart it is from particle 1. Particle 2 will instantly respond to the state of particle 1, even if it is on the other side of the universe.

At any instant measurement of particle 1 is performed, particle 2, which may be thousands of miles away, will acquire a definite spin—"up" or "down" if a vertical axis, "left" or "right" and a horizontal axis is chosen. How does particle 2 know which axis was chosen? There is no time for it to receive that information by any conventional signal (Capra, 1982, 85).

Wolfgang Pauli approached the entangled quantum system as follows:

> One must differentiate between different layers of reality, an R that contains all the information that can be obtained from measurements on 1 and one (deducible from R) r which only contains the information which can be obtained by measurements on 1 alone.
>
> A complete description would have to assign characteristics to the state of the particle 1 which must already contain all those properties of 1 which — according to possible measurements on 2, which do not interfere with 1 — can be predicted with certainty. (Pauli in a letter to Heisenberg, 1935, in Meyenn, 1985, 404)

6.1 A Proposal for a Space that Enables Schrödinger's "Entanglement"

The effect of the environment through decoherence becomes more accessible in the artistic interpretation of a higher-dimensional framework based on the Penrose tiling shown above:

Let us assume that for any interconnected system of two or more measurable quantum objects of higher order, say qubits (no matter how far they are separated), the "miracle" of a "minimal distance effect" occurs in a spatial framework that exhibits principles of the aperiodic Penrose tiling. For this, Conway's theorem holds that for any circular region of diameter d, the distance from the perimeter of the "home town to the perimeter" of the duplicate town is never more than d times half of the cube of the golden ratio, or 2.11 times d.

> If you walk in the right direction, you need not go more than that distance to find yourself inside an exact copy of your home town. The theorem also applies to the universe in which you live. Every large circular pattern (there is an infinity of different ones) can be reached by walking a distance in some direction that is certainly less than about twice the diameter of the pattern and more likely about the same distance as the diameter. On a Penrose pattern you are always very close to a duplicate of home. (Gardner, 1989, 10)

Following this assumption, we derive the 3D representation of the entanglement picture as interlocked spaces of two unit cells of the 5-dimensional space, supposedly what Plato called a "figure of marriage". In his *symposium*, Plato let Socrates quote his teacher Diotima, "Love is the medium".

6.2   *Entangled States in Quantum Computing*

The strikingly nonclassical properties of entangled states lead to more and more novel demonstrations of fundamental properties of quantum mechanical systems (Greenberger-Horne-Zeilinger, 1993). The young field of quantum information exploits such entangled quantum states for new types of information transmission and information processing (Bennett, 1995).

In 1995, the Austrian physicist Anton Zeilinger and co-workers reported the first experimental realization of quantum communication, verifying the increased capacity of a quantum information channel by "quantum dense coding" (Mattle, Zeilinger et al., 1996). The applied scheme was theoretically proposed by Bennett and Wiesner and utilizes two entangled two-state systems (Bennett & Wiesner, 1992).

The output of a quantum information processor is, by convention, measured in the basis states with Boolean values $|0\rangle$ and $|1\rangle$ denoted as A and B, often referred to as Alice and Bob. Any controlled unitary gate can be realized using at most two CNOT gates and four single-qubit gates. This acts on the state of two qubits, known as the control qubit, C, and the target qubit, T (Barnett, 2009, 146).

Typically, four specific two-qubit states with the maximal value of $2/\sqrt{2}$ are designated as Bell states. These Bell pairs represent all the ways that can be chosen to have correlations or anticorrelations between the $|0\rangle$ and $|1\rangle$ states of the computational basis and the $|+\rangle$ and $|-\rangle$ states of the X basis.

These four states form a complete basis for two qubits. So any state, entangled or not, can be expressed as a superposition of Bell states. They are known as the four maximally entangled two-qubit Bell states and form a maximally entangled basis known as the Bell basis of the 4-dimensional Hilbert space for two qubits.

From the perspective of quantum information theory, any two-state quantum system (represented by $a_0 |0\rangle + a_1 |1\rangle$) can be used to provide a physical implementation of a qubit like the bit in classical information theory. Examples that have been realized in the laboratory include the two orthogonal polarization states of a photon, the orientation of a spin-half particle, a pair of electronic energy levels in an atom, ion, or quantum dot, and a pair of paths in an interferometer. However, it is convenient to treat a qubit as though it were a spin-half particle and to describe it by Pauli operators (σ_x, σ_y, σ_z) (Barnett, 2009, 46).

6.3   A Space Composed of Cells

In the 1950s, the physicist Léon Brillouin, who applied information theory to physics, introduced space cells on the fundamental level of reality. Before he made major contributions to crystallography by finding the elementary, irreducible zone, the first Brillouin zone reduced by all symmetries in the point group of the lattice. He sketched a "miscellaneous" cell on the foundations of the quantum realm with information content per element of data (Brioullin, 1952, 72). The term "miscellaneous" stands for a compound of elements in movement which also applies to the 6-dimensional space lattice and inherent symplectic manifolds in movement. We envisage a compound of dynamic elements (points and straight lines) forming 5-simplices, 3D representation of the Penrose Kites and Darts tiling (E±). All parts of these unit cells of the 5-dimensional space, its points, edges and faces have to be thought of as being in constant movement, forming different angles and compositions. We imagine iterations of the size of the lattice forming simplices with partitions of circle decoration (cf. Conway decoration of the Penrose kites and darts tiling) forming wavelike structures in movement. Their traces would, according to Plato, lead to the formation of matter. We imagine that a particle as expressed by Schrödinger's wave equation is the result of permutations of space cells according to symmetry operations in accordance with group theory.

Our Platonic approach seems to meet symplectic Clifford algebra as interpreted by Basil Hiley: The basis is not matter or fields in interaction in space–time, but a notion of "structure process" from which the geometry of space–time and its relationship to matter emerge together, providing a way that could underpin general relativity and quantum theory (Hiley, 2016). All the quantum properties of Schrödinger and Pauli particles can be described entirely from within a Clifford algebra taken over the reals, there is no need to resort to any "wave function" (Hiley & Callaghan, 2010).

Such a space in motion would not permit the measurement of the momentum and position of a particle, and provides a logical geometrical framework for the Heisenberg uncertainty principle: Higher-dimensional space, as we know it from visualizations of hypercubes, is in motion.

In this respect, the dynamic structure of the infinite 5-dimensional space may be a visualization of the "pattern that connects" as coined in cybernetics by Gregory Bateson. Only by means of 3D animated geometry can we visualize the idea of dynamic building blocks for systems like "Jumbled Boxes", as Gordon Pask drew them (Clarke, 2009, 50).

Various Unified Theories can be visualized by means of this 3D representation and its composite configurations, such as of the 26-dimensional space continuum arising from inside the 4-dimensional dodecahedron, identified as Poincaré's dodecahedral space (Quehenberger, 2018).

However, we do not expect that this model will be accepted anytime soon since the physics community remains agnostic about an underlying space. In Quantum Gravity emergent space–time scenarios, space and time are denied at the fundamental level of reality (Oriti, 2006). In our model "time" is represented in the 4th dimension of reality on the globe, while if we apply the same model to quaternionic quantum mechanics we get quantum space–time crystals, as proposed by Nobel Laureate Frank Wilczek (2012). This system, which is periodic in time and also in space, was realized experimentally by confining ions in a ring-shaped trapping potential with a static magnetic field soon thereafter.

7 Hidden Parameters: Never Neglect a Woman (Grete Hermann)

Progress in quantum mechanics could have been achieved much faster if John von Neumann's *Mathematical Foundations of Quantum Mechanics* (1932) would have been less determining. His proof that no hidden variables can exist held for decades until it was finally refuted by John Bell's with his non-locality theory on hidden variables (Bell, 1966). Everybody continued to cite the von Neumann proof until John Bell (Bell, 1964) rediscovered that von Neumann's no hidden variables proof was based on an assumption that "can only be described as silly […] so silly […]" (Mermin, 1993).

But back in 1935, the German philosopher and mathematician Grete Herman already disproved John von Neumann's proof that no hidden variables can exist. She pointed out a glaring deficiency in his argument, however, she seems to have been entirely ignored. When the Einstein–Podolsky–Rosen paper was published, Wolfgang Pauli asked his German colleague Werner Heisenberg to write a "pedagogic response" concerning the determinism in quantum mechanics to the "old boys" (Pauli in a letter to Heisenberg, Zurich, 15 June, 1935 in Meyenn, 1985, 412).

Heisenberg replied that Grete Hermann already had already written a treatise which he found "overall reasonable (despite a philosophical rational tendency)" (letter from Heisenberg to Pauli, Leipzig, 2 July 1935, Meyenn, 1985, 408).

Grete Hermann was the first physicist who asked for the ontologic grounds of observables. With her philosophical request to find other characteristics which determine the path of a particle, we should not stop searching for underlying causes. She called for an improvement of the quantum mechanical formalism (Herzenberg, 2008):

> We need to find the causations of the measurement results and predictions — there is only one sufficient reason to stop looking for it — that you know already the source. (Hermann, 1935)

Grete Hermann distinguished between causality and predictability, and emphasized the fact that they are not identical:

> The fact that quantum mechanics assumes and pursues a causal account also for unpredictable occurrences proves that an identification of these two concepts is based on a confusion. (Hermann, 1935)

Grete Hermann showed that causality was retained in the sense that after an interaction, causes could be assigned for a particular effect. This allows for the possibility that physical processes may be strictly determined even though the exact prediction is not possible (Lenzen, 1969).

Her work with Werner Heisenberg, Carl Friedrich von Weizsäcker and other prominent physicists during the early 1930s led Hermann to the idea of a relational interpretation of quantum mechanics. This term refers to an interpretation of quantum theory that eliminates the concepts of the absolute state of a system, which was only recently developed more fully by Carlo Rovelli. Niels Bohr arrived considerably later at the relational concept of quantum mechanical states by considering the Einstein–Podolsky–Rosen thought experiment, and he became known for this approach, while Grete Hermann's work in introducing it has been all but forgotten (Jammer, 1974, 208).

Hermann has studied in Göttingen under Emmy Noether, one of the iconic figures of twentieth century mathematics, who revealed the very important general connection between symmetries and conservation laws in physics, known as Noether's theorem.

In her thesis, Grete Hermann demonstrated that Noether's proof of the Lasker–Noether theorem could be turned into polynomial ideals in Noetherian rings. This was the first efficient algorithm for computing primary decompositions for polynomial rings over a field of characteristic 0 (Hermann, 1926).

Hermann completed her work before the availability of computers, and even before the idea of an effective procedure had been formalized. Algorithms that compute primary decompositions are very useful nowadays for quantum algorithms.

7.1   More Ignorance and Paradoxes

Quantum computing is ironically built upon a paradox: Although "hidden parameters" were "forbidden" for decades, they became the foundations of Quantum Information.

Two decades after Grete Hermann, David Bohm found that in principle, these "hidden" variables determine the precise results of each individual measurement process. In practice, however, in measurements that we now know how to carry out, the observing apparatus disturbs the observed system in an unpredictable and uncontrollable way, so that the uncertainty principle is obtained as a practical limitation on the possible precision of measurements. Although widely applied, John von Neumann's description of measurement is insufficiently general for the simple reason that most performed observations are not of this type. The real world is noisy, and this ensures that our observations will include errors (Barnett, 2009, 92).

David Bohm pointed out that von Neumann's proof that quantum theory is not consistent with hidden variables and does not apply to his interpretation, because the hidden variables he speaks of depend both, on the state of the measuring apparatus and the observed system and so his theory would go beyond von Neumann's assumptions (Bohm, 1952).

Ironically, Bohm's hidden variable theory and Bohmian mechanics were falsely declared as deterministic and was even "verboten" (forbidden) to be taught in recent decades (Basil Hiley in QC-interview, Quehenberger, 2012a). In its original form, the mathematical formalism of the de Broglie-Bohm theory refers to hypothetical ontic elements, such as the quantum potential comparable with John Bell's "beables". John Bell was inspired by his teacher David Bohm, and his idea of an underlying reality that we can imagine like a pattern in movement (Bohm, 2004, 92).

"Bell's inequality" and separately the Kochen–Specker theorem by Simon Kochen and Ernst Specker (1967) first demanded contextuality to be a feature of quantum phenomenology was a correction of the history of quantum mechanics that opened up a new field of research, namely quantum information—and Bell's non-locality theorem became the basic feature of quantum computation.

Also, the Kochen–Specker theorem demonstrates that it is not possible to reproduce the predictions of quantum theory in terms of a hidden variable model where the hidden variables assign a value to every projector deterministically and non-contextually (Kunjwal & Spekkens, 2015).

When John Bell started to work on the foundations of quantum mechanics, there was hardly any interest in such topics. Even worse, working on foundations was not considered to be a proper topic for a physicist, so the CERN employée wrote his seminal article during his sabbatical leave, as his then-colleague Reinhold Bertlmann tells: John Clauser, the first who had the courage to carry out an experiment on Bell inequalities in the 1970s had to struggle enormously to get the resources for doing the experiment. Only after the experiments of Alain Aspect in the 1980s the community slowly began to realize that there was something essential to Bell's s theorem. In the 1990s, the field of applications of entangled states and of Bell's theorem opened up, followed by experiments on quantum teleportation, quantum cryptography, long-distance quantum communication and the realization of some of the basic entanglement-based concepts in quantum computation (Bertlmann & Zeilinger, 2017, vi).

Aspect was following Bell's example by using entangled photons instead of electrons, which proves physical reality must be un-local and all results show the impossibility of a Local Realistic interpretation of quantum mechanics (Aspect et al., 1981). After that, the perfect anti-correlation of entangled pairs has been repeated many times since the easiest method of producing entangled states was achieved by Anton Zeilinger's group after their entangled photon experiment (Kwiat et al., 1995). The pairs of polarization-entangled photons were produced by type-II down-conversion in a nonlinear beta-barium-borate crystal that was incidentally found in the desk of their colleague, Ralph Höpfel. The angle of 0.72° in their experimental set-up seems to indicate the 72° of the 5-dimensional space grid.

The second obstacle seems to be the lack of realism attributed to the quantum world. How could we build technology on something the existence of which is questioned?

The realist quantum approaches often distinguish between ψ-ontic, ontological quantum theories and ψ-epistemic theories (Fraassen, 1972). "Ontic states" are what Terry Rudolph calls real states versus "the mist (quantum state) is just knowledge/information about the real state" referred to as "psi-ontic" and "psi-epistemic", respectively (Rudolph, 2017). In other words, type-I ψ-ontic interpretations are defined as those in which the probabilities of measurement outcomes are determined by intrinsic properties of the observed system. The other, ψ-epistemic, theories view the quantum state as merely representing knowledge of an underlying objective reality in a sense somewhat analogous to that in which a state in classical statistical mechanics assigns a probability distribution to points in phase space (Cabello, 2017).

Schrödinger was considering to assign a physical realism not to the wave function ψ itself but to the quadratic function of ψ (Schrödinger in a letter to H. A. Lorentz, June 6, 1926, in Meyenn, 2010, 254ff.).

Twenty-five years ago, David Bohm and Basil Hiley described the close resemblance between key properties of deterministic, hidden variable, pilot-wave theory and emergence theory in the book *The Undivided Universe* (1993). Their theory describes the emergent formation of ordered (i.e., negentropic) states in nonlinear, self-organizing systems, such as deterministic chaos (Walleczek et al., 2019).

David Bohm's philosophy of the undividable wholeness where there are "no separate whirlpools interacting" was inspired by the Fourier Universe where the subscendent space is a carpet-like pattern of frequencies. This enfolded [or implicate] order contains the general totality of what is called the "explicate" or "unfolded" order (Bohm, 1980, xv).

We believe that the fractal nature of the Penrose tilings and their 3D representations where each triangle is decorated with a fraction of a circle, as suggested by John H. Conway for the Kites and Darts tiling, is appropriate to visualize this pattern of frequencies. Similarly, like the triangles with curves, the underlying reality as conceptualized by String theorists consists of fractions of flat space with a string attached, named Dirichlet branes (Polchinski, 1995). This idea seems to meet our concept with the triangular faces of epitahedra with the 1-dimensional "string-like" circular decoration. Joseph Polchinski also found that the product of the electric and magnetic charges is a single Dirac unit, and that the quantum of charge takes the value required by string duality.

If we assign some realism to the straight lines and attribute electric force to them, (like Maxwell and Faraday's lines of force) and magnetic force to circles, we derive the image of the wavelike nature of reality suggested by Joseph Fourier and Bohm.

8 The Geometry of Quantum States

In their seminal book on Quantum Geometry, physicists Ingemar Bengtsson and Karol Zyczkowski pointed out that Quantum mechanics is a significant generalization of classical probability theory and show how quantum states can be represented by geometrical elements (Bengtsson & Zyczkowski, 2006).

Usually, the observables are pictured as points on the surface of a sphere, the Bloch sphere—it is easy to visualize, but in some respects, it is quite misleadingly simple (Bengtsson & Zyczkowski, 2006, 215).

It is a sphere of unit radius, with each point on its surface corresponding to a different pure state. A given observable consists of two antipodal points on the surface of the Bloch ball representing a pair of mutually orthogonal states two orthogonal pure states. The north and south poles correspond to the quantum states so that the four possible results 00, 01, 10 and 11 correspond to measurements of the two qubits as the Bell basis.

During the Quantum Cinema research project, we were trying to find a more detailed 4-dimensional representation of such a quantum system.

There are examples of quantum states depicted as compositions of nested Platonic solids: Partially transposed states occupy the remaining four corners of the unit cube, which is filled with two tetrahedra in an antipodal position. Two states are regarded as "equally entangled" if they differ only by a choice of basis in Hilbert spaces H^1 and H^2. The separable subset in 2×2 dimensions is the intersection of two tetrahedra and can be represented by an octahedron (Vollbrecht and Werner, 2000).

Another representation of probabilities of maximally entangled states by using an octahedron inside a tetrahedron is quite similar to our intuitive "entanglement" image above (Fig. 4), where two epitahedra form the dodecahedron as Boolean intersection. We may assume that this emerging dodecahedral space can be compared to the maximal Kús-Zyczkowski ball of absolutely separable states that lies within the double pyramid. It touches the faces of the pyramids at the four points which mark the closest separable states to the four Bell states. The entangled Weyl states are located in the four corners of the tetrahedron outside the double pyramid, extending from the separable states to the maximally entangled Bell states at the tip (Bertlmann-Friis-Bulusu, 2016).

8.1 A New Object in Quantum Geometry

There are several indicators for fivefold symmetries in the quantum system. Most famously Klyachko and co-workers assigned a 5-vector an orthogonality graph in the form of a pentagram, and in this way derive a Kochen–Specker inequality for spin 1 systems where the set of all functions forms a Vorob'ev-Keller cone (Klyachko et al., 2008). This 5-vector can be visualized by the 5 edges of the pyramidal cone of the 3D representation of the Penrose Kites and Darts tiling E±. We assume that E± is a golden 5-simplex that corresponds to the unique optimal configuration of C^5, of which it is said that it looks like an umbrella (Bharti, Cabello et al., 2019).

We maintain that Mermin's magic pentagram as well as several other contextuality proofs can be drawn back to higher-dimensional spatial structures, namely the two antipodal entangled simplices (Quehenberger, 2017). The here presented entangled simplex representation (2E±) (Fig. 5) also seems to conform to the Penrose Dodecahedron as contextuality proof, an image that goes back to Penrose's idea of 5

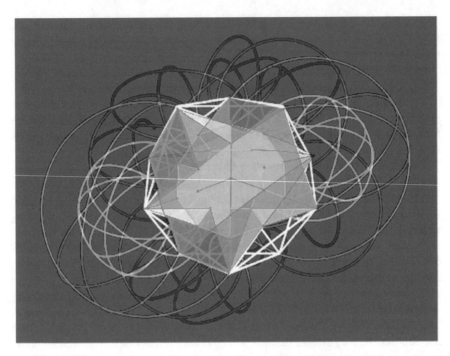

Fig. 4 "Entangled state" of two epitahedra with a dodecahedron in the centre; 3D animation of rotating faces result in full circles of the Conway circle decoration of the Penrose tiles, (image: Quantum Cinema/Renate Quehenberger, 2011© Bildrecht, Vienna 2021)

Fig. 5 3D Representation of a three-partite quantum system indicating spin states in the framework of the Poincaré sphere. The sphere is drawn with three great circles and labels for six basic polarization H (linear horizontal), V (linear vertical), D (linear diagonal), A (linear antidiagonal), R (right-hand circular) and L (left-hand circular) and images of the polarization vectors for each where the (i, 0) (image: Quantum Cinema/Renate Quehenberger, 2017© Bildrecht, Vienna 2021)

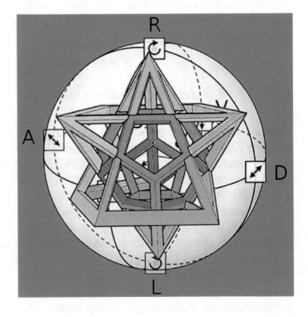

interlocked cubes (Zimba & Penrose, 1993). In our "entanglement figure", the apexes of these two irregular pyramids (E±) with dynamical edges from partitions of the pentagrid may be assigned to the spin up ↑/right and the spin down ↓/left, the other two measurements would then show 01, 10 (see Fig. 11).

We assume that it was possible to translate the *magic Mermin pentagram* from a finite projective geometry description by Michel Planat et al. (2006) into a hyper-Euclidean representation by means of the double epitahedron (2 E+). The deviated Pauli operators are in projective geometry represented by three points of PG(1, 2), their three "Jacobson" counterparts, and the four points whose both coordinates are zero-divisors.

These points on Mermin's magic pentagram correspond with the intersecting points of the entangled simplex figure.

Our 3D representation of the *magic pentagram* seems to correspond with the three different interlocked spaces associated with Stokes relations, summing up to zero. The coordinate system of Stokes vectors with components S_1, S_2 and S_3 (usually depicted as three vectors) appear in their spatial form. It appears that from these two entangled simplices (2E + , or better an indeterminate E± flipping from E+ to E-?) we derive the picture for a three-partite quantum system. Inside the Poincaré sphere, it illustrates the three spaces of possible polarizations of electromagnetic waves and their corpuscular behaviour. In this model, we may assign one direction of the spin in one epitahedron (E + beige), right (R), and another direction to the other (E + grey), left (L) (see Fig. 5), where the states |0⟩ and |1⟩ are the eigenstates (Quehenberger, 2017).

Entangled "state" of spin 1/2 particles can be assigned to the Pauli operators, a set of three 2×2 complex matrices (usually denoted as σ_x, σ_y, and σ_z ... by the Greek letter sigma (σ).

Here they are represented by the node points of the edges of the three intersecting space cells while partitions of the edges surrounding the dodecahedral centre perform the Clifford algebra of Pauli spin matrices as permuting elements. This construction, applicable for any number of degrees of freedom, reveals an interesting cyclic geometric structure that reminds us of a "quantum- merry-go-round" (cf. Uchida et al., 2015).

It looks as if this 8-dimensional representation of qubits (if we count two interlocked 4-dimensional systems) inside the Poincaré sphere could lead to a more detailed vision of quantum circuits. The advantage of the entangled E± figure, the two antipodal, spinning golden simplices, for the "conventional" Platonic structure is that it can be considered as a "real" partition of higher-dimensional space and the rational principles of algebraic structures may be applied in a dynamic way for a simulation of movement. Yet, it seems that until there is more discussion, abstract formalism will continue to prevail:

> I've always regarded the arrangement into a 5-pointed star of the ten operators in the eight-dimensional proof of the BKS theorem as nothing more than a useful mnemonic for what they are. All you have to remember is that the four going along one line of the star are XXX, XYY, YXY, and YYX.

> (N. David Mermin in an email to the author, 05/21/2015)

8.2   Quantum Gates

Charles Babbage first realized the idea of gates in his design of the Analytical Engine with barrels that might act on 50, 100 or more vertical systems of control levers with cranks, pivots and sliders (Bromley, 1982). These gates now became quantum gates—for example, the Hadamard gate, which is the one-qubit version of the quantum Fourier transform. Although Babbage knew Charles Fourier socially, Babbage did not seem to grasp the significance of his 1811 work on heat propagation, nor did he seem to know of Joule's endeavours with heat and mechanical energy. Both could not have had any idea that their works would merge in a quantum computer driven by Fourier amplitudes.

In quantum computation, it is sometimes useful to represent a quantum operation or protocol as a quantum circuit that is reminiscent of the equivalent circuits in electrical and electronic engineering.

Here, De Morgan's Laws are applied as powerful rules of Boolean algebra and set theory that relates the three basic set operations (union, intersection and complement) to each other. They denote a pair of transformation rules that are both valid rules of inference, providing an indispensable tool when simplifying digital circuits involving *and/or*, and *not* gates in quantum computing (Mano, 1988).

If A and B are subsets of a universal set U, de Morgan's laws state that.

$(A \cup B)' = A' \cap B'.$

$(A \cap B)' = A' \cup B'.$

where $\cup$ denotes the union (OR), $\cap$ denotes the intersection (AND) and A' denotes the set complement (NOT) of A in U, i.e., $A' = U \backslash A$ (Bernal, 2005).

In order to create a quantum algorithm, the CNOT and Hadamard gates followed by measurement in the computational basis constitute a Bell-state measurement. It is carried out on the first two qubits and leaves the last in one of the states $|\psi\rangle$-, $\sigma z|\psi|$, $\sigma x|\psi|$ and $\sigma y|\psi$. These four possible measurements of Bell states are associated with the binary numbers 00, 01, 10 and 11 (Barnett, 2009, 152).

The unitary operators on a single qubit, X, Y, Z correspond to the measurement of the spin along the x, y, and z axes, respectively. The X matrix is often known as the quantum gate, by analogy to the classical gate, while the Hadamard gate, which is also key to quantum computing, is the one-qubit version of the quantum Fourier transform.

The X and Z Pauli matrices are also sometimes referred to as the *bit flip* and *phase flip* matrices: the X matrix takes $|0\rangle$ to $|1\rangle$ and $|1\rangle$ to $|0\rangle$, earning the name *bit flip*; and the Z matrix leaves $|0\rangle$ invariant, and takes $|1\rangle$ to $-|1\rangle$, with the extra factor of -1 added known as a *phase factor*, so justifying the term *phase flip*. The Pauli gates X, Y and Z can be obtained by appropriately setting the phase shifts $\alpha = \pi/2$, $\beta 2 = \pi$, or $\alpha = \pi/4$ and $\beta 4 = -\pi/2$ (Hao et al., 2019).

The computation is performed step by step as follows:

Alice performs a Pauli X and Z gate to her half of the entangled pair, dependent on the message she would like to send. She then sends her half of the entangled pair in a string of N possible symbols through a CNOT gate to Bob, who then measures the two qubits in the Bell basis to infer the intended 2-bit classical message.

Given Alice's measurements, Bob performs one of four operations on his half of the EPR pair and recovers the original quantum state. If Alice performs a measurement and obtains the result ∣00 then Bob's system will be in the state ∣ψ (Nielsen and Chuang, 2010, 27).

The secret key determines a second transformation, this time on the encrypted message. This transformation is known as decryption, the inverse of encryption, allowing Alice to recover the original message (Nielsen and Chuang, 2010, 641).

8.3 Quantum Lattices—Why Not Use a 5-Dimensional Cubic Grid?

Quantum computing and lattices have had a close relationship ever since the work of Regev in 2004. He showed that the unique shortest vector problem is equivalent to bounded distance decoding the hidden subgroup problem on dihedral groups (Regev, 2004). The ordered lattices of all classical crystals are derived from three Platonic solids: the tetrahedron, the cube and the octahedron. Most lattice theories in quantum electrodynamics (QED) also rely on cubic lattices up to the 25-dimensional Leech lattice, which was discussed in the rise of higher-dimensional String Theories.

The grid of the 5-dimensional space formed by the edges of 12 E± (named epita-dodecahedron) is also actually a cubic lattice forming 24 subspaces and two counter-movements, totaling up to 26 dimensions if we count all subspaces and movements (Quehenberger, 2013, 2014a, 2014b, 2014c). The issue of its dimensionality was discussed with Gerard't Hooft, who first counted the 12 polyhedra forming the epita-dodecahedron and considered the epita-dodecahedron as 12-dimensional. On realizing that they form another 12 partitions from their intersection in the centre, he also saw 24 dimensions (in a personal conversation during the *Time and Matter* conference, Venice, 2012).

The essential ideas in quantum computation require operations upon quantum stuff as a user-defined circuit mapped onto space–time. In the space–time computer framework, interactions are taking place in discrete time intervals defined by a discrete set of space–time points, each labelling a setting-outcome pair. Any given quantum computation may be expressed as "tomographically complete" sets of quantum circuits. The structure of setting-outcome pairs at discrete events is the computational lattice (Hardy & Lewis, 2019, 1).

The problem with this model is, that "space–time points" are not sufficient to describe a continuum.

Here we suggest using a 6-dimensional space–time framework for one particle and the same six dimensions for the real world—if we are prepared to restructure the epistemic philosophical weltbild. The latter includes time in the 4th dimension and light as electromagnetic turbulence in the 5th dimension (Quehenberger, 2012b).

For quantum computing, a procedure to construct new lattices also enables the labelling of the signals belonging to particular signal constellations (Vieira et al., 2006). Signal constellations are considered to consist of the barycentres of the regular hyperbolic polygons with 4 g edges, denoted by F4g (Carvalho, 2001).

Hyperbolic polygons are partitions of some hyperbolic Poincaré half-plane models imagined as the signal space, where the noise is modelled as a hyperbolic Gaussian random variable.

Here we depict the hyperbolic space as hemispheres, defining the hypersphere forming the 4-dimensional space in imaginary movement. These hemispheres may also represent Boolean sets A and B as shown in Fig. 6.

The transformation matrices associated with the generators of Γ4g, with m = 1,..., 2 g and.

$$C = \begin{pmatrix} e^{\frac{i\pi}{4g}} & 0 \\ 0 & e^{\frac{-i\pi}{4g}} \end{pmatrix} \tag{1}$$

is the matrix corresponding to the elliptic transformation with order 4 g.

Our symplectic representation is composed of golden 5-cells named epitahedra (E±). Here we identify the edges of the trapezoid basis of E± as polygons with 4 g.

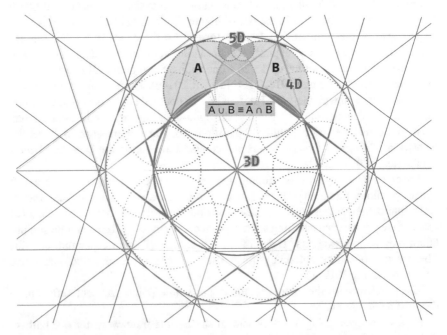

Fig. 6 Cubic Lattice of the five-dimensional space and the 6-dimensional space–time model with Venn diagram of subsets AB visualizing de Morgans's law (image: Quantum Cinema/Renate Quehenberger 2021© Bildrecht, Vienna 2021)

Formula (1) above poses a challenge to our imagination—how can we bring together the term e $^{i\pi}$ as an infinite logarithmic wavelike phenomenon with the space–time grid?

Within the context of the design of digital communications systems in homogeneous spaces, in particular in hyperbolic spaces, it is necessary to establish a systematic procedure for the construction of lattices (order of a quaternion algebra) as the basic entity to the design of signal constellations (quotient of an order by a proper ideal) to be employed in the transmission of digital information.

The computational lattice x = (lΔx, xΔt), composed of "wired" events is a tessellation of octagons and squares. Octagons are convenient because they adjoin to form "causal-diamond-like structures". The two-qubit gates situated on the octagons form a regular lattice, the gate lattice.

The choice of "causal-diamond-like structures" is the simplest construction of a rhomb by drawing two circles with the same radius, where the centrepoint of one circle lies on the circumference of the second. The same rhomb is forming the pattern of the golden mean, known as rhombic Penrose tiling and the lattice for quasicrystals, which we use for visualization of quantum phenomena.

By proposing the pentagrid as a model for a quantum lattice we also think of the infinite 5-dimensional space with angles in 72° (Fig. 7). Its inverse number, as well as its value behind the comma, seem to indicate a relationship to the experimental value for the Bell parameter function that was experimentally determined to be S_{exp} = 2.73± = 0.02 (Weihs et al., 1998)—named "awkward Irish angles" by John Bell (Bertmann and Friis, 2008).

9   The Hypersphere S^3

The hypersphere S^3 serves as a model for the Universe as well as for many features in quantum physics. Despite its importance, usually in literature, for the understanding of the topology of space, the depiction of the 3-sphere (S^3) remains to date more or less a question mark. There is no model for the 4-dimensional space and the hypersphere (S^3) is declared as "unimaginable".

For the hyper-Euclidean model of the hypersphere, which may also serve as a representation of the earth's complex system, we take an ordinary 3-dimensional sphere (in mathematics denoted as S^2: a 2-sphere, because it takes only the surface into account) and attach hemispheres on each point of its surface (Quehenberger, Rubin et al., 2022).

So we arrive at a complex 3-dimensional representation of a sphere with a 4-dimensional "real" Euclidean "tangent space" consisting of interconnected, hemispheres in imaginary movement. The model satisfies William Rowan Hamilton's mathematical foundations as played out in his *Algebra as the Science of Pure Time* (Hamilton, 1837), which appears to have been widely neglected. In his *Lectures on Quaternions* (1853), Hamilton noted that complex numbers (quaternions) are a composite of real and imaginary numbers ($z = x + iy \in C$ where $i = \sqrt{-1}$) living

Fig. 7 Epita-dodecahedron, composed of 12 E± forming a 4-dimensional dodecahedron, representing the infinite 5-dimensional space, **b** centre of the epita-dodecahedron, with Platonic and Archimedean solids; (see also: Epita-dodecahedron (2014), QC-film presented at the ICM Seoul (KR) 2014 (image: Quantum Cinema/Renate Quehenberger 2012© Bildrecht, Vienna 2021)

in 4-dimensional space—the piece of the curved space surrounding a sphere. In this way, the 4-dimensional space becomes visually accessible. General relativity avoids Hamiltonian formulations but recognizes that time is part of the physical system which quantum gravity is trying to quantize (Wüthrich, 2014). The confusion can be avoided by assigning time to the 4th dimension of space in imaginary movement instead of dealing with arbitrary points of space–time (Fig. 8).

In the 3-dimensional representation, we can visualize the embedding space $\mathbb{R}^4$—which is also the complex vector space C^2 (Bengtsson & Zyczkowski, 2006, 82). The visualization of a Clifford translation of a G + element, particularly a state, is usually assigned to appear along the big circle of the S^3 sphere. The circle is an intersection of the sphere with the plane of the Hamiltonian lift (Soiguine, 2017).

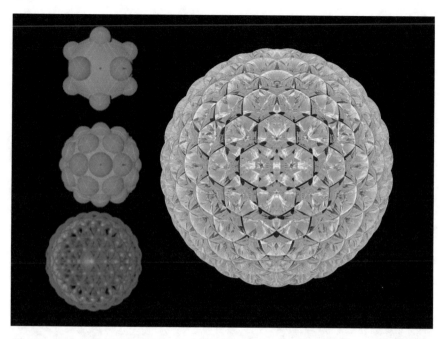

Fig. 8 Construction of the hypersphere: Attaching hemispheres on the ball started with icosahedral symmetry. More and more hemispheres are added until the centre of one is in the radius of its neighbour; the hypersphere fully packed with hemispheres exhibits hexagonal hyperbolic cells on the surface of S^3 (image: Quantum Cinema/Renate Quehenberger 2012© Bildrecht, Vienna 2021)

The currently still most quoted construction of a hypersphere is called Hopf fibration, which is a bijection between the points on a sphere resulting in sets of intertwining great circles on the surface of a sphere (Hopf, 1931).

The idea of fibered spaces goes back to the nineteenth century pedagogue Johann Friedrich Herbart who perceived the world as continua of fibre bundles. Herbart inspired the mathematician Bernhard Riemann to conceptualize 1-dimensional manifolds. However, these spaghetti-like spaces are counter-intuitive for an artist's mind. Nevertheless, we find these 1-dimensional lines as circles below (Fig. 9) and as wavelike logarithmic fibre spirals (Fig. 10).

Also, the Heegaard splitting is a decomposition of a compact oriented 3 manifold that results from dividing it into two handle bodies and does not lead to a representation of S^3. In topology S^1 is a circle, S^2 is a hyperbolic circle, two of which form the sphere S^2. However, it is not possible to glue two spheres together in order to create a hypersphere S^3. Consequently, a full picture of the hypersphere remains absent (Bengtsson & Zyczkowski, 2006, 425). Regardless of the lack of a proper imagination, the 3-sphere is an example of the 3-manifolds which play an essential role in topological quantum computing. Topological methods are used to assist in the performance of quantum error correction (Nielson and Chuang, 2010, 499).

Fig. 9 Construction Epita-dodecahedron, 3D Kites and Darts Penrose tilings with circle decoration in motion exhibit the spherical shape of a 3-sphere. (image: Quantum Cinema/Renate Quehenberger 2012© Bildrecht, Vienna 2021)

Topology speaks of handlebody-knots embedded in the 3-sphere S^3. Please note, that we get a torus if we cut the 4-dimensional space around the compact 3-dimensional space of the sphere in the centre.

In the context of 3-manifolds, some authors use Hausdorff topological spaces with a group-theoretical substitute for 3-manifolds, the fundamental group of a knot complement also known as knot group (Asselmeyer-Maluga, 2021). In our construction of the hypersphere, the knot is the fundamental space-generating element: It starts not with a knot in 1-dimensional string but with the knot of a 2-dimensional plane forming a pentagon. The triangular partitions of this pentagon undergo another knot-like enfolding that results in a double 3D pentagon, which we have named the epitahedron (E±). Twelve of these two 5-cells, form the epita-dodecahedron (Fig. 7), whose circles in movement form spherical shapes homeomorphic to the 3-sphere, as shown below in Fig. 9.

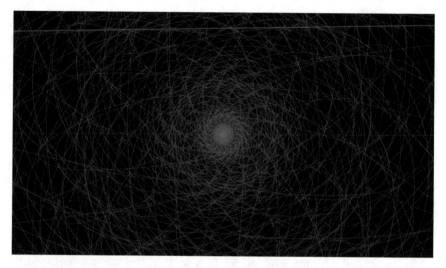

Fig. 10 The *4π evolution of e* (vimeo.com/50142342) evolution of the imaginary number e (image: Quantum Cinema/Renate Quehenberger 2012© Bildrecht, Vienna 2021)

The problem is that 3-manifolds require a 5-dimensional space like $\mathbb{R}^5$ as an embedding space for which they do not have any visual representation since the visualizations of the Quantum Cinema project are not yet widely discussed.

Today the main source for 3-manifolds remains mathematician William Thurston, who explains that the 3-sphere S^3 could be easily visualized as $\mathbb{R}^3$, together with one point at infinity (Thurston[, 1980], 2002), 1). However, Thurston's description of a horosphere, as mentioned by Carl Friedrich Gauss and Nikolai I. Lobachevsky in the early 1800s, seems more suitable for the hyper-Euclidean visualization method. It is a Euclidean sphere tangent to the unit sphere where the point X of tangency is the centre of the horosphere (Thurston, [1980] 2002, 38). We interpret the horosphere as attached to the sphere S^2.

When working on the foundations of topology, Henri Poincare investigated topological properties that characterized a sphere. He was able to show that the Poincaré sphere had a fundamental group of order 120, while the 3-sphere had a trivial fundamental group.

Our intuitively applied construction method for the hypersphere S^3—the full covering of the sphere S^2 with hemispheres where each centre of one lies in the radius of its neighbour—requires starting with icosahedral symmetry, as we see above (Fig. 8). We assume that Poincaré must have had a similar picture in mind when he suggested the icosahedral symmetries for the dodecahedral space. Surprisingly, we find hyperbolic hexagonal shapes resulting from the intersecting hemispheres on the 2-sphere as the outer surface of S^3.

We animated the epita-dodecahedron according to the description of the Poincaré dodecahedron space by William Threlfall and Herbert Seifert in their first joint paper (1931). It is characterized by the use of a dodecahedron, the opposite faces of which are identified after a turn by $\pi/5$ in order to construct a "spherical dodecahedron space". Analogous to their description, we use the 3-dimensional pentagons E±, a golden pentachoron (aka 5-cell or 4-simplex), to visualize the principle of the counter-movements of the opposite polyhedra E + instead of the pentagons and a spherical dodecahedron only appears inside from the red and green circle decorations. The same geometry of the dodecahedral space and its dual hypersphere was suggested to be responsible for the Earth' intrinsic dynamics (dC Rubin et al., 2021).

Hence, we get a 3-manifold that is composed according to Henri Poincaré's call for a hypergéométrie in the visualization of the space of 5 dimensions.

— Please note the relationship to Poincaré's famous question, "Is it possible that a manifold with a vanishing fundamental group is not homeomorphic to the 3-sphere?" (Poincaré, 1904, 46).

Poincaré assumed all these manifolds are simply connected, i.e., homeomorphic to hyper-spheres (Poincaré, 1895, 277). William Thurston established the following weak version of the Poincaré conjecture: If a simply connected 3-manifold M is a cyclic branched covering space of S^3, then M is in fact homeomorphic to S^3. We may well recognize this "cyclic branched covering space" as full circles of the epita-dodecahedron (Fig. 9).

With its circles in motion, this infinite 5-dimensional space in the shape of a dodec-ahedron reveals the same shape as the hypersphere. This picture of a 4-dimensional dodecahedron, the epita-dodecahedron, composed of the 3D representation of the Penrose Kites and Darts and animated according to the description of the Poincaré homology sphere gives rise to the claim that is homeomorphic to the hypersphere. However, we claim that Ricci flow surgery and Ricci calculus is relying on a system of point transformations and not considering Poincaré's special group transforma-tions. Moreover, any proof of the Poincaré conjecture cannot be void of any statement about the shape of the hypersphere (Quehenberger, 2021).

10   Quantum Algorithms and Fourier Transform

Fourier transform, or better quantum Fourier transform, became the dominant tool in quantum computing. Fourier transform may be used to solve the hidden subgroup problem, a generalization of the phase estimation and order-finding problems that has among its special cases an efficient quantum algorithm for the discrete logarithm problem, another problem thought to be intractable on a classical computer (Nielson and Chuang, 2011, 217).

In quantum computing the Fourier transform is the key to a general procedure known as phase estimation—also described as black box operation—which is, in turn, the key for many quantum algorithms.

In mathematics, the discrete Fourier transform over an arbitrary ring generalizes the discrete Fourier transform of a function whose values are complex numbers, i.e., performing the transform in the integers mod N for some integer N (Schönhage & Strassen, 1971). The algorithm uses recursive Fast Fourier transforms in rings with

2^{n+1} elements, a specific type of number theoretic transform (Knuth, 1997, 305). Quantum algorithms can be summarized as follows: First, reduce the question at hand to something resembling a hidden subgroup problem, then extend the existing hidden subgroup machinery to give a quantum algorithm for this particular variant. The key step in solving a hidden subgroup problem is the ability to compute the quantum Fourier transform over a group with exponentially many elements. So the relationship between the Fourier transforms over different cyclic groups of an arbitrary fixed superposition (Hales & Hallgren, 2000).

Suppose a unitary operator U has an eigenvector |u > with eigenvalue $e^{2\pi i\phi}$, where the value of ϕ is unknown. The product representation allows the construction of an efficient quantum circuit computing the Fourier transform and provides insight into algorithms based upon the quantum Fourier transform (Nielson and Chuang, 2011, 217).

A classical computer requires exponentially more operations to compute the Fourier transform than a quantum computer. The quantum circuit construction of the quantum Fourier transform apparently requires logic gates of exponential precision in the number of qubits used. Fourier analysis involves expanding periodic functions on the unit circle written in terms of a series of sines and cosines functions which are referred to as *harmonics*. Considering higher-dimensional analogues of the harmonics on the unit n-sphere, one arrives at the spherical harmonics—which we imagine as logarithmic spirals around the unit sphere.

In our visualization of the *4π evolution of e* (Fig. 10), the natural algorithm e is depicted by means of logarithmic spirals emerging from the unit sphere and curling around it. This construction is similar to the idea of fibered spaces; we coloured the spiral on the left red, and spirals on the right blue. Uneven numbers of logarithmic spirals mix, while even numbered spirals appear in bundles of the same colour. This is certainly another topic that requires deeper investigation.

At least it provides a suggestion for the wave state of a quantum particle. It reminds us of the measurement of the 4π value of a neutron by Helmut Rauch, which we discussed with him in depth during the Quantum Cinema project.

This wave bundle figure is reminiscent of a charged-particle motion (factors of $4\pi/c$) and Maxwell's equations describing a curl in an electric field that implies a change in the magnetic field equal to the current density, occurring at the speed of light. The given distribution of electric current density J is derived from the magnetic field B simply as $(c/4\pi i)\nabla \times B$ (Vasyliunas, 2005). This reminds us of Schrödinger's dream of a visual interpretation of the matrix elements "as amplitudes of partial oscillations of the electrical moment of the atom", which he had to abandon (Meyenn, 2010, 204).

According to Heisenberg's theory, the Schrödinger equation of a light quantum has a value of the order of 1 (Heisenberg in a letter to Pauli, May 30, 1935, in Meyenn, 1985, 397).

However, we may imagine the quantum object, the ideal quantum Fourier transform on n qubits, results if the controlled-Rk gates are performed to a precision $\Delta r = 1/p(n)$ for some polynomial p(n).

The prime state made of n qubits that corresponds to the quantum superposition of all prime numbers is less than 2^n (we take n > 1 so that 2^n is not a prime), where each prime number can be expressed in binary form $p = p_0 2^0 + p_1 2^1 + ... + p_{n-1} 2^{n-1}$.

All the states in the sum are orthogonal and the normalization of the state is related to the squared root of the number of primes less than 2^n, namely $\pi(2^n)$. In the case of, for example, n = 3,

$|P3\rangle = 1/\sqrt{4} \, (|2\rangle + |3\rangle + |5\rangle + |7\rangle)$

$= 1/2 \, (|\uparrow\downarrow\uparrow + |\uparrow\downarrow\downarrow + |\downarrow\uparrow\downarrow + |\downarrow\downarrow\downarrow\rangle)$ where the qubits $|0\rangle$ and $|1\rangle$ are described by the spin polarized states $\uparrow$ and $\downarrow$ of a spin 1/2 particle (Latorre & Sierra, 2013).

Applying the controlled-R2 gate produces the state of the qubit system.

At most n/2 swaps are required, and each swap can be accomplished using three controlled-CNOT gates. Accordingly, this circuit provides a $\Theta(n^2)$ algorithm for performing the quantum Fourier transform.

The ability to tune the control frequency of a qubit within the silicon chip by engineering its atomic configuration is extremely important. The molecules can be created with different resonance frequencies. This means that controlling the spin of one qubit will not affect the spin of the neighbouring qubit, leading to fewer errors—an essential requirement for the development of a full-scale quantum computer (Voisin, 2020).

10.1   Quantum Algorithms and Fuchsian Groups

Poincaré's research on the fundamental polyhedron has special importance for quantum computation since quantum gates are elements of the fundamental group represented as SU(2) matrices. In quantum computation, the arithmetic Fuchsian groups play an important role as a basis for code construction and signal constellation design (Cavalcante, Palazzo et al., 2005).

The arithmetic Fuchsian group is a subgroup of PSL(2\R) and derived from a quaternion algebra. The term "Fuchsian groups" for a class of groups that are important in the theory of modular functions was coined by Henri Poincaré because he was inspired by the work on inverse functions by Lazarus I. Fuchs (Fuchs, 1880).

Poincaré used them as a fundamental concept in the development of the theory of automorphic functions and created a completely new area of mathematics (Poincaré, 1882). Discrete and continuous transformations also play an important part in the analysis of differential equations carried out by Felix Klein. Which led to the bestowal of the name "Fuchsian" causing difficulties between Poincaré and Felix Klein. When Klein complained about the name because he was also working on the same algebraic structures, Poincaré countered in German by quoting Goethe's *Faust,* "Name ist Schall und Rauch" (a name is sound and smoke) in a letter to Klein, 4 April 1882 (Verhulst, 2012, 43).

Poincaré was building his research on a fundamental polyhedron in Group theory on the so-called Fuchsian groups, where each group is associated with a fundamental polygon with edges identified by certain motions and closed paths in a 3-manifold defined by a polyhedral region with faces identified by certain geometric transformations (Poincaré, 1892).

In this context, he was pondering certain shapes suitable as a fundamental region, like a cube, a pyramid or a torus belonging to a family of 3-dimensional manifolds and obtained as quotients of $\mathbb{R}^3$ by certain groups.

He assumed all these manifolds were simply connected, i.e., homeomorphic to hyper-spheres (Poincaré, 1904, 277). The question as to whether his conjecture was accurate or not became one of the famous millennium problems (Milnor, 2000, 2003).

Because of their 4-dimensional character, these polygons related to Fuchsian groups are usually thought of as being part of the hyperbolic plane. Fuchsian groups are associated with a regular tessellation and given edge-pairing transformations tessellation whose regular hyperbolic polygon generates an oriented compact surface with genus $g > 2$. This algorithm can be implemented in Maple or Mathematica (Benedito et al., 2016).

Distinct from the usual projective geometry interpretation that the quaternion space with Fuchsian groups can be treated as a tessellation of the sphere, we assume the quaternionic space is spanned over orthogonally attached vectors as real tangent space surrounding the sphere, forming the 4-dimensional space as shown in Fig. 6 above in the area of the yellow circles. The two intersecting circles may also represent the space of real states, with disjoint regions (cf. Jevtic & Rudolph, 2015).

Accordingly, our construction of the hyperspace is composed of hyperplanes for the upper-half plane $H = \{z \in C: \text{Im}(z) > 0\}$ attached on each point of the sphere (S^2) and constructs a hyperbolic space which is a real hemisphere on top of the hyperplane instead of the usual projective models of the hyperbolic plane, known as Poincaré disk or Beltrami–Klein disk model.

This gives us the images for the nodes as above, and the hyperbolic hexagon is only the basis of the closed mirroring self-dual epitahedra (E±) that is the fundamental polyhedron of the icosahedral/dodecahedral space we can compose of it. This image, which naturally contains the lattice Z^5 from the edges of the polyhedra (E±), enables lattices to be imagined that are associated with signal constellations in our presumed real hyperspace.

10.2 Factorization of Prime Numbers

A quantum computer can factor a number exponentially faster than the best known classical algorithms. The most spectacular discovery in quantum computing to date is that quantum computers can efficiently perform some tasks which are not feasible on a classical computer.

For example, finding the prime factorization of an n-bit integer is thought to require $\exp(\Theta(n^{1/3} \log^{2/3} n))$ operations using the best classical algorithm known at the time of writing, the so-called *number field sieve* (Nielsen and Chuang, 2010, 216).

Prime numbers are central objects in Mathematics and Computer Science. They appeared dramatically in Quantum Computation through Shor's algorithm, which converts the hard problem of factorization into a polynomial one using quantum interference (Shor, 1997). The entanglement properties of the Prime state remain to be explored in more detail.

The Prime state is highly entangled, and its entanglement measures encode number theoretical functions, such as the distribution of twin primes.

10.3 Bernoulli Numbers and the Riemann Hypothesis

Bernoulli numbers are characterized by the three most useful features, namely, a recursive equation, an explicit formula and a generating function. They were first applied to computer science in the first conception for the first published complex computer program by Ada Lovelace in note G on the Analytical Engine, where she describes an algorithm for generating Bernoulli numbers with Babbage's machine (Lovelace, 1842). Bernoulli factory algorithms sample the probability $f(\lambda)$. There is an analogy to the Bernoulli factory problem called the quantum Bernoulli factory, with the same goal of simulating functions of unknown probabilities. For every odd $n > 1$, $B_n = 0$. For every even $n > 0$, B_n is negative if n is divisible by 4 and positive otherwise.

Bernoulli numbers can also be expressed in terms of the famous Riemann zeta (ξe) function, an extension of Euler's zeta (ξ) function.

The Swiss mathematician Leonard Euler made pioneering and influential discoveries in many other branches of mathematics, such as analytic number theory, complex analysis, and infinitesimal calculus. He was invited to Russia by Empress Catherine II thanks to his good relations with the Bernoulli brothers and succeeded Daniel Bernoulli as professor of mathematics at the Academy of St. Petersburg in 1733. Euler discovered a formula relating $\xi(2k)$ to the Bernoulli numbers, yielding results such as $\xi(2) = \pi^2/6$ and $\xi(4) = \pi^4/90$. The values of the zeta function are expressed as powers of π and depend on these Bernoulli numbers.

Bernhard Riemann noted that his zeta function had trivial zeros at $-2, -4, -6, \ldots$ and that all non-trivial zeros were symmetric about the line $\mathrm{Re}(s) = 1/2$. He calculated the first six non-trivial zeros of the function and observed that they were all on the same straight line. In a report published in 1859, Riemann stated that this might very well be a general fact (Riemann, 1859).

In 1900, David Hilbert listed proving or disproving the Riemann hypothesis as one of the most important unsolved problems confronting modern mathematics.

Since then numerous attempts to solve this conjecture have been published. It is central to understanding the overall distribution of the primes. The Hilbert-Pólya Conjecture proposes a spectral interpretation of the non-trivial zeros of the Riemann zeta function. Hilbert called the set of eigenvalues of an operator the "spectrum", by analogy with the spectral lines produced by the radiation frequencies of atoms. Prime numbers are also related to the harmonic mean as formulated by the Pythagoreans.

For example, in 1901, von Koch showed that the Riemann hypothesis was equivalent to the logarithmic expression similar to the one Euler found (von Koch, 1901).

or(x) = Li (x) + 0 (x$^{1/2}$ ln x)

Jacques Hadamard and de la Vallee Poussin proved the prime number theorem, by showing that.

ad(x) = Li (x) + 0 (xe $^{-a\sqrt{\ln x}}$)

for some positive constant a, and they did this by bounding the real part of the zeros in the critical strip away from 0 and 1. The error term is directly dependent on what was known about the zero-free region within the critical strip.

The function f satisfies the equation f (1 − s) = f(s) for the complex argument s ≡ σ + iτ. Or, as the British mathematician Godfrey Harold Hardy put it, ξ (1/2 + it) and proved that the Riemann zeta function has infinitely many real zeros. Hardy also showed that infinite many non-trivial zeros lie on the critical line Re(s) = 1/2 (Hardy, 1914).

The Austro-Hungarian mathematician Marcel Riesz showed in 1916 that the Riemann hypothesis was equivalent to the claim that it is true for any e larger than 1/4 if we had, x tending to infinity on the positive real axis (Riesz, 1916). Also, all points where f vanishes are located on the critical line σ = 1/2 and all lines of constant phase of f corresponding to ±π, ±2π, ±3π... merge with the critical line (Schleich et al., 2018). This π-periodicity is visualized in Fig. 11, where after every π -turn the prime numbers appear on one line with the most beautiful pattern of "entangled" prime numbers on the y-axis visualized by small blue circles.

It seems to be a visual confirmation of Godfrey H. Hardy and John E. Littlewood's Twin Prime Conjectures, (Hardy & Littlewood, 1921). An analytical explanation of these periodical structures obtained is an open problem worthy of investigation by the experts in the analytical number theory (Machadoa & Luchkob, 2021).

In our visualization, the numbers 1, 2, 3, 4 are lying inside the radius of the sphere, while number 5 is already the first number on the spiral arm crossing the x-axis above the north pole of the sphere, representing the imaginary part of the system just like the red and blue bundles of lines in Fig. 10. By placing all numbers consecutively on the ±x and ±y-axis, we arrive at all even numbers on the x-axis and all odd numbers on the y-axis. Furthermore, we place the imaginary numbers i on a logarithmic spiral originating on the surface of the unit sphere at point 1/2 of a Cartesian coordinate system, which is the half diameter of the unit sphere that represents the real numbers. This implies that in our visualization, the so-called "critical line" with non-trivial zeros is a geodesic on the real number space unit sphere and the imaginary numbers on the logarithmic spiral form the 4-dimensional space around a central unit sphere which can be identified as Hamiltonian/quaternionic/Lorentzian space–time.

For quantum computation, Jose I. Latorre and German Sierra showed how to construct the quantum prime state, a quantum sequence state based on the sequence of prime numbers with an error below the bound that allows for verification of

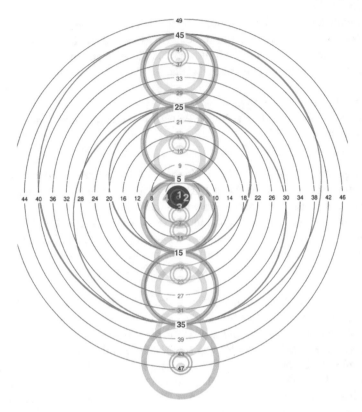

Fig. 11 The prime numbers in a logarithmic 4πi solution, all prime numbers on the y-axis, produce a series of twin prime numbers which are depicted by small blue circles, seemingly "entangled" (image: R. Quehenberger 2012© Bildrecht, Vienna 2021)

the Riemann hypothesis. This algorithm can be further combined with the quantum Fourier transform to yield an estimate of the prime counting function more efficiently than any classical algorithm (Latorre & Sierra, 2013). They assume it is likely that the quantum correlations emerging from the Prime state are profoundly related to theorems in Number Theory. Their claim is simple: All non-trivial zeros have a real part of 1/2.

With the help of computers, more than ten billion zeros of the zeta function lying on the critical, clearly highlighting Riemann's extraordinary intuition, have been calculated to date (Bayer, 2018). However, calculations ad infinitum do not provide sufficient proof (van de Lune et al., 1986).

Finally, we call for the reestablishment of the authority of geometry for proving algebraic operations so that we could claim the million dollars—at least for the Poincaré conjecture since, despite many claims to have done so, Henri Poincaré's question has never really been answered in the way we have shown here.

Although purely imaginative, geometry may provide a visual correspondence to algebraic descriptions of quantum systems. Moreover, we theorize that the much discussed "hidden parameters" are to be found in the above described dynamic 5-dimensional space.

John Bell did not agree that: "In science, whatever is provable must not be believed without proof." (Dedekind, 1888), but in all cases, the process of confirmation—and not the proof—remains the ultimate source from which knowledge derives its authority; it is the "experience of truth" (Bell, 2000).

In this context, we may recall Plato's theory of forms, which states: "...in any case there are forms although not perceivable but only accessible through the mind" (Plato, 97). In this tradition of Platonic thought, John Bell quoted Hermann Weyl, who shared his lifelong conviction that intuition, or insight—rather than proof—furnishes the ultimate foundation of mathematical knowledge (Bell, 2000).

> Transparent penetrable space, the purest image of my knowing, cannot be inspected but must be seen intuitively, and within it my inspecting itself is so seen. The light is not outside of me, but rather in me. (Hermann Weyl, as quoted by Bell, 2000)

References

Abbott, E. A. (1884). *Flatland: A romance of many dimensions.* Seeley & Co.

Aspect, A., Grangier, P., & Roger, G. (1981). Experimental tests of realistic local theories via Bell's theorem. *Physical Review Letters, 47*, 460–463.

Babbage, C. (1838). *The ninth Bridgewater treatise. A fragment.* J. Murray.

Baez, J. C. (2017). From the icosahedron to E8. arXiv:1712.06436v2

Barnett, S. (2009) *Quantum information* (Oxford Master Series in Physics: Atomic, Optical, and Laser Physics). Oxford University Press.

Bayer, P. (2018). The Riemann hypothesis: The great pending challenge. *Mèthode Science Studies Journal 8*, 34-41. https://doi.org/10.7203/metode.8.8903

Bell, J. S. (1964). On the Einstein Podolsky Rosen paradox. *Physics Physique Fizika, 1*, 195–200.

Bell, J. S. (1966). On the problem of hidden variables in quantum theory. *Reviews of Modern Physics, 38*, 447–452.

Bell, J. S. (2000). Hemann Weyl on intuition and the continuum. *Philosophia Mathematica, 8(3)*, 259–273.

Bellman, R. E. (1957). *Rand corporation dynamic programming.* Princeton University Press.

Benedito, C. W. D. O., Palazzo, R., & Interlando, J. C. (2016). An algorithm to construct arithmetic Fuchsian groups derived from quaternion algebras and the corresponding hyperbolic lattices. *Journal of Pure and Applied Algebra, 220(5)*, 1902–1923.

Bengtsson, I., & Zyczkowski, K. (2006). *Geometry of quantum states: An introduction to quantum entanglement.* Cambridge University Press.

Bennett, C. H. (1995). Quantum information and computation. *Physics Today, 48(10)*, 24.

Bennett, C. H., & Wiesner, S. J. (1992). Communication via one- and two-particle operators on Einstein-Podolsky-Rosen states. *Physical Review Letters, 69*, 2881.

Bharti, K., Ray, M., Varvitsiotis, A., Kwek, L.-C., & Cabello, A. (2019). Local certification of programmable quantum devices of arbitrary high dimensionality. arXiv:1911.09448v1

Bohm, D. (1952). A suggested interpretation of the quantum theory in terms of "Hidden" variables. I *Physical Review, 85*, 166 and Part. II *Physical Review, 85*, 180.

Bohm, D. (1982b). *The holographic paradigm and other paradoxes, exploring the leading edge of science*. In: K. Wilber (Ed.). Shambala.

Bohm, D. (2004). *On creativity*. In: L. Nichol (Ed.). Routledge.

Bohm, D., & Hiley, B. (1993). *The undivided universe*. Routledge.

Bernal, R. A. (2005). De Morgan's laws revisited: To be AND/OR NOT to be, paper PO25, PharmaSUG.

Bertlmann, R. A., & Friis, N. (2008). Theoretical physics T2 quantum mechanics course of lectures, T2–script of Sommersemester 2008 by Reinhold A. Bertlmann.

Bertlmann, R., & Zeilinger, A. (Eds.). (2017). *Quantum [Un]Speakables II*. Springer International Publishing Switzerland.

Brezina, C. (2006). *Al-Khwarizmi: The inventor of algebra*. The Rosen Publishing Group.

Brillouin, L. (1956). *Science and information theory*. Academic Press.

Brody, D. C., & Gräfe, E.-M. (2011). Six-dimensional space-time from quaternionic quantum mechanics. *Physical Review D, 84*, 125016. arXiv:1105.3604

de Broglie, L. (1925). Recherches sur la Théorie des Quanta, Thesis 1924. *Annales de Physique, 10*e série, t. III.

de Broglie, L. (1927). L'univers à cinq dimensions et la mécanique ondulatoire. *Journal De Physique Et Le Radium, 8*, 66–73.

de Broglie, L. (1943). Die Elementarteilchen, Individualität und Wechselwirkung, H Coverts; orig. (1941). Continu et discontinu en physique moderne (Continuous and discontinuous in Modern Physics), Paris: Albin Michel.

Bromley, A. G. (1982). Charles Babbage's analytical engine, 1838. *Annals, 4*(3).

Brouwer, L. E. J. (1913). Über den natürlichen Dimensionsbegriff. *Reine Und Angewandte Mathematik, J, 42*, 146–215.

Cabello, A. (2017). Interpretations of quantum theory: A map of madness. In: O. Lombardi, S. Fortin, F. Holik, C. López (Eds.), *What is quantum information?* Cambridge University Press.

Cavalcante, R. G., Lazari, H., Lima, J. D., & Palazzo, R. Jr. (2005). A new approach to the design of digital communication systems. In: A. Ashikhimin & A. Barg (Eds.), *Discrete mathematics and theoretical computer science* (Vol. 68, pp. 145–177). DIMACS Series. American Mathematical Society.

Carvalho, E. D. (2001). Construction and labeling of geometrically uniform signal constellations in Euclidean and hyperbolic spaces. Doctoral Dissertation, FEEC-UNICAMP.

Capra, F. (1982). *The turning point*. Simon and Schuster.

Clarke, B. (2009). Heinz von Foerster's Demons. In: B. Clarke, M. B. N. Hansen (Eds.), *Emergence and embodiment: New essays on second-order systems theory*. Duke University Press.

Dakić, B., & Brukner, Č. (2016). The classical limit of a physical theory and the dimensionality of space. *Fundamental Theory Physics, 181*, 249–282.

Deleuze, G. (1989). Part 2: Cinema 2, University of Minnesota Press. Here quoted in German from: Transl. Klaus Englert, (1991) Das Zeit-Bild, Kino 2, Frankfurt/Main: Suhrkamp.

Descartes, R. ([1637] 1902). Discours de la Méthode. Adam et Tannery.

Dirac, P. A. M. (1930). *The principles of quantum mechanics*. Clarendon Press.

de Bruijn, N. G. (1981). Algebraic Theory of Penrose's Non-periodic Tilings of the Plane I and II, Proc Koninklijke Nederlandse Akademie van Wetenschapen, 84, 39–52 and 53–66.

Einstein, A. (1916). Die Grundlage der allgemeine Relativitätstheorie. *Annalen Der Physik, 49*, 769–822.

Einstein, A., Podolsky, B., & Rosen, N. (1935). Can quantum-mechanical description of physical reality be considered complete? *Physical Review, 47*, 777–780.

Faulhaber, J. (1631). Academia Algebrae, darinnen di miraculo'sische, Inventiones zu den höchsten Costen weiters continuirt u. profitiert werden, Augsburg: Johann Remelins.

Feynman, R. (1982). Simulating physics with computers. *International Journal of Theoretical Physics, 21*, 6/7, 467–488.

Fourier, J. B. J. (1822). *Theorie analytique de la chaleur*. Firmin Didot.

van Fraassen, B. (1972). A formal approach to the philosophy of science in paradigms and paradoxes: The philosophical challenge of the quantum domain. In R. G. Colodny (Ed.), *Paradigms and paradoxes: The philosophical challenge of the quantum domain* (pp. 303–366). University of Pittsburgh Press.

Fuchs, L. (1880). Über die Verallgemeinerung des Kehrungsproblems. *Journal De Crelle, 89,* 151–169.

Gardner, M. (1989). *Penrose tiles to trapdoor ciphers.* H. Freeman & Co.

Greenberger, D. M., Horne, M. A., & Zeilinger, A. (1993). Multiparticle interferometry and the superposition. *Physics Today, 46*(8), 22.

Grünbaum, B., & Shephard, G. C. (1987). *Tilings and patterns.* W. H. Freeman and Company.

Grössing, G. (2003). Zum Bilderstreit in der Quantentheorie, http://phaidon.philo.at/~nulleins/arc hiv/archiv.html

Hales, L., & Hallgren, S. (2000) An improved quantum Fourier transform algorithm and applications. In: Proceedings of the 41st Annual Symposium on Foundations of Computer Science (pp. 45–46).

Hao, L., Wang, Z., Yang, J., et al. (2019). Anomalous magnetoresistance due to longitudinal spin fluctuations in a Jeff $= 1/2$ Mott semiconductor. *Nature Communications, 10,* 5301. https://doi.org/10.1038/s41467-019-13271-6

Hardy, G. H. (1914). Sur les zeros de la fonction zeta $\zeta(s)$. *Comptes Rendus De L'académie Des Sciences, 158,* 1012–1014.

Hardy, G. H., & Littlewood, J. E. (1921). The zeros of Riemann's Zeta-function on the critical line. *Mathematische Zeitschrift, 10*(3–4), 283–317. https://doi.org/10.1007/bf01211614

Hardy, L., & Lewis, A. G. M. (2019) Quantum computation with machine-learning-controlled quantum stuff. arXiv:1911.13282v1

Hermann, G. (1926). Die Frage der endlich vielen Schritte in der Theorie der Polynomideale. *Mathematische Annalen, 95,* 736–788.

Hermann, G. (1935). Die naturphilosophischen Grundlagen der Quantenmechanik. *Abhandlungen Der Fries'schen Schule, 6*(2), 69–152.

Heisenberg, W. (1927) Über den anschaulichen Inhalt der quantentheoretischen Kinematik und Mechanik (On the Perceptual Content of Quantum Theoretical Kinematics and Mechanics), Copenhagen, Institut for theoretical Physics at the Kopenhagen University, March 23 and Zeitschrift für Physik 43, 172–198.

Heisenberg, W. (1989). *Encounters with Einstein: And other essays on people, places, and particles.* Princeton University Press.

Heisenberg, W. (2011). Quantenphysik und Philosophie, Stuttgart: Philip Reclam jun. (reprint from 1979).

Herzenberg, C. L. (2008) Grete Hermann: An early contributor to quantum theory. arxiv.org/pdf/0812.3986.

Hiley, B. J., & Callaghan, R. E. (2010). The Clifford algebra approach to quantum mechanics B: The Dirac particle and its relation to the Bohm approach. arXiv:1011.4033v1

Hiley, B. J.(2016). Structure process, weak values and local momentum. *Journal of Physics: Conference Series, 701*(1). Article id. 012010

Hopf, H. (1931). Über die Abbildungen der dreidimensionalen Sphäre auf die Kugelfäche, Mathematische, *Annalen, 104*(1), 637–665.

Holmes, R. (2015). Computer Science: Enchantress of abstraction. *Nature, 525,* 30–32.

Toole, B. A. (2010) Ada, the enchantress of numbers: Poetical science. Critical Connection.

Jammer, M. (1974). *The philosophy of quantum mechanics: The interpretations of quantum mechanics in historical perspective.* Wiley.

Jevtic, S., & Rudolph, T. (2015). How Einstein and/or Schrödinger should have discovered Bell's theorem in 1936. *Journal of the Optical Society of America B - Optical Physics, 32,* A50-A55. ISSN 0740-3224

Jarry, A. ([1911] 1996). Exploits and Opinions of Dr. Faustroll, Pataphysician, Boston: Exact Change.

Jones, W. (1794). On the philosophy of the Asiatics. *Asiatic Researches, IV*, 164.
Jones, W. (1794). On the philosophy of the Asiatics. *Asiatic Researches*, IV, 164.
Klein, F. (1872). Vergleichende Betrachtungen über neuere geometrische Forschungen, Programm zum Eintritt in die philosophische Fakultät und den Senat der Friedrich-Alexander-Universität zu Erlangen, Erlangen: Deichert.
Klein, F. (1892). Vorlesungen über Nicht-Euklidische Geometrie. In Klein, Felix (1928) Vorlesungen über Nicht-Euklidische Geometrie, Rosemann, W. (Ed.) Berlin: Springer.
Klyachko, A., Can, M. A., Binicioğlu, S., & Shumovsky, A. S. (2008). Simple test for hidden variables in spin-1 systems. *Physical Review Letters, 101*, 020403.
Knuth, D. E. (1997). *The art of computer programming, volume 2: Seminumerical algorithms*, 3rd edn. Addison-Wesley Professional.
von Koch, H. (1901). Sur la distribution des nombres premiers. *Acta Mathematica, 24*, 159–182.
Kochen, S., & Specker, E. P. (1967). The problem of hidden variables in quantum mechanics. *Journal of Mathematics and Mechanics, 17*(1967), 59–87.
Kunjwal, R., Spekkens, R. W. (2015). From the Kochen-Specker theorem to noncontextuality iinequalities without assuming determinism, submitted to QPL. www.cs.ox.ac.uk/qpl2015/pre proceedings/50.pdf
Kwiat, P. G., Mattle, K., Weinfurter, H., Zeilinger, A., Sergienko, A. V., & Shih, Y. H. (1995). New high-intensity source of polarization-entangled photon pairs. *Physical Review Letters, 75*, 4337–4341.
Latorre, J. I., & Sierra, G. (2013) Quantum computation of prime number functions. *Quantum Information & Computation, 14*(7). arXiv:1302.6245v3
Lenzen, V. F. (1969). Concepts and reality in quantum mechanics. *Philosophy of Science, 16*(4), 279–286.
Leonardo da Vinci. (1877). *Treatise on painting* (J. F. Rigaud, Trans.) George Bell & Sons.
Lovelace, A. (1842). Notes to "A Sketch of the Analytical Engine invented by Charles Babbage by LF Menabrea", translated with a detailed commentary Ada Augusta, Countess of Lovelace, Bibliothèque Universelle de Genève, No. 82, October 1842. In: R. Taylor, (Ed.), *Scientific memoirs, selections from transactions of foreign academies and learned societies and from foreign journals* (Vol. 3, Article XXIX). FSA (1843).
van de Lune, J., te Riele, H. J. J., & Winter, D. T. (1986). On the zeros of the Riemann zeta function in the critical strip, iv. *Mathematics of Computation, 46*(174), 667–681.
Lück, R. (2000). Dürer-Kepler-Penrose, the development of pentagon tilings. *Materials Science and Engineering A*, 294–296.
Machadoa, T., & Luchkob, Y. (2021). Multidimensional scaling and visualization of patterns in distribution of nontrivial zeros of the zeta-function. *Communications in Nonlinear Science and Numerical Simulation, 102*, 105924.
Mackay, A. L. (1990). Crystals and fivefold symmetry. In: I. Hargittai (Ed.), *Quasicrystals, networks, and molecules of fivefold symmetry*. VCH.
Mackay, A. L. (1981). De nive Quinquangula — On the Pentagonal Snowflake, Kristallogafiya 26, 1981, 910–919. *Soviet Physics Crystallography, 26*(1981), 517–522.
Mandolesi, A. L. G. (2019). Quantum fractionalism: The Born rule as a consequence of the complex Pythagorean theorem. 1905.08429. *Physics Letters A, 384*(28), 126725.
Mattle, K., Weinfurter, H., Kwiat, P. G., & Zeilinger, A. (1996). Dense Coding in Experimental Quantum Communication, *Physical Review Letters, 76*, 4656.
Mano, M. M. (1988). *Computer engineering: Hardware design*. Prentice Hall.
Mermin, N. D. (1990). Simple unified form for the major no-hidden-variables theorems. *Physics Review Letters, 65*(27), 3373–3376.
Mermin, N. D. (1993). Hidden variables and the two theorems of John Bell. *Reviews of Modern Physics, 65*, 803–815.
von Meyenn, K. (1985) Wolfgang Pauli: Wissenschaftlicher Briefwechsel mit Bohr, Einstein, Heisenberg u.a. Band II: 1930–1939. Springer.

von Meyenn, K. (2010). *Eine Entdeckung von ganz außerordentlicher Tragweite: Schrödingers Briefwechsel zur Wellenmechanik und zum Katzenparadoxon*. Springer.

Miller, A. I. (1996). *Insights of genius, imagery and creativity in science and art*. Springer.

Milnor, J. (2000). The Poincaré conjecture. https://www.claymath.org/sites/default/files/poincare.pdf

Milnor, J. (2003). Towards the Poincaré conjecture and the classification of 3-manifolds. *Notices of the AMS, 50*(50), 1226–1233.

Musk, E. (1990) Neuralink, An integrated brain-machine interface platform with thousands of channels. bioRxiv: 703801.https://doi.org/10.1101/70380

von Neumann, J. (1932). *Mathematische Grundlagen der Quantenmechanik*. Julius Spinger.

Nielsen, M. A., & Chuang, I. L. (2010). *Quantum computation and quantum information*, 10th Anniversary Edition, CUP. Cambridge University Press.

Oriti, D. (2006). Disappearance and emergence of space and time in quantum gravity. In: D. Rickles (Ed.), *Studies in history and philosophy of modern physics*. Special issue "Quantum Gravity". arXiv:1302.2849.

Padua, S. (2015). *The thrilling adventures of Lovelace and Babbage—The (Mostly) true story of the first computer*. Pantheon.

Pauli, W. (1955). Die Wissenschaft und das abendländische Denken, Wolfgang Pauli Nachlass im CERN-ARCH-PMC-06-291, WB 4/IVA, letter [2707]. http://cds.cern.ch/record/96387/files/CERN-ARCH-PMC-06-291.pdf?version=1.

Pauka, S., et al. (2021). A cryogenic CMOS chip for generating control signals for multiple qubits. *Nature Electronics, 4*, 64–70. https://doi.org/10.1038/s41928-020-00528-y

Penrose, R. (1974). The role of aesthetics in pure and applied mathematical research, southend-on-sea. *Bulletin of Mathematical Analysis and Applications, 10*, 266–271.

Plato. (2003). *Timaios*. Reclam.

Poincaré, H. (1882). Theorie des Groupes Fuchsiens. *Acta Mathematica, 1*, 1–63.

Poincaré, H. (1895). Analysis Situs, Journal de l'École Polytechnique, 11e Série—Premier Cahier, Paris. (Transl. by J. Stillwell: Papers on Topology, Analysis Situs and Its Five Supplements, Henri Poincaré, 2009).

Poincaré, H. (1892). Sur l'Analysis situs, note de Henri Poincare. *Comptes Rendus Des Seances De L'academie Des Sciences, 115*, 633–636.

Poincaré, H. (1900). Second supplement to analysis situs. *Proceedings of the London Mathematical Society, 32*, 277–308.

Poincaré, H. (1904). Cinquième complément à l'analysis situs, Rend. Circ. Mat.: Palermo, 18, 45–110.

Poincaré, H. (1912). Pourquoi l'espace á trois dimensions. *Revue De Métaphysique Et De Morale, 4*, 483–504.

Polchinski, J. (1995). Dirichlet-Branes and Ramond-Ramond charges. arxiv.org/abs/hep-th/9510017.

Pool, R. (2020). A new kind of quantum. https://spie.org/news/photonics-focus/novdec-2020/a-new-kind-of-quantum?SSO=1.

Price, H. (2013). Taming the quantum spooks. https://aeon.co/essays/can-retrocausality-solve-the-puzzle-of-action-at-a-distance.

Pusey, M. F., Barrett, J., & Rudolph, T. (2012). On the reality of the quantum state. arXiv:1111.3328

Quehenberger, R. (2011). QC expert talk with Gerard 't Hooft (Utrecht University) 12 November 2011, Vienna. http://vimeo.com/34081638

Quehenberger, R., Weibel, P., Rauch, H., Katzgraber, H., & Friemel, R. (2012). A new digital 3-D dynamic geometry for the visualization of complex number space. In: *Proceedings of International Conference on Computational Electronics and Nanotechnology (ICOCENT-2012) Conference*, March 2012. Jaipur (pp. 287–292).

Quehenberger, R. (2012a). Quantum cinema expert-talk with Basil Hiley and Helmut Rauch. vimeo.com/50374353.

Quehenberger, R., & Weibel, P. (2012). Recording of "Quantum Fluxus Dada Duett" with voice of Erwin Schrödinger (recording "Do electrons think?" by courtesy of Ruth Braunitzer) Live stream with Zabelka, M. (violin) as ARTs Birthday present 2012, Ö1 Kunstradio. http://www.kunstradio.at/2012A/22_01_12.html

Quehenberger, R. (2012b). A reflection on theories of light. In: A. Khrennikov, H. Atmanspacher, A. Migdall, & S. Polyakov (Eds.), *Quantum theory: Reconsideration of foundations* 6. AIP *Conference Proceedings 1508* (pp. 459–463).

Quehenberger, R. (2013). Quantum information traced back to ancient Egyptian mysteries. *Technoetic Arts: A Journal of Speculative Research, 11*(3), 319–334.

Quehenberger, R. (2014a). A newly found golden heptahedron named epitahedron. In: G. Darvas (Ed.) *Journal Symmetry: Culture and Science*. Symmetrion no. 25, 177–192.

Quehenberger, R. (2014b). A new visualization of the homology sphere. ICM 2014. In: *International Congress of Mathematicians*, Seoul (KR) Poster presentation, P5 #29.

Quehenberger, R. (2014c). Epitadodecahedron, film presented at IMAGINARY at ICM 2014 & International Conference of Mathematicians, ICM Seoul 2014. www.imaginary.org/film/the-epita-dodecahedron-visualizing-poincares-dodecahedral-space

Quehenberger, R. (2016). Description of the cover-images: The Epitahedron. In: G. Darvas (Ed.), *Symmetry: Culture and science* (Vol. 27(1), pp. 56–58). Symmetrion.

Quehenberger, R. (2017) A Proposal for a Psi-ontological model based on 5-dimensional geometry. In: *QCQMB Workshop: Quantum Contextuality in Quantum Mechanics and Beyond*, Prague 2017, http://www.psych.purdue.edu/~ehtibar/workshop/files/Poster_prag_Q CQMB%202017_RCZQ_A0_corr.png

Quehenberger, R. (2018). A quest for an epistemic reset in higher dimensional space, *Entropy, Best Poster Awards at Linnaeus Conference: Towards Ultimate Quantum Theory (UQT)*. https://res.mdpi.com/data/1st-place-r.c.-z.-quehenberger.pdf

Quehenberger, R. (2019) Enfolding the 5th dimension: With origami to the 3D representation of the Penrose kites & darts pattern, symmetry: art and science, 2019. In: *11th Congress and Exhibition of SIS Special Theme: Tradition and Innovation in Symmetry*, Katachi 形 Kanazawa, Japan, November 25–30, 2019.

Quehenberger, R., Kern L., & Tirelli, D. dC. Rubin, S. (2022). GAIA 5.0—A five-dimensional geometry for the 3D visualization of Earth' climate complexity. In: T. Lanza (Ed.) S. Illingworth et al., (Eds.) *Special issue: Five years of earth sciences and art at the EGU (2015–2019)*. Geoscience Communication, Copernicus, 2020–2021.

Quehenberger, R. (2021). Interactive comment on GAIA 5.0—A five-dimensional geometry for the 3D visualization of Earth' climate complexity. Geoscience Communication Discussion. https://doi.org/10.5194/gc-2020-27-AC3

Quehenberger, R. (2022, forthcoming) Kepler & die 5. Dimension, (Transl.: Kepler & the 5th Dimension, forthcoming) Aichbergiana, Handreiche zum Alltag, No. 27.

Riemann, G. F. B. (1859). Über die Anzahl der Primzahlen unter einer gegebenen Grösse. Monatsberichte der Berliner Akademie, 671–680.

Rauch, H., Zeilinger, A., Badurek, G., Wilfing, A., Bauspiess, W., & Bonse, U. (1975). Verification of coherent spinor rotation of fermions. *Physics Letters A, 54*(6), 425–427.

Regev, O. (2004). Quantum computation and lattice problems. *SIAM Journal on Computing, 33*(3), 738–760.

Riesz, M. (1916). Sur l'hypothèse de Riemann. *Acta Mathematica, 40*, 185–190.

Rudolph, T. (2017). *Q is for quantum*. Kindle Edition by Terry Rudolph.

Rupprich, H. (Ed.) (1969). Dürer. Schriftlicher Nachlass, Berlin: Deutscher Verlag für Kunstwissenschaft, 1956/1966/1969, 3rd vol. 1969.

Schleich, W. P., Bezděková, I., Kim, M. B., Abbott, P. C., Maier, H., Montgomery, H. L., & Neuberger, J. W. (2018). Equivalent formulations of the Riemann hypothesis based on lines of constant phase. *Physica Scripta, 93*(6), [065201]. https://doi.org/10.1088/1402-4896/aabca9

Schopenhauer, A. (1819). Die Welt als Wille und Vorstellung (Engl.: The World as Will and Representation), in: Arthur Schopenhauers sämtliche Werke in sechs Bänden (Ed. Eduard Grisebach), Verlag von Philipp Reclam jun., 1892.

Schönhage, A., & Strassen, V. (1971). Schnelle Multiplikation grosser Zahlen. *Computing, 7*, 281–292.

Schrödinger, E. (1935). Die gegenwärtige Situation in der Quantenmechanik, Naturwissenschaften, 23:48, 807–812, 823–828; 844–849.

Schrödinger in a letter to Born, 14 April 1926, in Meyenn, 2010, p. 201) "... sich irgendwie mit der Atommechanik von Heisenberg, Jordan und mir vereinigen lassen. Ich kann mir natürlich ungefähr denken, wie das geht, denn partielle Differentialgleichungen sind ganz allgemein äquivalent mit quadratischen Formen von unendlich vielen Variabeln; aber ich wüsste doch gern genauer, wie es zusammenhängt und wäre Ihnen sehr dankbar, wenn..."

Shannon, C. E (1948). *A Mathematical Theory of Communication, 27*, 379–423 & 623–656.

Shechtman, D., Blech, I., Gratias, D., & Cahn, J. W. (1984). Metallic phase with long-range orientaional order and no translational symmetry. *Physical Review Letters, 53*(20), 1951–1953.

Shor, P. W. (1997). Polynomial-time algorithms for prime factorization and discrete logarithms on a quantum computer. *SIAM Journal on Computing, 26*(5), 1484–1509.

Soiguine, A. (2017). Quantum computing in geometric algebra formalism: Light Beam Guide Implementation Future Technologies Conference (FTC) 29–30 November 2017, Vancouver, Canada

Svozil, K. (2000). The information interpretation of quantum mechanics. arxiv.org/abs/quant-ph/0006033v2.

Threlfall, W. & Seifert, H. (1931).Topologische Untersuchung der Diskontinuitätsbereiche endlicher Bewegungsgruppen des dreidimensionalen sphärischen Raumes, *Mathematische Annalen 104*(1), 1–70.

Thurston, W. P. ([1980] 2002). The geometry and topology of three-manifolds, notes distributed by Princeton University, Electronic version 1.1, March 2002. http://www.msri.org/publications/books/gt3m/

Toole, B. A. (2010). Ada, the Enchantress of Numbers: Poetical Science, Sausalito CA: Critical Connection.

Tydecks, W. (2020). Re-entry nach Spencer-Brown, presentation at formlabor, Berlin, 31 January 2020.

Uchida, G., Bertlmann, R. A., & Hiesmayr, B. C. (2015) Entangled entanglement: The geometry of GHZ states. *Physics Letters A, 379*, 2698, arxiv.org/abs/1410.7145.

Vasyliunas, V. M. (2005). Relation between magnetic fields and electric currents in plasmas. *Annales Geophysicae, 23*(7), 2589–2597.

Vieira, L. V., Palazzo, R., & Faria, M. B. (2006). On the arithmetic Fuchsian groups derived from quaternion orders VI. In: *International Telecommunications Symposium (ITS2006)*. https://doi.org/10.1109/ITS.2006.4433342

Verhulst, F. (2012). Henri Poincaré: Impatient Genius, Springer US.

Voisin, B., et al. (2020). Valley interference and spin exchange at the atomic scale in silicon. *Nature Communications*. https://doi.org/10.1038/s41467-020-19835-1

Walleczek, J., Grössing, G., Pylkkänen, P., & Hiley, B. (2019). Emergent quantum mechanics: David Bohm centennial perspectives. *Entropy, 21*(2), 113. https://doi.org/10.3390/e21020113

Weibel, P. (2005). Time slot—Geschichte und Zukunft der apparativen Wahrnehmung vom Phenakistiskop bis zum Quantenkino, Cologne: Walther König.

Weihs, G., Jennewein, T., Simon, C., Weinfurter, H., & Zeilinger, A. (1998). Violation of Bell's inequality under strict Einstein locality conditions. *Physical Review Letters, 81*, 5039–5043.

Wheeler, J. A. (1983) Law without law. In: J. A. Wheeler & W. H. Zurek (Eds.). *Quantum theory and measurement*. Princeton University Press.

Wheeler, J. A. (1989). Information, physics, quantum: the search for links. In: *Proceedings III International Symposium on Foundations of Quantum Mechanics* (pp. 354–368). Tokyo.

Wilczek, F. (2012). Quantum time crystals. *Physical Review Letters, 109*(16): 160401. arXiv:1202. 2539

Wüthrich, C. (2014). Raiders of the lost spacetime. In: D. Lehmkuhl (Ed.), *Towards a theory of spacetime theories*. Birkhäuser. arXiv:1405.5552v1

Yau, S.-T., & Nadis, S. (2010). *The shape of inner space, string theory and the geometry of the universe's hidden dimensions*. Basic Books.

Zeilinger, A. (1999). A foundational principle for quantum mechanics. *Foundations of Physics, 29*, 631–643.

Zeilinger, A., Bernstein, H. J., & Horne, M. A. (1994). Information transfer with two-state, two-particle quantum systems. *Journal of Modern Optics, 41*, 2375–2381.

Zimba, J., & Penrose, R. (1993). On Bell non-locality without probabilities: More curious geometry. *Studies in History and Philosophy of Science.* SHPSB524

How to Make Qubits Speak

Bob Coecke, Giovanni de Felice, Konstantinos Meichanetzidis, and Alexis Toumi

Abstract This is a story about making quantum computers speak, and doing so in a quantum-native, compositional and meaning-aware manner. Recently we did question-answering with an actual quantum computer. We explain what we did, stress that this was all done in terms of pictures, and provide many pointers to the related literature. In fact, besides natural language, many other things can be implemented in a quantum-native, compositional and meaning-aware manner, and we provide the reader with some indications of that broader pictorial landscape, including our account on the notion of compositionality. We also provide some guidance for the actual execution, so that the reader can give it a go as well.

Keywords Quantum natural language processing · Compositionality · ZX-calculus · String diagrams · DisCoPy software

1 Networks of Words

What is a word?

In particular, what is the meaning of a word? This could be what you find in a dictionary. Then again, before there were dictionaries people also used words, and so did people that could not read. In fact, most words we know we didn't look up in a dictionary. We probably learned them by hearing them used a lot. This idea of meaning is closer to how machines learn meanings of words today: they learn them

B. Coecke (✉) · G. de Felice · K. Meichanetzidis · A. Toumi
Cambridge Quantum/Quantinuum, 17 Beaumont St, Oxford OX1 2NA, UK
e-mail: bob.coecke@cambridgequantum.com

G. de Felice
e-mail: giovanni.defelice@cambridgequantum.com

K. Meichanetzidis
e-mail: k.mei@cambridgequantum.com

A. Toumi
e-mail: alexis.toumi@cambridgequantum.com

© Springer Nature Switzerland AG 2022
E. R. Miranda (ed.), *Quantum Computing in the Arts and Humanities*,
https://doi.org/10.1007/978-3-030-95538-0_8

from the context in which they are used, an idea coined by Wittgenstein as "meaning is use", and expressed by Firth's derived dictum (Firth, 1957):

"You shall know a word by the company it keeps".

What is a sentence?

In particular, what is the meaning of a sentence? A sentence is not just a "bag of words",[1] but rather, a kind of network in which words interact in a particular fashion. In fact, in a very particular fashion, given that when we hear a sentence that we never heard before, provided we do understand the words that occur in it, then we surely also understand that sentence. That's exactly why we don't have dictionaries for sentences—besides the fact a 'sentence dictionary' would take up an entire library, if not more. There is an important academic question here:

● How do we deduce the meaning of a sentence given the meanings of its words?

We can also ask the converse question:

● Can we infer the meanings of words in sentences, from the meanings of sentences?

Both turn out to also be essential practical questions.

Some 10 years ago, in (Coecke et al., 2010), BC Mehrnoosh Sadrzadeh and Steve Clark started to draw networks in order to address the first of these questions. These networks look like this:

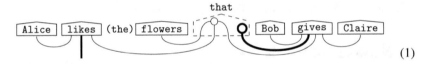

$$(1)$$

The usual technical term for these networks is "string diagram" (see e.g. Baez & Stay 2011). These string diagrams provided a straight-forward manner for how meanings of words combine in order to produce the meaning of a sentence, even resulting in a cover-heading feature in New Scientist.[2]

In order to better see how these string diagrams do so, let's consider a simpler example:

$$(2)$$

The idea here is that the boxes represent the meanings of words and that the wires are channels through which these meanings can be transmitted. So, in the example above, the subject Alice and the object Bob are both sent to the verb hates, and

---

[1] Here, 'bag' is a widely used technical term, referring to the fact that in many 'natural language processing' applications grammatical structure has been entirely ignored for many years (Harris, 1954).

[2] https://www.newscientist.com/article/mg20827903-200-quantum-links-let-computers-understand-language/.

together they then make up the meaning of the sentence. This idea scales to much bigger sentences like the one depicted above, and even to large text made up of multiple sentences (Coecke, 2019).

This flow of words in sentences can be traced back to work originally started in the 1930s by Adjukiewicz (1935) and later by Bar-Hillel (1953), Chomsky (1956, 1957) and Lambek (1958, 2008), among others. What they did was unifying grammatical structures across different languages as a single mathematical structure. An overview of these developments from the perspective of string diagrams is in Coecke (2013).

2 Language Is Quantum-Native

One particularly interesting aspect of this graphical framework for linguistics was that the string diagrams were inherited from previous work that provided quantum theory with a network-like language (Coecke, 2005). This work was accumulated eventually in a 900-page book (Coecke & Kissinger, 2017).

In summary, a direct correspondence was established between, on the one hand, the meanings of words and quantum states, and on the other hand, grammatical structures and quantum measurements. This is illustrated in Fig. 1.

Obviously this led to the question: Can one make quantum computers handle natural language? This was first proposed in a paper by Will Zeng and BC in 2016 (Zeng & Coecke, 2016), creating a new paradigm for natural language processing (NLP) in a quantum computing context.

The idea of making quantum computers process natural language is just not only incredibly cool but also a very natural thing to do for the reasons indicated above. More researchers started to take an interest, and at some point Intel supported a basic attempt to develop some of the ideas contained in (Zeng & Coecke, 2016) on their quantum simulator (Intel News Release, 2019; O'Riordan et al., 2020).

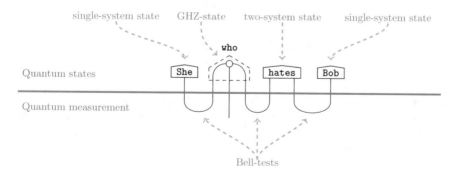

Fig. 1 Illustration of how word meanings can be interpreted as quantum states, and grammatical structure in terms of quantum measurement

However, there were some significant challenges to the proposal. Most importantly, there weren't any sufficiently capable quantum computers that could implement the NLP tasks proposed. Additionally, an assumption was made that one could encode word meanings on the quantum computer using quantum random access memory (QRAM) (Giovannetti et al., 2008), which to this day, and despite theoretical progress and experimental proposals, remains a distant possibility.

3   Our New Attempt

In the past year, we have been examining ways to use existing NISQ (= Noisy Intermediate Scale Quantum) devices, in the first instance one of IBM's quantum devices, for NLP.

The string diagrams as depicted above can't be interpreted directly by IBM's machine, which instead, needs something in the form of a 'quantum circuit'. Natural language when we map it to a 'quantum circuit skeleton' now looks like this (Coecke et al., 2020):

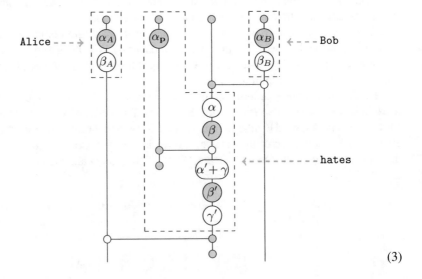

$$(3)$$

In this picture as well as in our experiments we used the 'ZX-calculus' (Coecke & Duncan, 2011) for drawing quantum circuits, which is again part of the same string diagram language of quantum theory—we give the translation of ZX-calculus circuit components in standard circuit gates in Fig. 2.

X-preparation	X-phase gate	Z-phase gate	CNOT gate	X-measurement
$\begin{pmatrix} 1 \\ 0 \end{pmatrix}$	$\begin{pmatrix} 1 & 0 \\ 0 & e^{i\alpha} \end{pmatrix}$	$H \circ \begin{pmatrix} 1 & 0 \\ 0 & e^{i\alpha} \end{pmatrix} \circ H$	$\begin{pmatrix} 1 & 0 & 0 & 0 \\ 0 & 1 & 0 & 0 \\ 0 & 0 & 0 & 1 \\ 0 & 0 & 1 & 0 \end{pmatrix}$	$\begin{pmatrix} 1 & 0 \end{pmatrix}$

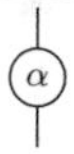

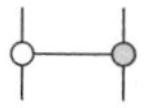

Fig. 2 Translation of ZX-calculus circuit components in standard circuit gates. For a short tutorial see (Coecke & Duncan, 2012), or for a more extensive ones see (Coecke & Kissinger, 2017; van de Wetering, 2020)

The building blocks of the ZX-language are these 'spiders':

$$(4)$$

and rules for computing and reasoning look as follows:

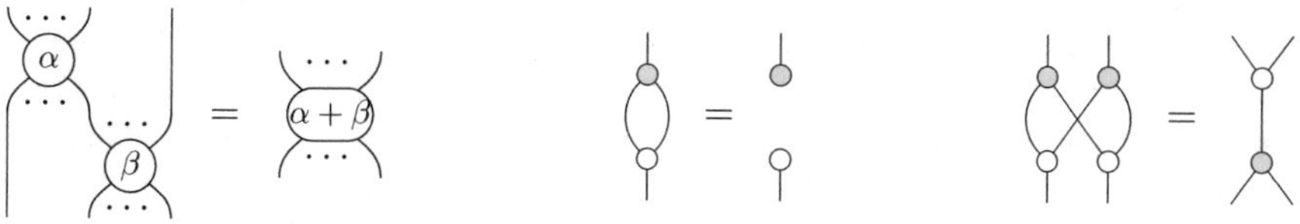

$$(5)$$

So we are genuinely dealing with an entirely pictorial language, and in fact, recently it was shown in (Hadzihasanovic, 2018) that all the equations that can be derived using the usual quantum mechanical formalism, can also be derived using only pictures!

The key part of the passage from a sentence diagram representing grammatical structure like (2) to a quantum circuit uses the ZX-calculus in a fundamental way. For example, the fact that the CNOT-gate arises by sticking together two spiders as follows:

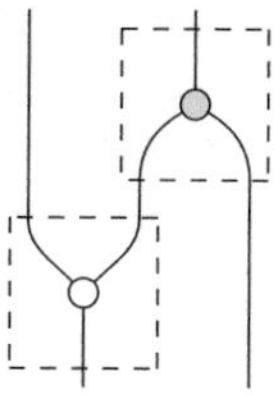

$$(6)$$

is essential. Using this decomposition, the circuit arises as follows from (2):

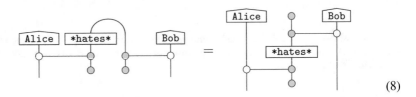

$$(7)$$

We could then also decide to reduce the number of qubits as follows:

$$(8)$$

From this (3) arises when parametrising the word meanings by gates. A detailed discussion of the entire passage is in (Coecke et al., 2020).

In the form (3), natural language can be implemented on NISQ devices, and of course, will still work well as these devices scale in terms of size and performance. Crucially, our solution provides a way forward in the absence of QRAM. By employing quantum machine learning we do not directly encode the meanings of words, but instead quantum gates—those in the circuit (3) carrying Greek letters—learn their meanings directly from text (Meichanetzidis et al., 2020). By way of analogy with classical machine learning, in quantum machine learning we can indeed use quantum circuits instead of classical neural networks in order to learn patterns from data (Ma et al., 2019; Benedetti et al., 2019). Interestingly, neural network architectures are the state-of-the-art in classical NLP, but the majority of methods do not take advantage of grammatical structures. In contrast, we saw that our approach to natural language naturally accommodates both grammar and meaning.

Using our framework, once the meanings of words and phrases are encoded as quantum gates, we are able to encode the meaning of grammatical sentences on quantum hardware. Posing a question to the quantum computer, constructed by the vocabulary and grammar the quantum computer has learned, it returns the answer— as illustrated in Fig. 3.[3]

Naturally, we next turned our attention toward the design and execution of an experiment that is non-trivial, not least of all since our design is predicated on the program being scalable. This means that the dimension of the meaning space grows significantly with the number of qubits available whilst the size of the circuits dictated by the grammar does not grow too large with the size of the sentence.

---

[3] Earlier theoretical work by our team towards question-answering includes (Coecke et al., 2018; Felice et al., 2019).

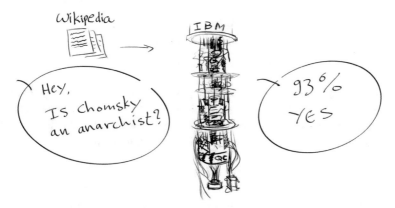

Fig. 3 A quantum computer putting an end to social media disputes

4 The Actual Experiment

Once a quantum circuit is created from a sentence, it needs to be evaluated in order to compute the meaning of that sentence. We may choose to perform this evaluation on a classical computer, where we employ state-of-the-art methods for performing the costly task of multiplying exponentially big matrices, or, we may choose to implement the circuit on a quantum computer. This is of course what we decided to do. A schematic presentation of the experiment on IBM quantum hardware is in Fig. 4.

As we see in the quantum circuit (3), each of the parts of speech (subject, object, verb) is in the quantum circuit a function of some parameters. For example, there are sets of parameter values $\alpha_A, \beta_A, \alpha_B, \beta_B, \alpha, \beta, \ldots$ such that:

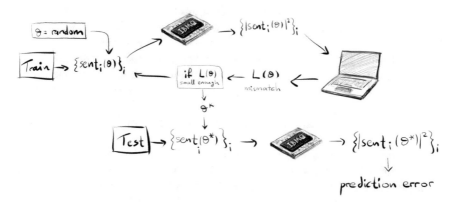

Fig. 4 A schematic presentation of the actual experiment on IBM quantum hardware, depicting both the training and testing parts, with $\theta = \{\alpha_A, \beta_A, \alpha_B, \beta_B, \alpha, \beta, \ldots\}$

$$\texttt{subject}(\alpha_A, \beta_A) = \texttt{Alice}$$
$$\texttt{object}(\alpha_B, \beta_B) = \texttt{Bob}$$
$$\texttt{verb}(\alpha, \beta, \ldots) = \texttt{hates}$$

The values are determined empirically by a text corpus and are then used to answer questions about the corpus. In order to ensure that our experiment can be executed effectively on near-term NISQ devices, but at the same time be complex enough to be interesting, we chose a vocabulary of a few words, for example:

$$\{\texttt{Alice, Bob, loves, hates, rich, silly}\}$$

and generated not some, but all grammatical sentences from their combinations. From these sentences, we created their corresponding parameterised circuits. Moreover, we interpret the language diagrams such that the sentence space is one-dimensional, i.e. just a number indicating the truth-value of the sentence:

$$0 \le \texttt{sentence}(\alpha_A, \beta_A, \alpha_B, \beta_B, \alpha, \beta, \ldots) \le 1$$

A value close to 1 represents "*true*" and a value close to 0 indicates "*false*".
The labeled toy corpus would look like:

$Corpus = \{(\texttt{Alice loves Bob}, false), (\texttt{Bob is silly}, true), (\texttt{Alice is rich}, true), \ldots\}$

Now that we have our corpus of sentences we split the corpus:

$$Corpus = Train \cup Test$$

in a training set $Train$ and a test set $Test$. Sentences in the training set $Train$ are used to do supervised quantum machine learning in order to learn the parameters that result in the correct measurement of the truth-labels. In this way, the parameters for the circuits that prepare the meaning states for nouns {Alice, Bob}, verbs {is, loves, hates}, and adjectives {rich, silly}, are learned.

The scheme for learning the parameters for words in sentences in the training set is as follows. The circuit of a sentence in the training set is evaluated for the current set of parameters on the quantum computer. By sampling measurement outcomes we estimate:

$$|\texttt{sentence}(\alpha_A, \beta_A, \alpha_B, \beta_B, \alpha, \beta, \ldots)|^2$$

This number is read by a classical computer that checks how well this matches the desired truth label of the sentence. If there is a mismatch, our system updates the parameters so that the quantum computer may evaluate an updated circuit. Iterating this procedure until the parameters converge and all truth labels are reproduced for the sentences in the training set.

After training, sentences in $Test$ are used to estimate how well the truth labels of new sentences, i.e. not in $Train$, are inferred. These new sentences share the same vocabulary as the sentences used for training, but they are grammatically and semantically different.

Note that finding the optimal sequence of updates is, in general, a so-called 'hard optimization problem', so it is important that our quantum meaning space is well designed and allows the learning to be as tractable as possible. This design feature is critical.

With this learning framework, we can now ask questions of the quantum computer, as long as the questions are grammatical sentences expressed in terms of the vocabulary and grammar used during training. We are pleased to add that in our experiment, questions can, in fact, be posed as compositions of already learned sentences. For example, we can use the relative pronoun who (Sadrzadeh et al., 2013) (which we model in terms of CNOT gates within a quantum circuit) and ask:

Does Bob who is silly love Alice who is rich?

This is the same as asking whether:

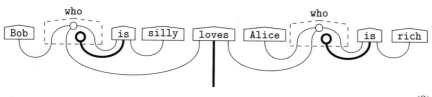

$$(9)$$

is $true$. This amounts to evaluating a bigger quantum circuit than the one that the model has been trained on. However, because the model was trained on the same grammar and vocabulary as used to express the question, we get the expected truth label, false in this case.

One critical enabling feature of our experiment is making computers understand the pictures that we have been drawing. For this we used GdF and AT's DisCoPy Python library (Felice et al., 2020). Another critical feature is effective and efficient compiling and optimization. In order to run a circuit on a quantum device, it needed to be compiled. Compiling entails morphing the circuit such that quantum operations are expressed in terms of device-native operations, as well as accommodating for the quantum processor's restricted connectivity. For this task, we used Cambridge Quantum Computing's quantum software development platform, t|ket⟩ (Sivarajah et al., 2020) which again crucially makes use of ZX-calculus.

5 QNLP

We've done it, it exists, so let's give the kid a name: 'quantum natural language processing', or QNLP in short. A much more detailed account on QNLP as well as the data for a larger-scale experiment can be found in (Coecke et al., 2020; Meichanetzidis et al., 2020; Lorenz et al., 2021). With the successful execution of QNLP on quantum hardware, and the great promise for the future, Cambridge Quantum Computing has now resourced an Oxford based team dedicated to QNLP, and structure-driven quantum AI more generally, evidently including each of us. Let's single out a few special features of QNLP.

Firstly, as already indicated above, QNLP is quantum-native. What we mean by that is that the model of natural language that we employ is a quantum model (Clark et al., 2014). Not just because we think that having a quantum model is particularly cool, but because it was effectively the case that the most economic manner for bringing language meaning and grammatical structure together was provided by the 'categorical quantum mechanics' framework (Abramsky & Coecke, 2004; Coecke, 2005; Coecke & Kissinger, 2017).[4] The immediate consequence of this is that when we try to 'simulate' QNLP on a classical computer, it would be exponentially expensive, just as it is the case for simulation quantum systems on a classical computer. Hence, QNLP truly loves being on quantum hardware.

Secondly, it is fair to say that it is meaning-aware, at the very least, it is definitely more meaning-aware than the current deep learning based NLP implementations. Indeed, the language pictures clearly indicate how the meanings flow in a meaningful way. We say a bit more about this below in Sect. 9.

Thirdly, as already indicated in (Zeng & Coecke, 2016), there are many more advantages to using quantum computers, most notably, we get quantum speed-up for a large variety of NLP tasks, like question-answering. This space of opportunities is in fact still largely unexploited, and the mission of our QNLP team is to do just that, in as many possible manners as we can.

6 Beyond Language

We are now well on the way to make qubits speak. But can we make them do other things as well? We are pretty sure we can, a first thing could be to make qubits represent all of the inputs that we get through our fleshy physical embodiment, like taste, smell, vision, hearing. The technical term for the kinds of spaces that represent these sensory modes is 'conceptual spaces'. They were introduced by Peter Gärdenfors (Gärdenfors, 2000, 2014). There already are existing frameworks that represent the input of these senses in a manner that exactly matches our language

---

[4] Categorical quantum mechanics is really just a fancy name for doing quantum mechanics in terms of pictures.

diagrams, and hence allow for these sensory inputs to interact with each other (Bolt et al., 2018).

These conceptual spaces are convex, and fit well with some other work within the framework of our natural language diagrams, namely on the use of the quantum mechanical concept of density matrices, which also form convex spaces (Ludwig, 1985; Barrett, 2007). This additional convex structure allows one to bring many more linguistic features into play, like word ambiguity. For example, the word queen is clearly ambiguous, as it can be a royal, a rock band, a bee, a chess piece, a playing card, etc. In order to represent that ambiguity, we simply add all the different meanings in order to form a density matrix (Piedeleu et al., 2015):

$$
\begin{aligned}
\rho_{\text{queen}} = &\ |\text{queen-royal}\rangle\langle\text{queen-royal}| \\
&+ |\text{queen-band}\rangle\langle\text{queen-band}| \\
&+ |\text{queen-bee}\rangle\langle\text{queen-bee}| \\
&+ |\text{queen-chess}\rangle\langle\text{queen-chess}| \\
&+ |\text{queen-card}\rangle\langle\text{queen-card}| \\
&+ \ldots
\end{aligned}
$$

Alternatively, these density matrices can also be used to express hierarchical relationships between words (Bankova et al., 2019), for example, lion, tiger and cheeta, are examples of big cat, which further generalises to mammal, which generalises to vertebrate, etc. Using density matrices we encode this hierarchy as follows:

$$
\begin{aligned}
\rho_{\text{lion}} &= |0\rangle\langle 0| \\
\rho_{\text{tiger}} &= |+\rangle\langle +| \\
\rho_{\text{cheeta}} &= |-\rangle\langle -| \\
\rho_{\text{big cat}} &= |0\rangle\langle 0| + |1\rangle\langle 1| \\
\rho_{\text{mammal}} &= |0\rangle\langle 0| + |1\rangle\langle 1| + |2\rangle\langle 2| \\
\rho_{\text{vertebrate}} &= |0\rangle\langle 0| + |1\rangle\langle 1| + |2\rangle\langle 2| + |3\rangle\langle 3|
\end{aligned}
$$

In (Coecke & Meichanetzidis, 2020; Lewis, 2019a, b; De las Cuevas et al., 2020; Meyer & Lewis, 2020) this use of density matrices in language is further elaborated upon.

7 String Diagrams Are Everywhere

Sticking with the theme of big cats, what's the difference between a tiger and a lion? In particular, why does a tiger have stripes, while a lion doesn't (Fig. 5).

A more traditional scientist will tell you that these days we understand all of that perfectly well. Before, if we dissected these animals, we found exactly the same

Fig. 5 Two close relatives with very different coats

organs. If we do some smaller scale animal chopping, we end up with cells that again have exactly the same structure. However, digging even further, and hitting the molecular level, we encounter DNA, and then we truly understand the difference.

Really? Some obscure humongously large code that is impossible to grasp by any human 'explains' the difference? It explains as much the difference between these two animals as the codes of two computer programs written in terms of 0's and 1's explain the difference between accounting software and the operating system of your smart phone.

That said, most of the exact sciences have adopted this perspective of understanding things by breaking them down to smaller things, like elementary particles in particle physics, and the elements of a set in set-theory. In contrast, in the arts and humanities things really get their meanings by putting them in a context. For example, in the social sciences properties are mostly thought of in terms of behaviours, rather than by trying to build a map of the inside of someone's brain, and most modern art has very little meaning without a context.

We think that we can all agree that for our tiger and our lion, what explains the difference is really the hunting process which each of them needs to do in order to survive, and their respective coats provide the best camouflage for doing so, given they hunt in very different environments. While of course these differences are encoded in the genes, it is the process of evolution that makes our tiger and our lion look as they do. The role of DNA in explaining the stripes of tigers is similar to the role of our parameters α_A, β_A, α_B, ... in computing the meaning of sentences. Indeed, learning the optimal parameters is an evolution-like process that depends on the environment of the words, i.e. on the training data that we feed the machine.

So the two views explained above are really about either focussing on components vs. context, and the latter is very much in line with Wittgenstein's "meaning is use". We are saying all of this because the language diagrams that we saw above are the mathematical incarnation of this way of thinking, for example:

$$(10)$$

Indeed, this way of thinking requires a completely different kind of mathematics, be it in the form of our diagrams, or its symbolic counterpart, so-called monoidal categories.

This is not the place to go into this new kind of mathematics, but the interested reader may want to check out the book (Coecke & Kissinger, 2017), which is probably the most comprehensive account on these language diagrams, and their communality with quantum theory. Shorter more informal accounts can be found in (Coecke, 2017). For the braver reader, the symbolic counterpart of all of this exists within the realm of category theory, and tensor categories in particular (Coecke & Paquette, 2011; Selinger, 2011). Meanwhile, string diagrams have started to pop up in a wide variety of disciplines, for example, in computer science (Abramsky, 1996; Pavlovic, 2013), where they have been for a while, in engineering (Bonchi et al., 2014; Baez & Fong, 2015), in economy and game theory (Ghani et al., 2018; Felice et al., 2020), and even in cognition (Tull & Kleiner, 2020; Signorelli et al., 2020).

We believe that this is the kind of mathematics that would also serve the social sciences very well. As it is quite new, most mathematics that one finds in the social sciences is statistics, which, in all fairness, is what one uses in absence of anything better.

With new mathematics also come new software tools. For this we are using the new very flexible DisCoPy software (Felice et al., 2020) which we already mentioned above, and continues to be further developed by AT, GdF and collaborators. We're inviting everyone to play around with this toolbox, and even better, help us out further developing DisCoPy—we may even reward you with a pie. For ZX-calculus specifically there is now also the PyZX software (Kissinger & van de Wetering, 2019), which communicates directly with DisCoPy. PyZX also helps us out with our QNLP experiments, by making the circuits that we feed into the quantum computer as simple as possible, a task for which ZX-calculus is now setting the state-of-the-art (Duncan et al., 2019; Cowtan et al., 2019; Beaudrap et al., 2020).

8 Compositionality

Above we saw how we can turn language diagrams into circuits. A typical thing about circuits is, of course, that you can compose different circuits together for forming bigger circuits. This is the principle of "compositionality", and in fact, string diagrams themselves are a result of this way of thinking. We can now, once we moved to circuits, compose several sentences together in order to produce larger text:

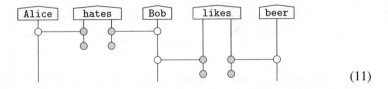

$$(11)$$

But why do we have to go to quantum circuits to do so? Indeed, it goes without saying that in absence of quantum circuits we can also compose sentences in order to form larger text, so there should be manner for doing so without having to dive in the realm of the quantum world.

This is indeed the case, when stopping halfway in the middle when passing from language diagrams to quantum circuits (Coecke, 2019). There we also find circuits that can be composed, and interestingly, some of the features specific to different languages (e.g. specific orderings of words like subject-verb-object in English) are gone (Coecke & Wang, 2020). So in a manner we end up with something that is more universal than language diagrams, and, for example, stands in a particularly close relationship with visual representations (Coecke, 2019). Here's the language circuit:

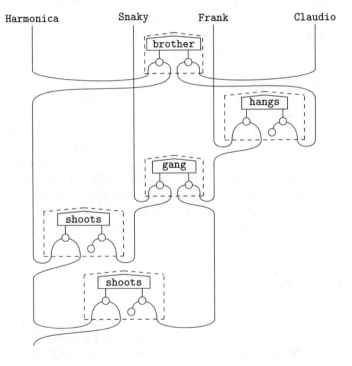

$$(12)$$

and here's the corresponding visual representation:

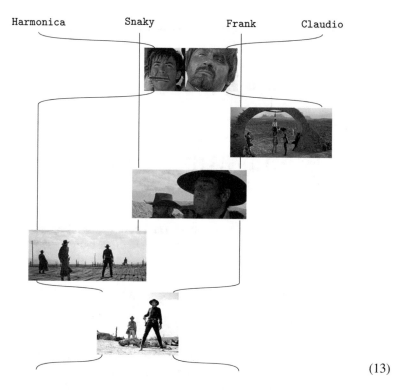

Harmonica   Snaky   Frank   Claudio

(13)

Maybe there is something deeper here. Maybe the origin of language should be looked for in the visual world, rather than in some symbolic abstract realm. The passage from these visual circuits to the more complex grammatical structure may actually boil down to forcing something that wants to live in two (or more) dimensions on a line, and force additional bureaucracy (or, bureau-crazy?) like the ordering of words, which ends up being different for different languages. Why forced into one-dimension? Since it seems to be hard for us humans to communicate verbally in any other way than in terms of one-dimensional strings of sounds.

In fact, the very success of string diagrams is that it allows things to live in two (or more) dimensions. Here's a demonstration of this fact. Consider the operations of parallel composition $\otimes$ (a.k.a. 'while') and sequential composition $\circ$ (a.k.a. 'after'). Typically, in textbooks, this structure is represented by one-dimensional strings of algebraic symbols, and then, equations are needed to express their interaction (a.k.a. bifunctoriality (Mac Lane, 1998)):

$$(g_1 \otimes g_2) \circ (f_1 \otimes f_2) = (g_1 \circ f_1) \otimes (g_2 \circ f_2)$$

On the other hand, representing it in terms of two-dimensional diagrams we have:

$$\left(\boxed{g_1} \otimes \boxed{g_2}\right) \circ \left(\boxed{f_1} \otimes \boxed{f_2}\right) = \left(\boxed{g_1}\ \boxed{g_2}\right) \circ \left(\boxed{f_1}\ \boxed{f_2}\right) = \begin{array}{cc}\boxed{f_1} & \boxed{f_2} \\ \boxed{g_1} & \boxed{g_2}\end{array}$$

$$\left(\boxed{g_1} \circ \boxed{f_1}\right) \otimes \left(\boxed{g_2} \circ \boxed{f_2}\right) = \left(\begin{array}{c}\boxed{f_1}\\\boxed{g_1}\end{array}\right) \otimes \left(\begin{array}{c}\boxed{f_2}\\\boxed{g_2}\end{array}\right) = \begin{array}{cc}\boxed{f_1} & \boxed{f_2}\\\boxed{g_1} & \boxed{g_2}\end{array}$$

That is, by using the two-dimensional format of diagrams, the symbolic equation is always satisfied, so it is not needed anymore. In other words, the equation was a piece of bureaucracy due to the one dimensional symbolic representation.

9 Meaning-Aware Relational Wholism

Let's end with a bit of philosophy, which puts our language diagrams within a broader context of different approaches that one may take towards machine learning. We'll again do this by means of a metaphor. Assume one wants to design a home. One way to do so would be to train a machine using many different images of homes, possibly with some indications of appreciation. Then, let the machine be creative and design a home based on all of that input.

And it comes up with an amazing design. But what dit it really do? It kind of carved everything out of one big block of stuff, and there will be forks and plates on the table, chairs at the table, pictures on the wall, since that's were they were in all of the training images. However, the machine is unaware of what the use is of each of these objects, and in particular that there is a very special relationship between the plate and the fork, and also with the chair at the table (and less so with the picture on the wall), and even more so, that there is a person who happens to be not in the picture, but who wants to exploit this special relationship with respect to the food they are cooking in the kitchen.

On the other hand, our language diagrams, when applied to a broader context, can capture these relationships, in contrast to the black-box approach that is taken within machine learning:

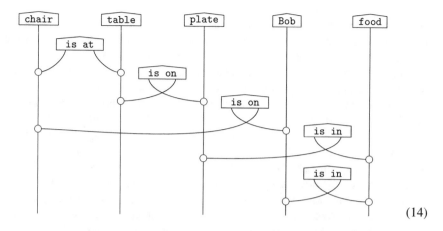

$$(14)$$

The kind of thinking encoded in our language diagrams, which is fundamentally wholistic while still aiming to grasp relationships between the parts, is what we mean by the title of this section.

What our language diagrams do is express how words interact with each other. Once we have established that structure, one thing that we can do is compute the meaning of a sentence from the meanings of words. The more interesting thing however, is to derive the meanings of words from the meaning of sentences. This is exactly what we did in our QNLP experiment: from the knowledge that the sentences in $Train$ were true, we figured out the optimal parameters $\alpha_A, \beta_A, \alpha_B, \beta_B, \alpha, \beta, \ldots$ which determine the meanings of the words.

This is really what this paper was all about, instantiated in order to make qubits speak. After this philosophy interlude...

10 ...Let's Get Playing!

First, we need some data to play with. For this we have two options: (1) we generate an artificial dataset by designing the grammar ourselves, this is the simplest option and the one we have implemented in (Coecke et al., 2020) or (2) we take some real-world text data (e.g. every sentence from Alice in Wonderland) and use a parser to construct the diagrams automatically for us. In any case, we begin our experiment with a collection of diagrams for sentences, with some annotation telling us whether each sentence is true or false (or whether or not it comes from Alice in Wonderland).

Next, we need to choose an ansatz for each type of word, i.e. an appropriate circuit shape. Instantaneous quantum polynomials (IQP) (Shepherd & Bremner, 2009) are a reasonable choice, since they are shallow circuits but are still believed to be hard to simulate for classical computers. Once we've chosen the circuit shape for each type, this defines a *landscape*, the total number of parameters defining our model: the sum of the parameters defining each word. For each point in this landscape, i.e. for each combination of parameters, we can run the circuits corresponding to each sentence in our dataset and check how far away we are from the correct label. The sum of the distance between our model's prediction and the correct label is called the *loss function*, this is the objective that we want to minimise.

Once we have defined the landscape and the loss, we can feed them as input to the optimisation algorithm of our choice. The easiest choice, and the one we implemented in (Coecke et al., 2020), is to pick a black box optimiser such as simultaneous perturbation stochastic approximation (SPSA) (Spall, 1992). A more advanced solution is to go fully diagrammatic and compute the gradient of our loss directly from the shape of the diagrams for sentences (Toumi et al., 2021). When the process has converged to some (locally) optimal parameters, tadam! We have successfully trained our quantum natural language model. We may now evaluate it on some testing data, typically a subset of the sentences that we withhold from the training set. Note that crucially, even though our model has never seen these sentences during training, hopefully it has already learnt the meanings for the words, and it knows how to combine these meanings in new ways.

Thus, a typical QNLP experiment may be summarised as follows: (1) we draw the diagrams for each sentences, (2) we pick a circuit shape for each word type, (3) we define a loss function, (4) we find the optimal parameters and (5) we test the model's prediction. These five steps fit in a few lines of Python using the tools from the DisCoPy library, the result can then be presented in the form of a Jupyter notebook, see the documentation for plenty of examples:

https://discopy.readthedocs.io/en/main/notebooks.qnlp.html

Also, here is a more detailed tutorial as well:

https://discopy.readthedocs.io/en/main/notebooks/qnlp-tutorial.html

11   Outlook

So what's next? Well, as we speak the editor of this volume is himself getting to play, and obviously, we will also embark on a musical adventure. Evidently, we will also further develop QNLP bringing in more conceptual depth, for the specific case of language as well as towards broader AI features. Experimentally, we keep pushing along as hardware improves.

All together this is just the beginning of an adventure that, we hope, on the one hand, will create the next generation of AI, and on the other hand, will further the quest towards better understanding the human mind and its interaction with the world around us.

References

Abramsky, S. (1996). Retracing some paths in process algebra. In *CONCUR'96: concurrency theory*, vol. 1119 of *Lecture notes in computer science* (pp. 1–17). Springer.

Abramsky, S., & Coecke, B. (2004). A categorical semantics of quantum protocols. In *Proceedings of the 19th annual IEEE symposium on logic in computer science (LICS)* (pp. 415–425). arXiv:quant-ph/0402130.

Adjukiewicz, K. (1935) Die syntaktische Konnexität. *Studia Philosophica*.

Baez, J. C., & Fong, B. (2015). A compositional framework for passive linear networks. arXiv:1504.05625

Baez, J. C., & Stay, M. (2011). Physics, topology, logic and computation: a Rosetta stone. In B. Coecke (ed.), *New structures for physics, lecture notes in physics* (pp. 95–172). Springer.

Bankova, D., Coecke, B., Lewis, M., & Marsden, D. (2019). Graded hyponymy for compositional distributional semantics. *Journal of Language Modelling, 6*(2), 225–260.

Bar-Hillel, Y. (1953). A quasiarithmetical notation for syntactic description. *Language, 29*, 47–58.

Barrett, J. (2007). Information processing in generalized probabilistic theories. *Physical Review A, 75*, 032304.

Benedetti, M., Lloyd, E., Sack, S., & Fiorentini, M. (2019). Parameterized quantum circuits as machine learning models. *Quantum Science and Technology, 4*, 043001.

Bolt, J., Coecke, B., Genovese, F., Lewis, M., Marsden, D., & Piedeleu, R. (2018). Interacting conceptual spaces I: Grammatical composition of concepts. In M. Kaipainen, A. Hautamäki, P. Gärdenfors, & F. Zenker (eds.), *Concepts and their applications, synthese library, studies in epistemology, logic, methodology, and philosophy of science*. Springer.

Bonchi, F., Sobocinski, P., & Zanasi, F. (2014). A categorical semantics of signal flow graphs. In *CONCUR'14: concurrency theory. Lecture notes in computer science* (vol. 8704, pp. 435–450). Springer.

Chomsky, N. (1956). Tree models for the description of language. *I.R.E. Transactions on Information Theory, IT, 2*, 113–124.

Chomsky, N. (1957). *Syntactic Structures*. Mouton.

Clark, S., Coecke, B., Grefenstette, E., Pulman, S., & Sadrzadeh, M. (2014). A quantum teleportation inspired algorithm produces sentence meaning from word meaning and grammatical structure. *Malaysian Journal of Mathematical Sciences, 8*, 15–25. arXiv:1305.0556.

Coecke, B. (2005). Kindergarten quantum mechanics. In A. Khrennikov (ed.), *Quantum theory: Reconsiderations of the foundations III* (pp. 81–98). AIP Press. arXiv:quant-ph/0510032

Coecke, B. (2013). An alternative Gospel of structure: Order, composition, processes. In C. Heunen, M. Sadrzadeh, & E. Grefenstette (eds.), *Quantum physics and linguistics. A compositional, diagrammatic discourse* (pp. 1 – 22). Oxford University Press. arXiv:1307.4038

Coecke, B. (2017). From quantum foundations via natural language meaning to a theory of everything. In S. B. Cooper & M. I. Soskova (eds.), *The incomputable: Journeys beyond the turing barrier, theory and applications of computability* (pp. 63–80). Springer International Publishing. arXiv:1602.07618

Coecke, B. (2019). The mathematics of text structure. arXiv:1904.03478

Coecke, B., De Felice, G., Marsden, D., & Toumi, A. (2018). Towards compositional distributional discourse analysis. In M. Lewis, B. Coecke, J. Hedges, D. Kartsaklis, & D. Marsden (eds.), *Proceedings of the 2018 workshop on compositional approaches in physics, NLP, and social sciences. Electronic proceedings in theoretical computer science* (vol. 283, pp. 1–12)

Coecke, B., de Felice, G., Meichanetzidis, K., & Toumi, A. (2020). Foundations for near-term quantum natural language processing. arXiv preprint arXiv:2012.03755

Coecke, B., & Duncan, R. (2011). Interacting quantum observables: Categorical algebra and diagrammatics. *New Journal of Physics, 13*, 043016. arXiv:quant-ph/09064725

Coecke, B., & Duncan, R. (2012). Tutorial: Graphical calculus for quantum circuits. In *International workshop on reversible computation* (pp. 1–13). Springer

Coecke, B., & Kissinger, A. (2017). *Picturing quantum processes. A first course in quantum theory and diagrammatic reasoning.* Cambridge University Press.

Coecke, B., & Meichanetzidis, K. (2020). Meaning updating of density matrices. arXiv:2001.00862

Coecke, B., & Paquette, É. O. (2011). Categories for the practicing physicist. In B. Coecke (ed.), *New structures for physics, lecture notes in physics* (pp. 167–271.) Springer. arXiv:0905.3010

Coecke, B., Sadrzadeh, M., & Clark, S. (2010). Mathematical foundations for a compositional distributional model of meaning. In J. van Benthem, M. Moortgat, & W. Buszkowski (eds.), *A festschrift for Jim Lambek. Linguistic analysis* (vol. 36, pp. 345–384). arxiv:1003.4394

Coecke, B., & Wang, V. (2020). De-humanising grammar for improved machine use. *Draft paper.*

Cowtan, A., Dilkes, S., Duncan, R., Simmons, W., & Sivarajah, S. (2019). Phase gadget synthesis for shallow circuits. arXiv preprint arXiv:1906.01734

de Beaudrap, N., Bian, X., & Wang, Q. (2020). Fast and effective techniques for t-count reduction via spider nest identities. In *15th conference on the theory of quantum computation, communication and cryptography (TQC 2020).* Schloss Dagstuhl-Leibniz-Zentrum für Informatik. arXiv:2004.05164

de Felice, G., Di Lavore, E., Román, M., Toumi, A. (2020). Functorial language games for question answering. arXiv preprint arXiv:2005.09439

de Felice, G., Meichanetzidis, K., & Toumi, A. (2019). Functorial question answering. arXiv preprint arXiv:1905.07408

de Felice, G., Toumi, A., & Coecke, B. (2020). DisCoPy: Monoidal categories in python. *Accepted for ACT 2020.* arXiv preprint arXiv:2005.02975

De las Cuevas, G., Klingler, A., Lewis, M., & Netzer, T. (2020). Cats climb entails mammals move: preserving hyponymy in compositional distributional semantics. arXiv:2005.14134 [quant-ph]

Duncan, R., Kissinger, A., Pedrix, S., & van de Wetering, J. (2019). Graph-theoretic simplification of quantum circuits with the zx-calculus. arXiv preprint arXiv:1902.03178

Firth, J. R. (1957). A synopsis of linguistic theory, 1930–1955. *Studies in Linguistic Analysis,* 1–32.

Gärdenfors, P. (2000). *Conceptual spaces: The geometry of thought.* MIT Press.

Gärdenfors, P. (2014). *The geometry of meaning: Semantics based on conceptual spaces.* MIT Press.

Ghani, N., Hedges, J., Winschel, V., & Zahn, P. (2018). Compositional game theory. In *Proceedings of the 33rd annual ACM/IEEE symposium on logic in computer science,* (pp. 472–481). ACM.

Giovannetti, V., Lloyd, S., & Maccone, L. (2008). Quantum random access memory. *Physical review letters, 100*(16), 160501.

Hadzihasanovic, A., Ng, K. F., & Wang, Q. (2018). Two complete axiomatisations of pure-state qubit quantum computing. In *Proceedings of the 33rd annual ACM/IEEE symposium on logic in computer science* (pp. 502–511). ACM.

Harris, Z. S. (1954). Distributional structure. *Word, 10*(2–3), 146–162.

Kissinger, A., & van de Wetering, J. (2019). Pyzx: Large scale automated diagrammatic reasoning. arXiv preprint arXiv:1904.04735

Lambek, J. (1958). The mathematics of sentence structure. *American Mathematics Monthly, 65.*

Lambek, J. (2008). *From word to sentence.* Milan: Polimetrica.

Lewis, M. (2019a). *Compositional hyponymy with positive operators* (pp. 638–647). Varna, Bulgaria: In Proceedings of Recent Advances in Natural Language Processing.

Lewis, M. (2019b). Towards negation in discocat. In: Proceedings of SemSpace.

Lorenz, R., Pearson, A., Meichanetzidis, K., Kartsalkis, D., & Coecke, B. (2021). Qnlp in practice: Running compositional models of meaning on a quantum computer. arXiv preprint arXiv:2102.12846.

Ludwig, G. (1985). *An axiomatic basis of quantum mechanics* (vol. 1). Berlin Heidelberg: Derivation of Hilbert Space. Springer-Verlag.

Ma, Y., Tresp, V., Zhao, L., & Wang, Y. (2019). Variational quantum circuit model for knowledge graph embedding. *Advanced Quantum Technologies,* 1800078.

Mac Lane, S. (1998). *Categories for the working mathematician.* Springer.

Meichanetzidis, K., Gogioso, S., de Felice, G., Toumi, A., Chiappori, N., & Coecke, B. (2020). Quantum natural language processing on near-term quantum computers. *Accepted for QPL 2020.* arXiv preprint arXiv:2005.04147

Meichanetzidis, K., Toumi, A., de Felice, G., & Coecke, B. (2020). Grammar-aware question-answering on quantum computers. arXiv:2012.03756

Meyer, F., & Lewis, M. (2020). Modelling lexical ambiguity with density matrices. In *Proceedings of the 24th Conference on Computational Natural Language Learning* (pp. 276–290). arXiv:2010.05670 [cs.CL]

O'Riordan, L. J., Doyle, M., Baruffa, F., & Kannan, V. (2020). A hybrid classical-quantum workflow for natural language processing. arXiv preprint arXiv:2004.06800

Pavlovic, D. (2013). Monoidal computer I: Basic computability by string diagrams. *Information and computation, 226*, 94–116.

Piedeleu, R., Kartsaklis, D., Coecke, B., & Sadrzadeh, M. (2015). Open system categorical quantum semantics in natural language processing. In *6th conference on algebra and coalgebra in computer science (CALCO 2015)*. Schloss Dagstuhl-Leibniz-Zentrum fuer Informatik.

Intel News Release. (2019). *Intel to support the Irish Centre for High End Computing on new collaborative quantum computing project*, March 6, 2019. https://newsroom.intel.ie/news-releases/intel-to-support-the-irish-centre-for-high-end-computing-on-new-collaborative-quantum-computing-project/

Sadrzadeh, M., Clark, S., & Coecke, B. (2013). The Frobenius anatomy of word meanings I: Subject and object relative pronouns. *Journal of Logic and Computation, 23*, 1293–1317. arXiv:1404.5278

Selinger, P. (2011). A survey of graphical languages for monoidal categories. In B. Coecke (ed.), *New structures for physics, Lecture notes in physics* (pp. 275–337). Springer. arXiv:0908.3347

Shepherd, D., & Bremner, M. J. (2009). Temporally unstructured quantum computation. *Proceedings of the Royal Society A: Mathematical, Physical and Engineering Sciences, 465*(2105), 1413–1439.

Signorelli, C. M., Wang, Q., & Khan, I. (2020). A compositional model of consciousness based on consciousness-only. arXiv preprint arXiv:2007.16138

Sivarajah, S., Dilkes, S., Cowtan, A., Simmons, W., Edgington, A., & Duncan, R. (2020). t|ket>: A retargetable compiler for NISQ devices. arXiv preprint arXiv:2003.10611

Spall, J. C., et al. (1992). Multivariate stochastic approximation using a simultaneous perturbation gradient approximation. *IEEE Transactions on Automatic Control, 37*(3), 332–341.

Toumi, A., Yeung, R., & de Felice, G. (2021). Diagrammatic differentiation for quantum machine learning. arXiv preprint arXiv:2103.07960

Tull, S., & Kleiner, J. (2020). Integrated information in process theories. arXiv preprint arXiv:2002.07654

van de Wetering, J. (2020). ZX-calculus for the working quantum computer scientist. arXiv:2012.13966

Zeng, W., & Coecke, B. (2016). Quantum algorithms for compositional natural language processing. *Electronic Proceedings in Theoretical Computer Science, 221*. arXiv:1608.01406

A Quantum Computing Approach to Harnessing the Logic of the Mind for Brain–Computer Interfacing

Eduardo Reck Miranda

Abstract This chapter presents a new quantum computing-based method to study and harness neural correlates of mental activity for the development of Brain–Computer Interface (BCI) systems. It introduces the notion of logic of the mind, whereby neurophysiological data are encoded as logical expressions representing mental activity. These representations then become instructions for BCI control. Long complex logical expressions are intractable for classical computers. Hence, a quantum computing algorithm is proposed. Despite the fact that quantum hardware for running the proposed algorithm is not yet available commercially, we provide a simulation, which demonstrates it plausibility. An attempt at running the system on a quantum computer is also detailed, accompanied with an explanation of current limitations. In addition to building BCI systems, we argue that our quantum computing-based approach to study brain functioning and neural correlates of the mind has the potential to impact medical research, including diagnosing brain disorder.

1 Introduction

The human brain is allegedly the most complex object known to mankind: it has circa one hundred billion neurones forming a network of quadrillions of connections. The amount of information that circulates through this network is immense.

There is a school of thought, called Dualism, which considers the mind and the brain as separate entities (Rozemond, 1988). What is more, it has been suggested that minds would not even need brains to exist. Although the separation between mind and brain enjoys some currency in philosophical circles, it is generally agreed nowadays that the mind results from the functioning of the brain.

The scientific community does not have yet a clear understanding of how brain activity gives rise to the mind. Even though the behaviour of individual neurones is fairly well understood nowadays, the way in which they cooperate in ensembles of

E. R. Miranda (✉)
Interdisciplinary Centre for Computer Music Research (ICCMR), University of Plymouth, Plymouth, UK
e-mail: eduardo.miranda@plymouth.ac.uk

© Springer Nature Switzerland AG 2022
E. R. Miranda (ed.), *Quantum Computing in the Arts and Humanities*,
https://doi.org/10.1007/978-3-030-95538-0_9

millions has been difficult to unpack. Yet, this is of paramount importance to fathom how the brain creates the mind.

The advent of increasingly sophisticated brain scanning technology has been enabling a plethora of research activity to comprehend the neuronal correlates of mental activities (Kraft et al., 2008; Vartanian et al., 2013). This comprehension is paramount for philosophy, psychology, medicine and engineering. Indeed, emerging technology that enables users to control systems with their mind banks on such developments.

A Brain–Computer Interface, or BCI, is a piece of equipment that enables users to control systems with their mind. It reads signals from the brain and harnesses them for communicating with a computer; for example, to control a mechanical device, such as a robotic arm, a musical instrument or a wheelchair (Fig. 1).

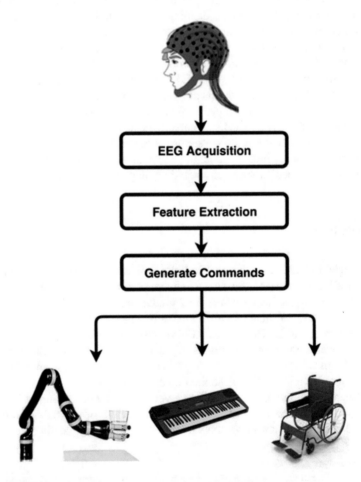

Fig. 1 A BCI harness brain signals to communicate with devices

This chapter proposes a novel approach to study and harness neuronal correlates of mental activity for the development of BCI systems. It introduces the notion of a *logic of the mind*, where neurophysiological data are encoded as logical expressions representing mental activity. These representations then become instructions for BCI control. Ultimately, the logic of the mind is tool for studying the neuronal correlates of mental activities.

Effective logical expressions are likely to be extensive and complex, involving dozens of variables. Large expressions require considerable computational power to be processed. This is problematic for BCI applications because brain data need to be processed rapidly in order to execute sequences of commands on the fly.

Quantum computers hold much promise in terms of processing speed for a number of problems that are intractable by classical computers, including those involving evaluation of logical expressions. Hence, the rational for using quantum computers to process the logic of the mind.

Quantum computers are fundamentally different from the typical computer as we know it. The speed-up prophecy depends, amongst other things, on algorithms that cleverly exploit fundamental properties of quantum physics, such as *superposition*, *entanglement* and *interference*. As it will be demonstrated below, the proposed quantum algorithm to process logic of the mind expressions takes advantage of those properties.

As a proof-of-concept, the chapter presents a system that reads the electroencephalogram of a person at given lapses of time and builds logic expressions encoding brain activity. The system generates respective quantum circuits and submits them to a quantum computer. The computer then checks the satisfiability of the expressions. Those expressions that are satisfied trigger commands to control a robotic arm.

As an additional example, we introduce an application of a BCI to make music. In this case, the system converts the satisfiability results into sounds. The BCI becomes a musical instrument controlled by the mind of the player.

The paper begins with a brief introduction to BCI and the electroencephalogram (EEG) which is the neurophysiological signal that is normally used in BCI systems, including the one presented in this chapter. Then, it briefly discusses how the EEG corresponds to mental states, and presents the logic of the mind concept. Next, there is a short introduction to quantum computing. Here, we focus on the basics deemed necessary to understand the proposed quantum algorithm that deals with logic of the mind expressions. Then, the paper shows how those expressions are processed and describes the proof-of-concept BCI systems. The limitations of current quantum computing hardware technology are also discussed.

2 Brain–Computer Interfacing

By and large, BCI research is concerned with the development of assistive technology for persons with severe motor impairment. However, BCI technology is also being developed for other applications, such as computer games (Hasan & Gan, 2012), biometrics (Palaniappan, 2008) and cursor control (Wilson & Palaniappan, 2011). And this author has developed BCI for making music (Miranda, 2006; Miranda et al., 2011).

There are different types of brain signals and respective sensors to read them. The type of signal that is most commonly used for BCI technology is the electroencephalogram.

2.1 The Electroencephalogram (EEG)

Neurones communicate with one another through electrical impulses. Neuronal electrical activity can be recorded with electrodes placed on the scalp (Fig. 2). This recording is called the electroencephalogram, or EEG. The EEG conveys the overall activity of millions of neurones in the brain in terms of electric current. This is measured as the voltage difference between two or more electrodes, one of which is taken as a reference. See (Marcuse et al., 2015) for more details.

It is also possible to record electrical brain activity with electrodes surgically implanted under the skull, on the surface of the cortex or deep inside the brain. This is often called electrocorticography (ECoG) or intracranial EEG (iEEG). Whereas implanted electrodes provide a far better signal to work with than surface ones, brain implants are not yet routinely used for BCI systems for health and safety reasons. Other technologies for brain scanning include functional Magnetic Resonance

Fig. 2 Cap furnished with electrodes relay EEG data to a computer for further processing

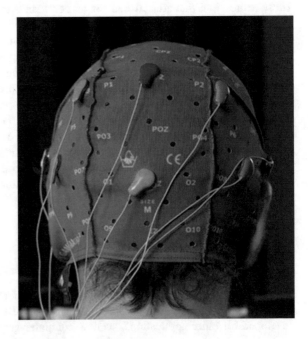

Fig. 3 Convention for placing electrodes on the scalp

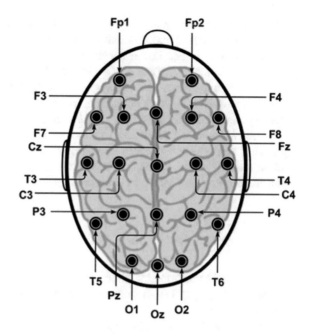

Imaging (fMRI), near-infrared spectroscopy (NIRS) and magnetoencephalography (MEG). However, these technologies are prohibitively expensive, less portable and/or offer inadequate time-resolution resolution for a BCI application (McFarland & Wolpaw, 2017).

For this project, we used a device manufactured by g.tec[1] to read the EEG. It consists of a cap furnished with electrodes and a transmitter that relays the EEG wirelessly to a computer for further processing.

Electrodes positioning on the head may vary depending on the purpose of the system or experiment. A commonly adopted convention is shown in Fig. 3. The terminology for referring to the positioning of electrodes uses letters to indicate a brain region and a number: Fp (for pre-frontal), F (for frontal), C (for central), T (for temporal), P (for parietal) and O (for occipital). Odd numbers are for electrodes on the left side of the head and even numbers are for those on the right side; the letter 'z' stands for the central region.

The EEG from scalp electrodes is a difficult signal to handle. It is filtered by the meninges (the membranes that separate the cortex from the skull), the skull and the skin before it reaches the electrodes. Moreover, the signal is weak: it is measured in terms of microvolts (μV). It needs to be amplified considerably in order to be useful for a BCI. But amplification invariably brings spurious signals. Thus, the EEG needs to be harnessed with signal processing and analysis methods in order to render it suitable for a BCI system.

---

[1] https://www.gtec.at/.

Power spectrum analysis is a popular method to harness the EEG. This method breaks the EEG signal into different frequency bands and reveals the distribution of power between them. Power spectrum analysis is useful because it can reveal patterns of brain activity, and a computer can be programmed to recognize and translate them into commands for a system. Although this chapter focuses on power spectrum analysis, it is worth noting that there are other EEG analysis methods for detecting mental activity as well; e.g. Hjorth analysis (Oh et al., 2014).

Typically, users must learn how to voluntarily produce specific patterns of EEG signals in order to be able to control something with a BCI. A fair amount of research is underway to develop lexicons of detectable EEG patterns, understand what they mean and develop methods to train users to produce them.

2.2 The Semantics of the EEG

The EEG conveys information about mental activity (Anderson & Sijercic, 1996; Petsche & Etlinger, 1998). In medicine, the EEG is an important tool for diagnosis of brain disorders.

There has been an increasing amount of research aimed at the identification of EEG correlates of all sorts of mental activities (Hinterberger et al., 2014; Jeunet et al., 2020; Nicolas-Alonso & Gomez-Gil, 2012; So et al., 2017; Yelamanchili, 2018). For instance, Giannitrapani (1985) identified EEG patterns associated with abstract intellectual tasks, such as doing arithmetic operations mentally. Guenter and Brumberg (2011) discovered neural correlates for the production of speech, which can be detected in the EEG. And more recently, Daly et al. (2018) detected EEG signatures correlated with emotions elicited while subjects were listening to music.

It is generally known that the distribution of power in the spectrum of the EEG can indicate different states of mind. For example, a spectrum with salient low-frequency components is often associated with a state of drowsiness, whereas a spectrum with salient high-frequencies is associated with a state of alertness. Giannitrapani (1985) linked the prominence of low-frequency components with a passive state of mind, as opposed to an active state, which is characterized by high-frequency spectral components.

Research exploring mental correlates of the EEG normally considers spectral components up to 40 Hz (Kropotov, 2008). There are four recognized spectral frequency bands, also known as EEG rhythms, each of which is associated with specific mental states (Table 1).

The exact boundaries of the bands listed in Table 1 and their respective mental states vary from one author to another.

The EEG that gives raise to mental states is highly dynamic. Figure 4 shows a snapshot of EEG mapped onto 2D and 3D topographic representations using the *OpenVibe* software (Renard et al., 2010). The line cursor on the 'Signal display' pane shows the precise moment of the snapshot. At this moment, prominent beta rhythms are detected at the lateral sides of the brain.

Table 1 Typical EEG rhythms and associated mental states. Frequency bands are in Hertz (Hz)

Bands	Rhythms	Mental states
$f < 4$	Delta	Sleep; can indicate of cerebral anomaly
$4 \leq f < 8$	Theta	Drowsiness; hypnotic state; can indicate cerebral anomaly
$8 \leq f < 15$	Alpha	Relaxed, meditative, unfocused, almost drowsy state of mind
$15 \leq f < 40$	Beta	High arousal, active thinking, consciously focusing state of mind

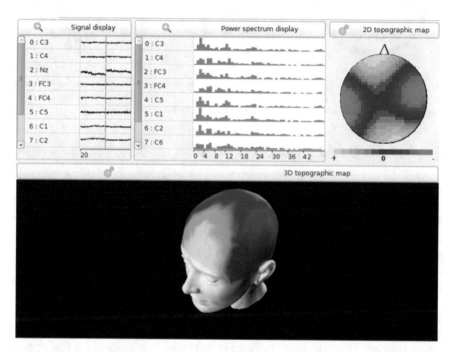

Fig. 4 Snapshot of EEG activity mapped onto 2D and 3D topographical representations (The electrodes' labelling arrangement here is slightly different from the convention introduced in Fig. 2.)

It is common knowledge that different regions of the brain have distinct roles; e.g. the visual cortex processes images, the auditory cortex processes sound, the motor cortex controls our limbs and so on (Squire et al., 2008). Assorted regions of the brain cooperate to perform mental tasks. Moreover, spectral amplitudes of scalp EEG are constantly changing over the skull surface. Therefore, states of mind cannot be accurately inferred simply by looking at a snapshot of the averaged EEG from the whole set of electrodes at once. Rather, one needs to look at the time-based interrelationships between spectral components of the EEG recorded simultaneously at different locations on the scalp.

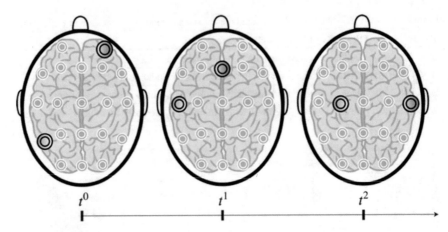

Fig. 5 Tracking the trajectories of two EEG rhythms in time

3   An EEG-Based Logic of the Mind

Petsche and Etlinger (1998) advocate that different bandwidths of the EEG spectrum are distinct windows to a dynamic landscape of electrical brain activity. It is suggested that mental states are correlated with interrelations between these windows. We follow this up by proposing a method to represent those interrelationships using logical expressions. Hence, the notion of an EEG-based logic of the mind.

Consider a system, which extracts information from the EEG of a number of electrodes at given lapses of time; e.g. periods lasting for 500 ms. For the sake of clarity, let us say that the system extracts beta and alpha rhythms from the EEG signals (Table 1). With this information, the system tracks the behaviour of these rhythms over the scalp as time progresses. For this example, it tracks the electrodes' locations where beta and alpha rhythms displayed most power. This is shown in Fig. 5: beta rhythms (denoted by the red electrodes) were detected prominently by electrode Fp2 at time t^0. Next, at time t^1 they were detected prominently by electrode Fz and then by electrode T4 at time t^2. And alpha rhythms (denoted by the blue electrodes) were detected prominently by electrodes T5, T3 and C3, respectively.

Next, at each time step t^n, the EEG information is encoded as a logic expression. The variables represent the activity of the respective electrodes.

In order to keep the example simple, let us consider only the following subset of electrodes: {Fp2, Fz, T3, C3, T4, T5}. In this case, the electrodes that detect the most prominent EEG signals are represented in the expression as 'True'. All the others are 'False'. Thus, the expressions for the time lapses in Fig. 5 could be written follows[2]:

---

[2] See (Smith, 2020) for an introduction to formal logic. The symbol ¬ is the negation operator (i.e. NOT), and ∧ stands for the logical conjunction operator (i.e. AND). Variables corresponding to a True electrode are in bold for clarity. These expressions could be stated in reduced form, but we leave them expanded here for the sake of intelligibility.

$$t^0 : \beta(\textbf{Fp2} \wedge \neg Fz \wedge \neg T3 \wedge \neg C3 \wedge \neg T4 \wedge \neg T5)$$
$$\wedge \, \alpha(\neg Fp2 \wedge \neg Fz \wedge \neg T3 \wedge \neg C3 \wedge \neg T4 \wedge \textbf{T5})$$

$$t^1 : \beta(\neg Fp2 \wedge \textbf{Fz} \wedge \neg T3 \wedge \neg C3 \wedge \neg T4 \wedge \neg T5)$$
$$\wedge \, \alpha(\neg Fp2 \wedge \neg Fz \wedge \textbf{T3} \wedge \neg C3 \wedge \neg T4 \wedge \neg T5)$$

$$t^2 : \beta(\neg Fp2 \wedge \neg Fz \wedge \neg T3 \wedge \neg C3 \wedge \textbf{T4} \wedge \neg T5)$$
$$\wedge \, \alpha(\neg Fp2 \wedge \neg Fz \wedge \neg T3 \wedge \textbf{C3} \wedge \neg T4 \wedge \neg T5)$$

The first term, inside the parenthesis on the left side of the conjunction operator, corresponds to the beta rhythms and the second term to the alpha ones.

As a matter of fact, the expressions above could have been written with other logical operators. For instance, t^0 could have been written with the logical disjunction operator[3] between the beta and alpha terms:

$$t^k : \beta(Fp2 \wedge \neg Fz \wedge \neg T3 \wedge \neg C3 \wedge \neg T4 \wedge \neg T5)$$
$$\vee \, \alpha(\neg Fp2 \wedge \neg Fz \wedge \neg T3 \wedge \neg C3 \wedge \neg T4 \wedge T5)$$

The forms of those logical expressions depend on what they are meant to represent. They might, for example, depend on the correlated mental states which they are supposed to stand for. For instance, it is generally agreed that increased beta activity in the pre-frontal cortex corresponds to 'focused attention in problem-solving' (Ligeza et al., 2015). The pre-frontal cortex is covered by electrodes Fp1 and Fp2, and to some extent F7, F3, Fz, F4 and F8. Thus, the expressions for t^0 and t^1 above would correspond to (or 'satisfy' in logic parlance) this mental state. Conversely, the expression for t^2 would not.

The problem of establishing which values for the variables of a logical expression can render the whole expression True, or satisfiable, is known as the Boolean satisfiability problem. As an illustration, consider this simple expression, with variables A and B: $(\neg A \wedge B)$. If A = False and B = True, then this expression is True. Hence, these values for A and B render the expression satisfiable. But if, say, B = False then the whole expression would be False.

The ability to represent EEG correlates in terms of logical expressions is useful for implementing BCI systems because they can be programmed to activate actions associated to specific expressions. As the EEG is acquired and analysed, a system would check when the information satisfies given logical expressions. Then, it would perform the respective actions for those expressions that return True. For example:

If at t^n, $\beta(a \wedge \neg(b \wedge c \wedge d \wedge e \wedge f)) \vee \alpha(\neg(a \wedge b \wedge c \wedge d \wedge e) \wedge f)) = $ True.

then { move robot arm to the left by 180°};

---

[3] The symbol for the logical disjunction operator (i.e. OR) is $\vee$.

If at t^n, $\beta(b \wedge \neg(a \wedge c \wedge d \wedge e \wedge f) \wedge \alpha(\neg(a \wedge b \wedge c \wedge d \wedge f) \wedge e)) =$ True.

then { move robot arm to the left by 45°};

If at t^n, $\beta((a \vee d) \wedge \neg(b \wedge c \wedge e \wedge f)) \wedge \alpha(\neg(a \wedge b \wedge c) \wedge (d \vee e \vee f)) =$ True.

then { move robot arm to the right by 90°};

and so on.

Satisfiable complex Boolean expressions may return True for different combinations of logic values. And such expressions can have a great number of variables and a variety of satisfactory combinations. These are properties that make the logic of the mind concept rather powerful for BCI.

In addition to building BCI systems, the proposed EEG-based logic of the mind is potentially useful for cataloguing mental states and EEG correlates. We envisage systems to automatically generate logical expressions while subjects are performing specific mental tasks. The hypothesis is that such systems are able to evolve sophisticated time-based ontologies of mental activity in unprecedented ways.

The caveat of developing a catalogue of mental states represented as logical expressions is that Boolean satisfiability problems are very demanding in terms of computation. Such expressions could easily grow to dozens of logical variables. Expressions with, say, 50 variables would require as many as 2^{50} combinations to be evaluated. A mid-range personal computer is capable of performing circa 2 billion operations per second. Thus, such a computer would take almost an entire week to evaluate an expression. Emerging quantum computing technology promises considerable speed-up for tasks such as these (Ball, 2014). Hence, the rationale for researching the potential of quantum computing for BCI.

4 A Brief Introduction to Quantum Computing and Logic Operations

4.1 The Basics of Quantum Computing

This section introduces the basics of quantum computing and logic operations. It focuses on the basics deemed necessary to contextualize and understand how to program quantum computers for solving logical expressions. For more detailed explanations of quantum computing, the reader is referred to (Bernhardt, 2019; Grumbling & Horowitz, 2019; Johnston et al., 2019; Mermin, 2007; Rieffel & Polak, 2011).

Classical computers manipulate information represented in terms of binary digits, each of which can value 1 or 0. They work with microprocessors made up of billions of tiny switches activated by electric signals. Values equal to 1 and 0 reflect the on and off states of the switches, respectively.

In contrast, a quantum computer deals with information in terms of quantum bits, or *qubits*. Qubits are subject to the laws of quantum mechanics because they operate at the subatomic level. At the subatomic level, a quantum object does not necessarily exist in a determined state. Its state is unknown until one observes it.

A qubit is a two-level quantum system where the two basis states are usually written as $|0\rangle$ and $|1\rangle$. In fact, a qubit can be in state $|0\rangle$, $|1\rangle$ or (unlike a classical bit) in a state of *superposition*, which is a linear combination of both. Superposition is the first fundamental property of quantum computing mentioned at the Introduction. To a greater extent, the art of programming a quantum computer involves manipulating qubits to perform operations while they are in such indeterminate state. This makes quantum computing fundamentally different from digital computing.

The difficulty with building quantum processors using superconducting technology is that they need to be well isolated from the environment in order to keep the qubits coherent to perform computations. Quantum processors rely on qubits being kept in superposition for as long as possible. However, keeping a qubit in superposition is like balancing a tiny thin coin upright on a floating surface: any movement, even the tiniest vibration, will cause it to fall to head or tail. And this gets harder with groups of entangled qubits (e.g. when the state of a qubit depends on the state of another). This fall is referred to as decoherence.

Still, total isolation is impossible, because one needs to access the qubits in order to read information. The very act of reading qubits can change their state, because it is an intervention from the external world.

In order to picture a qubit, imagine a transparent sphere with opposite poles. From its centre, a vector whose length is equal to the radius of the sphere can point to anywhere on the surface. In quantum mechanics, this sphere is known as the Bloch sphere. And the vector is referred to as a state vector. The opposite poles of the sphere are denoted by $|0\rangle$ and $|1\rangle$, which is the notation used to represent quantum states (Fig. 6).

A qubit's state vector can point at anywhere on the Bloch sphere's surface. Mathematically, this is described in terms of polar coordinates using two angles, θ and φ. The angle θ is the angle between the state vector and the z-axis (latitude) and the angle φ describes the vector's position in relation to the x-axis (longitude).

When a qubit is in a state of superposition of $|0\rangle$ and $|1\rangle$, the state vector could be pointing anywhere between the two. However, we cannot really know where exactly a state vector is pointing to until we read the qubit. In quantum computing terminology, the act of reading a qubit is called 'measurement'. Measuring the qubit will make the vector point to one of the poles and return either 0 or 1 as a result.

The state vector of a qubit in superposition is described as a linear combination of two vectors, $|0\rangle$ and $|1\rangle$, as follows:

$$|\Psi\rangle = \alpha|0\rangle + \beta|1\rangle, \text{ where} |\alpha|^2 + |\beta|^2 = 1.$$

The state vector $|\Psi\rangle$ is a superposition of vectors $|0\rangle$ and $|1\rangle$ in a two-dimensional complex space, referred to as Hilbert space, with amplitudes α and β. Consider the squared values of α and β as probability values representing the likelihood of the

Fig. 6 Bloch sphere (*Source* Smite-Meister, https://commons.wikimedia.org/w/index.php?curid=5829358)

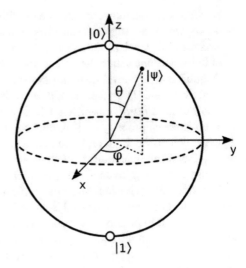

measurement return 0 or 1. For instance, let us assume the following:

$$|\Psi\rangle = \alpha|0\rangle + \beta|1\rangle, \text{ where } \alpha = \frac{1}{2} \text{ and } \beta = \frac{\sqrt{3}}{2}$$

In this case, $|\alpha^2| = 0.25$ and $|\beta^2| = 0.75$. Therefore, the measurement of the qubit has a 25% chance of returning 0 and a 75% chance of returning 1.

Quantum computers are programmed using sequences of commands, or quantum gates, that act on qubits. For instance, the 'not gate', performs a rotation of 180° around the x-axis. Hence, this gate is often called the X gate (Fig. 7). A more generic rotational $Rx(\vartheta)$ gate is available for quantum programming, where the angle for the rotation is specified. Therefore, $Rx(180)$ applied to $|0\rangle$ or $|1\rangle$ is equivalent to applying X to $|0\rangle$ or $|1\rangle$. In essence, all quantum gates perform qubit rotations, which change the amplitude distribution of the system.

An important gate is the Hadamard gate (referred to as the H gate). This gate puts the qubit into a superposition state consisting of an equal-weighted combination of two opposing states: $|\Psi\rangle = \alpha|0\rangle + \beta|1\rangle$, where $|\alpha|^2 = 0.5$ and $|\beta|^2 = 0.5$ (Fig. 8). For other gates, please consult the references given earlier.

Qubits in a program typically start in ground state $|0\rangle$, and then a sequence of gates are applied. Then, the qubits are read and the results are stored in standard digital memory, which are accessible for further handling. A quantum algorithm is often depicted as a circuit diagram of quantum gates, showing sequences of gate operations on the qubits (Fig. 9).

The vertical z-axis of the Bloch sphere forms the so-called *standard computational basis*. The x-axis forms the *conjugate computational basis* and y-axis for the *circular computational basis*. As its name suggest, the standard basis is the most commonly used, and it is the one adopted for the work presented in this chapter. A detailed

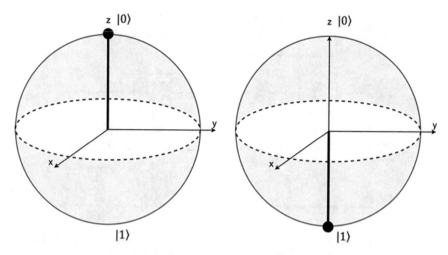

Fig. 7 The X gate rotates a qubit's state vector (pointing upwards on the figure on the left side) by 180° around the x-axis (pointing downwards on the figure on the right side)

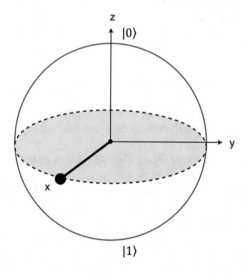

Fig. 8 The Hadamard gate puts the qubit into a superposition state halfway two opposing poles

explanation of these bases and their significance to computation can be found in (Bernhardt, 2019). What is important to bear in mind here is that changing the basis on which a quantum state is expressed, corresponds to changing the measurement performed to read the outcomes of the computations. It is important to note that changing the basis to express a state does not change anything physical per se.

What makes quantum computation interesting are gates that operate on multiple qubits, such as the conditional (or controlled) X gate; referred to as CX gate. The CX

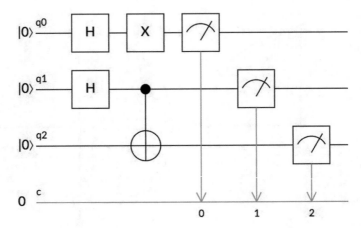

Fig. 9 A quantum algorithm depicted as a circuit of quantum gates. The squares with dials represent measurements, which are saved on classical registers represented at the bottom line

gate is an entangling gate. Depending on the input control state, it can put 2 qubits in *entanglement*, which is the second fundamental property of quantum computing mentioned at the introduction. Entangled qubits can no longer be thought of as independent units. They become one quantum entity described by a state vector of its own right.

The CX gate applies an X gate on a qubit only if the state of another qubit is $|1\rangle$. Thus, the CX gate establishes a dependency of the state of 1 qubit with the value of another. The schematic representation of the CX gate is shown in Fig. 10. In fact, any quantum gate can be built in controlled form. And entanglement can take place between more than 2 qubits.

The Bloch sphere is useful for visualizing what happens with a single qubit, but it is not suitable for multiple qubits, in particular when they are entangled. Entangled qubits can no longer be thought of as independent units. They become one quantum entity described by a state vector of its own right. Hence, from now on, we will have to use mathematics to represent quantum systems.

The notation used above to represent quantum states ($|\Psi\rangle$, $|0\rangle$, $|1\rangle$), is called Dirac notation. It provides an abbreviated way to represent vectors. For instance, $|0\rangle$ and $|1\rangle$ represent the following vectors, respectively:

Fig. 10 The CX gate creates a dependency of the state of 1 qubit with the state of another. In this case, q1 will be flipped only if q0 is $|1\rangle$

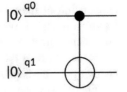

$$|0\rangle = \begin{bmatrix} 1 \\ 0 \end{bmatrix} \text{and} |1\rangle = \begin{bmatrix} 0 \\ 1 \end{bmatrix}$$

Mathematically, quantum gates are expressed as matrices. For instance, the X gate is represented as

$$X = \begin{bmatrix} 0 & 1 \\ 1 & 0 \end{bmatrix}$$

Therefore, quantum gate operations are expressed as matrix operations. Thus, the application of an X gate to $|0\rangle$ is the multiplication of a matrix (gate) by a vector (qubit state), which looks like this:

$$X(|0\rangle) = \begin{bmatrix} 0 & 1 \\ 1 & 0 \end{bmatrix} \times \begin{bmatrix} 1 \\ 0 \end{bmatrix} = \begin{bmatrix} 0 \\ 1 \end{bmatrix} = |1\rangle$$

Quantum processing with multiple qubits is represented by means of tensor vectors. A tensor vector is the result of the tensor product, represented by the symbol $\otimes$, of 2 or more vectors. A system of 2 qubits looks like this $|0\rangle \otimes |0\rangle$, but it is normally abbreviated to $|00\rangle$. It is useful to look at the expanded form of the tensor product to trace what it represents:

$$|00\rangle = |0\rangle \otimes |0\rangle = \begin{bmatrix} 1 \\ 0 \end{bmatrix} \otimes \begin{bmatrix} 1 \\ 0 \end{bmatrix} = \begin{bmatrix} 1 \times 1 \\ 1 \times 0 \\ 0 \times 1 \\ 0 \times 0 \end{bmatrix} = \begin{bmatrix} 1 \\ 0 \\ 0 \\ 0 \end{bmatrix}$$

The CX gate, for instance, is defined by the matrix:

$$CX = \begin{bmatrix} 1 & 0 & 0 & 0 \\ 0 & 1 & 0 & 0 \\ 0 & 0 & 0 & 1 \\ 0 & 0 & 1 & 0 \end{bmatrix}$$

Thus, application of CX to $|10\rangle$ is represented as

$$CX(|10\rangle) = \begin{bmatrix} 1 & 0 & 0 & 0 \\ 0 & 1 & 0 & 0 \\ 0 & 0 & 0 & 1 \\ 0 & 0 & 1 & 0 \end{bmatrix} \times \begin{bmatrix} 0 \\ 0 \\ 1 \\ 0 \end{bmatrix} = \begin{bmatrix} 0 \\ 0 \\ 0 \\ 1 \end{bmatrix}$$

Table 2 CX gate table, where q1 is flipped only if q0 is $|1\rangle$	Input	Result
	$|00\rangle$	$|00\rangle$
	$|01\rangle$	$|11\rangle$
	$|10\rangle$	$|10\rangle$
	$|11\rangle$	$|01\rangle$

The resulting vector is then abbreviated to $|11\rangle$ as shown below:

$$\begin{bmatrix} 0 \\ 0 \\ 0 \\ 1 \end{bmatrix} = \begin{bmatrix} 0 \\ 1 \end{bmatrix} \otimes \begin{bmatrix} 0 \\ 1 \end{bmatrix} = |1\rangle \otimes |1\rangle = |11\rangle$$

Table 2 shows the resulting quantum states of CX gate operations, where the second qubit flips only if the first qubit is $|1\rangle$. Note that in quantum computing, qubit strings are usually enumerated from the right end of the string to the left: e.g. $|q_2\rangle \otimes |q_1\rangle \otimes |q_0\rangle$. This is the norm adopted in this chapter from now on. Thus, the 'first qubit' is the rightmost one.

Another useful controlled gate is the multiple controlled form of the X gate, also known as the Toffoli gate (Aharonov, 2003). An example of a 3-qubit Toffoli gate (also known as CCX or CCNOT) in shown in Fig. 11.

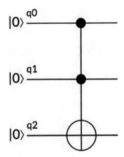

Fig. 11 The 3-qubit Toffoli gate creates a dependency of the state of 1 qubit with the state of 2 qubits. In this case, q2 flips only if both, q0 and q1 are $|1\rangle$)

Table 3 shows resulting quantum states of the 3-qubit Toffoli gate portrayed in Fig. 11.

4.2 Quantum Logic Operators

In digital logic, one can build any logic operator and entire Boolean expressions with just one basic NAND operator (Akerkar and Akerkar, 2004). Likewise, quantum

Table 3 The 3-qubit Toffoli gate table where q2 flips only if both, q0 and q1 are $|1\rangle$	Input	Result
	$|000\rangle$	$|000\rangle$
	$|001\rangle$	$|001\rangle$
	$|010\rangle$	$|010\rangle$
	$|011\rangle$	$|111\rangle$
	$|100\rangle$	$|100\rangle$
	$|101\rangle$	$|101\rangle$
	$|110\rangle$	$|110\rangle$
	$|111\rangle$	$|011\rangle$

logic operators and expressions can be built using only the X gate and its controlled forms. A few examples of such quantum logic operators are shown in Fig. 12.

In addition to the X gate, another useful basic gate with which we can build Boolean expressions is the $Rz(\vartheta)$ gate. This gate rotates the state vector of a qubit around the z-axis of the Bloch sphere by a given angle ϑ. In cases where the angle ϑ is equal to 180° then the $Rz(\vartheta)$ gate is often called as the Z gate (Fig. 13).

Rotations around the z-axis represent changes in the phase of a qubit. For a qubit in uniform superposition with an equal probability to be measured $|0\rangle$ or $|1\rangle$, the state vector will point to the equator line. The Z gate thus reverses the phase of the qubit, while maintaining its superposition.

Boolean gates built using controlled Z gates encode the results of operations in the phases of the qubits. This allows for the representation of multiple outcomes in superposition, which is something that cannot be done using digital bits. Hence, the beauty and processing power of quantum computing.

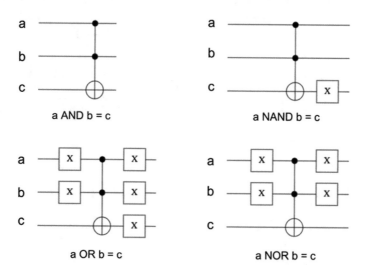

Fig. 12 Quantum logic operators built with X gates. It is assumed that qubit c is initialized to $|0\rangle$

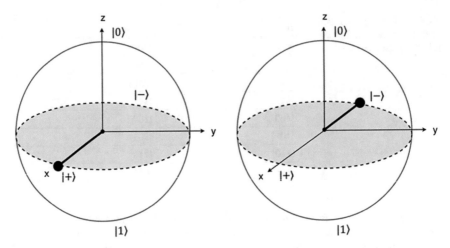

Fig. 13 The Z gate rotates a qubit's state vector by 180° around the z-axis. The opposite ends for this state vector are notated as $|+\rangle$ and $|-\rangle$

A controlled Z gate acts only when both, the control qubit(s) and the target qubit are pointing to $|1\rangle$. Therefore, the logic circuit will change the phases only of those qubits that satisfy the operations they represent. These changes are referred to as 'spin-marking' the possible outcomes. Examples of logic operators built with controlled Z gates are shown in Fig. 14.

Notice that one cannot simply link sequences logic statements using a combination of controlled X and Z gates in a circuit. Their 'inputs' and 'outputs' are incompatible because they operate on different computational basis.

Bear in mind that rotations on the z-axis do not affect the amplitudes of the state vector on the *standard computational basis* (z-axis of the Bloch sphere). And if one measures a phase logic operator on the standard basis, phase information is

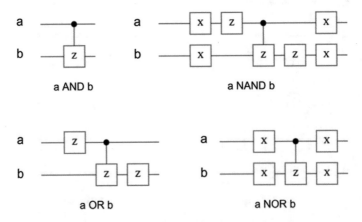

Fig. 14 Quantum phase logic operators using Z gates

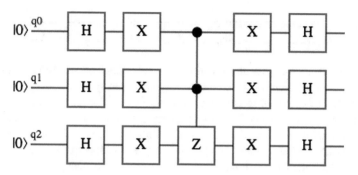

Fig. 15 The amplitude amplifier circuit[4]

lost. Hence, in order to be extracted, phase information needs to be converted into amplitude information. This is done by means of a technique known as amplitude amplification.

An amplitude amplifier is a device that increases the probability of revealing which qubits have been spin-marked by the Z gates in a logical operation. The combination of spin-marking and amplitude amplification illustrates the third fundamental property of quantum computing that makes it different from classical computing: *interference.* Figure 15 shows how the circuit for amplitude amplification looks like. In this case, the circuit is for three qubits. If more qubits are needed, then identical gate sequences are added for each additional one.

5 Building Logic Expressions from EEG Information

In order to explain how to build logical expressions from EEG information automatically, let us begin by defining an example of a scheme for representing EEG information as logic expressions with only three clauses and three logical variables (A, B, and C) as follows:

$$(Clause_1) \wedge (Clause_2) \wedge (Clause_3)$$

Each clause has two terms of the form $(Term_1 \vee Term_2)$. Below are examples of logic expressions in the proposed format:

$$(\neg A \vee C) \wedge (\neg B \vee \neg C) \wedge (A \vee C)$$
$$(A \vee B) \wedge (\neg B \vee C) \wedge (A \vee \neg C)$$
$$(A \vee B) \wedge (\neg B \vee \neg C) \wedge (A \vee C)$$

---

[4] In the case of controlled Z gates, the black dot and the boxed circuit representation have the same effect; sometimes only a vertical line of black dots are used.

Next, consider that the clauses represent EEG information captured by electrodes positioned at specific places on the head (Fig. 3), as shown in Table 4 and illustrated in Fig. 16.

In this example, we are dealing with a BCI system that banks on EEG beta rhythms in order to activate commands to control a hypothetical system.

The terms of the clauses are stipulated as follows: for each clause, the system selects the two electrodes with the two highest EEG amplitudes. As an example, let us pretend that at a certain moment in time the electrodes Fp1 and O1 registered the highest amplitudes. Then, the system analyses the spectrum of the EEG from these two electrodes and extracts the most prominent frequency component in the spectrum, for each electrode. Let us say that the strongest component for Fp1 was 33.18 Hz and for O1 was 23.61 Hz. These correspond to terms A and C of $Clause_1$; that is, A = 33.18 Hz and C = 23.61 Hz.

Next, the system checks if the frequencies conveyed by the respective terms are beta rhythms. If a frequency is equal to, or higher than, 15 Hz, then the respective term is True. Otherwise, it is False. In this case A = True and C = True. Therefore, these two terms form the clause (A ∨ C). Had Fp1 been equal to, say, 10.36 Hz, then this term would have been (¬A ∨ C) instead.

For this example, the EEG analysis for the nine electrodes resulted in the following expression: (A∨B)∧(¬B∨¬C)∧(A∨C). The next step is to verify if this expression is satisfiable. In other words, the system checks if those EEG values would render

Table 4 Specific electrodes are allocated to distinct A, B and C variables for each clause		A	B	C
	$Clause_1$	Fp1	T3	O1
	$Clause_2$	Fz	Cz	Oz
	$Clause_3$	Fp2	T4	O2

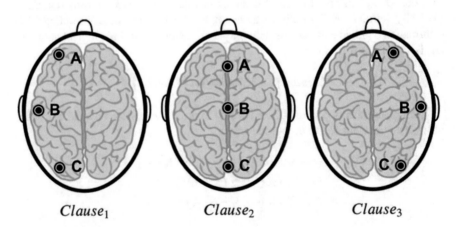

Fig. 16 Electrodes and respective logic variables allocations

this expression True. Let us examine how the system generates a quantum circuit to check the satisfiability of logical expressions.

6 Generating Quantum Circuits for Logic Satisfiability

The system uses X and Z gates, and their multiple-controlled forms, to build logical expressions. The one introduced above are formed by three clauses linked by conjunction operators. And each of the three clauses are formed by terms linked by disjunction operators.

The three disjunction clauses are built with X gates. And then, they are linked with conjunction operators implemented with a controlled Z gate. The circuit requires only 6 qubits: three of them (q0, q1 and q2) represent the logical variables A, B and C. The other three (q3, q4 and q5) are ancillary qubits, which are used to represent the results of the disjunction clauses. The full circuit is shown in Fig. 17 and the respective Quil[5] code in Code 1.

```
DECLARE ro BIT[3]            X 0                          CCNOT 0 1 3
H 0                          X 2                          X 0
H 1                          X 5                          X 1
H 2                          CONTROLLED CONTROLLED Z 3 4 5   H 0
X 0                          X 0                          H 1
X 1                          X 2                          H 2
CCNOT 0 1 3                  X 5                          X 0
X 0                          CCNOT 0 2 5                  X 1
X 1                          X 0                          X 2
X 3                          X 2                          CONTROLLED CONTROLLED Z 0 1 2
X 1                          X 1                          X 0
X 2                          X 2                          X 1
X 1                          X 1                          X 2
X 2                          X 2                          H 0
CCNOT 1 2 4                  X 4                          H 1
X 1                          CCNOT 1 2 4                  H 2
X 2                          X 1                          MEASURE 0 ro[0]
X 4                          X 2                          MEASURE 1 ro[1]
X 1                          X 1                          MEASURE 2 ro[2]
X 2                          X 2
X 0                          X 0
X 2                          X 1
CCNOT 0 2 5                  X 3
```

Code 1 Generated Quil code for the circuit in Figure 18   CCNOT stands for CCX, or Toffoli gate.

To begin with, all qubits are initialized to $|0\rangle$ and the ones representing the logical variables are put in uniform superposition with the H gate.

The disjunction clauses are specified, one after the other, as shown in dashed boxes at the top of Fig. 17. The results from each clause are held in ancillary qubits: N1 in q3, N2 in q4 and N3 in q5. Then, the tripartite conjunction operation is implemented with

---

[5] Quil is a quantum instruction language developed by Rigetti.

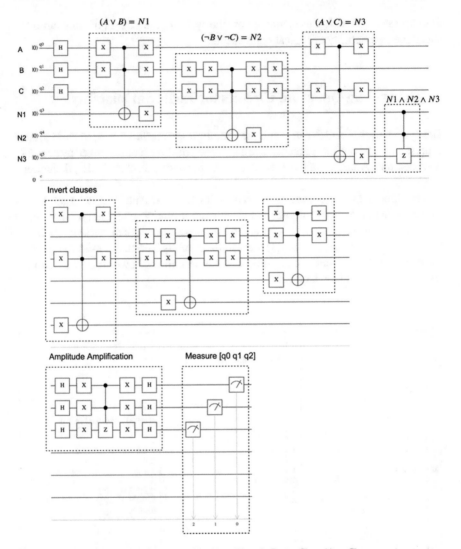

Fig. 17 Quantum circuit for the expression $(A \lor B) \land (\neg B \lor \neg C) \land (A \lor C)$

a 3-qubit controlled Z gate applied to the ancillary qubits: they will be spin-marked by the phase-logic AND operation.

Next, the three disjunction clauses need to be uncomputed in order to return the ancillary qubits to their initial states. Uncomputation is achieved by running the respective logic sub-circuits back to front; the 'Inverted clauses' in the middle section of Fig. 17. Then, amplitude amplification is applied to the logic variables to reveal the spin-marked qubits in terms of amplitudes. Finally, the qubits representing the logic values are measured to yield the result.

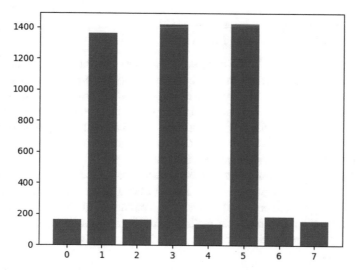

Fig. 18 Result yielded by running the circuit example for 5,000 shots. Binary numbers were converted to decimals for plotting on the horizontal axis. Vertical axis are the times each of the results were observed

The ancillary qubits are not part of the result. Therefore, they are not measured in the end. However, they can interfere with the measurements because they are entangled with the qubits that represent the logical variables. Hence, the ancillary qubits must be uncomputed to make them return to their unentangled state.[6]

In this example, the circuit spin-marks the quantum states $|001\rangle$, $|011\rangle$ and $|101\rangle$. Thus, each time the algorithm runs, it will output either $001, 011$ and 101 with much higher probability than any of the other possible values. The plot in Fig. 18 shows the times that each of the possible outcomes were observed after running and measuring the circuit for 5,000 times, or 5,000 'shots' in quantum computing terminology, on a Rigetti's Quantum Virtual Machine (QVM).[7] See Appendix 1 for more examples.

What does this result mean? Bearing in mind that in quantum computing qubit strings are enumerated from the right end of the string to the left (i.e. $|CBA\rangle$), the expression $(A \vee B) \wedge (\neg B \vee \neg C) \wedge (A \vee C)$ is satisfied when:

Case 1, $|001\rangle$: A = True, B = False and C = False.
Case 2, $|011\rangle$: A = True, B = True and C = False.
Case 3, $|101\rangle$: A = True, B = False and C = True.

In other words, a BCI command associated with this expression is triggered when significant beta rhythms are detected as follows:

(a)   by electrodes Fp1 and Fp2 (Case 1)

---

[6] Refer to (Johnston et al., 2019) for a didactic discussion on uncomputing ancillary qubits and implications to measurement.

[7] The Rigetti Quantum Virtual Machine is an implementation of the Quantum Abstract Machine described in (Smith et al., 2017). Noise simulation models to account for the effect of quantum hardware decoherence were not used in the examples discussed in this chapter.

(b)   by electrodes Fp1, Fp2 and T3 (Case 2)
(c)   by electrodes Fp1, Fp2 and O2 (Case 3)

In terms of mental states, this expression encodes a state of focused attention. It is well known that beta rhythms in the frontal regions of the brain are associated with this mental state (Berta et al., 2013). We programmed the proof-of-concept to move the robotic arm vertically, upwards and downwards, in response to detecting a state of focusing attention in the brain of the user. An additional example is the case of expression $(A \vee B) \wedge (B \vee A) \wedge (B \vee C)$, which is shown in Appendix 1. With this expression, the robotic arm moves horizontally.

Whereas an in-depth account of mental states associated to EEG is far beyond the scope of this chapter, the example above offers a glimpse into the potential for neurotechnology of the proposed logical of the mind with quantum computing.

The next section describes our method to render into sound the outcomes from running a quantum circuit to check the satisfiability of a logic expression, over a number of shots.

7   Technical and Practical Considerations

7.1   Reducing Outliers

In the example presented in Sect. 6, the system spin-marked the quantum states $|001\rangle$, $|011\rangle$ and $|101\rangle$. As shown in Fig. 18, it outputted 001, 011 and 101 significantly more times than any of the other possible values. But there are five outliers, whose probability of being observed were nearly 12% of the spin-marked ones. These outliers take place due to the very nature of the amplitude amplification algorithm (Johnston et al., 2019).

The probabilities of the outliers appearing could have been squashed further by repeating the circuit between the initial Hadamard gates and the measurement section a few times. What happens here is that interference can amplify the amplitudes (or 'probabilities') of spin-marked qubits and decrease the amplitudes of all others.

For the example above, repeating the circuit three times squashes the outliers to negligible levels, as shown in Fig. 19. However, one must exercise caution here. The number of repetitions needs to be specified carefully. In this particular case, if the circuit is repeated only twice, then the probabilities of the outliers would have increased instead. Moreover, a more important caveat is that such repetitions increase circuit's depth (i.e. the overall length of the circuit) considerably. On quantum hardware, this escalates significantly the likelihood of errors due to decoherence.

The problem of decoherence poses severe limitations to the quantity of successive gates that can be used in a circuit for a real quantum processor. The higher the number of gates sequenced one after the other, the higher the likelihood of decoherence to occur.

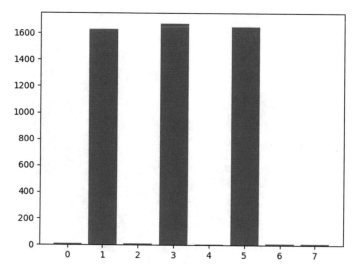

Fig. 19 Results yielded by running three copies of the circuit in Fig. 17 for 5,000 shots on a Rigetti's QVM. Binary numbers were converted to decimals for plotting on the horizontal axis. Vertical axis are the times each of the results were observed

7.2 Running on Quantum Hardware

At the time of writing, quantum processors struggle to maintain coherence for more than a dozen successive gates involving superposition and entanglement. Currently, coherence is assessed in terms of a few microseconds rather than seconds. However, research is progressing fast to improve this (Cho, 2020). In addition to improving hardware technology, there is much research activity to develop efficient quantum error correction methods.

In order to illustrate how challenging the problem of decoherence is, let us examine what happened when we ran Code 1 on the state of the art Rigetti's Aspen-8 quantum chip, with no added noise correction algorithm. Aspen-8 affords three-fold connectivity (i.e. supports 3-qubit gates) and coherence times of approximately 20 μs.

Figure 20 plots the outcomes after running the code for 2,048 shots on the quantum chip. The results do not match the ones obtained with the QVM simulator; they are all over the place. This is because the circuit is indeed too deep for this chip.

In truth, even the circuit for the first logic clause (A ∨ B) is problematic. The Quil code to examine the satisfiability of this simple clause is shown in Code 2. Table 5 shows the outcomes that satisfy this clause. Thus, one can predict that each time we run Code 2, it would output either 000, 101, 110, 111. Indeed, Fig. 21 shows the times that each of the possible outcomes were observed after running the circuit for 2,048 shots on Rigetti's QVM. They match the prediction perfectly. Yet, the outcomes from running exactly the same code on Aspen-8 are not accurate (Fig. 22). Why is this so?

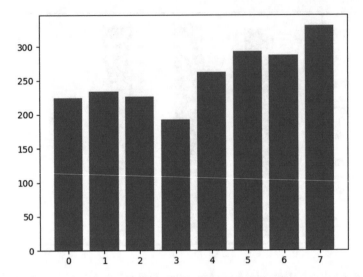

Fig. 20 Results from running the circuit shown in Fig. 17 for 2,048 shots on a Rigetti's Aspen-8 quantum chip. Binary numbers were converted to decimals for plotting on the horizontal axis. Vertical axis are the times each of the results were observed

Table 5 The outcomes that satisfy the logical clause $(A \vee B)$

$(A \vee B) = |q_2\rangle$	$B = |q_1\rangle$	$A = |q_0\rangle$	Outcome	Boolean
0	0	0	$000 = 0$	True
0	0	1	$001 = 1$	False
0	1	0	$010 = 2$	False
0	1	1	$011 = 3$	False
1	0	0	$100 = 4$	False
1	0	1	$101 = 5$	True
1	1	0	$110 = 6$	True
1	1	1	$111 = 7$	True

```
DECLARE ro BIT[3]
H 0
H 1
X 0
X 1
CCNOT 0 1 2
X 0
X 1
X 2
MEASURE 0 ro[0]
MEASURE 1 ro[1]
MEASURE 2 ro[2]
```

Code 2 Quil code for the $(A \vee B)$ portion of the circuit shown in Figure 17

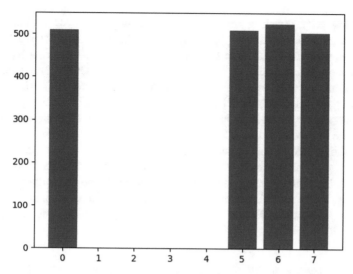

Fig. 21 Results from running the circuit shown in Code 2 for 2,048 shots on Rigetti's Quantum Virtual Machine (QVM). Binary numbers were converted to decimals for plotting on the horizontal axis. Vertical axis are the times each of the results were observed

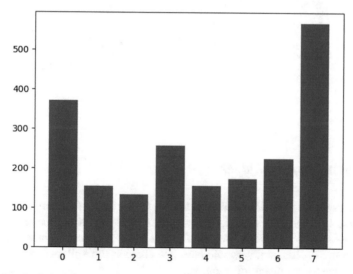

Fig. 22 Results from running the circuit shown in Code 2 for 2048 shots for 2048 trials on Rigetti's Aspen-8 quantum chip. Binary numbers were converted to decimals for plotting on the horizontal axis. Vertical axis are the times each of the results were observed

Fundamentally, quantum computing programming languages are built upon a handful of universal gates that physically act on quantum chips. Rigetti's Aspen-8 is a superconducting quantum chip, which can enact the following basic gates: Rx(ϑ), Rz(ϑ), CZ and XY(ϑ).[8] These are referred to as Quil's native gates. The measurements are done natively in the computational basis (z-axis). All other standard Quil gates (H, X, Z, Y, CNOT and so on) are built using native gates. Thus, before compilation, a standard Quil code needs to be transpiled[9] to native Quil. Code 3 shows the transpilation from Code 2.

Even though the circuit shown in Code 2 is 12 lines long and uses only eight standard gates, its transpilation results in 39 lines of code with as many as 35 native Quil gates. Unfortunately, this is too deep for Aspen-8. The real culprit here is the CCNOT, Toffoli gate. This gate is notoriously expensive in native gates.

Similar tests on an IBM's quantum computing resources can be found in Appendix 2.

```
DECLARE ro BIT[3]      RZ(-pi/4) 5        RX(pi/2) 6
RX(pi) 5               RX(-pi/2) 5        CZ 6 7
RZ(pi/2) 6             CZ 5 6             RX(pi/2) 6
RX(-pi/2) 6            RX(pi/2) 5         RZ(pi/4) 6
CZ 5 6                 RZ(pi/4) 5         RX(-pi/2) 6
RZ(-pi/2) 5            RX(-pi/2) 5        CZ 6 7
RX(pi/2) 5             CZ 7 5             RX(pi/2) 6
RZ(pi/4) 5             RX(pi/2) 5         RZ(-pi/2) 6
RX(-pi/2) 5            RZ(3*pi/4) 5       MEASURE 6 ro[1]
RZ(pi/2) 7             RX(pi/2) 5         RZ(-pi/4) 7
RX(-pi/2) 7            RZ(-pi/2) 5        RX(pi) 7
CZ 5 7                 MEASURE 5 ro[2]    MEASURE 7 ro[0]
RX(pi/2) 5             RZ(-3*pi/4) 6      HALT
```

Code 3 The transpiled native version of the code shown in Code 2

7.3   Quantum Advantage?

The expressions in Sect. 6 are limited to three logic variables. One may not even need a computer to check their satisfiability, even less so a quantum computer. Should the clauses have entailed a unique variable for each electrode, then the task of checking the expressions' satisfiability would have been harder. But still, not hard enough to justify the need for a quantum computer.

However, if the expressions encompassed the whole set of 20 electrodes shown in Fig. 3, each of which corresponding to a unique logic variable, then the satisfiability problem would become considerably harder. In this case a quantum computer could well be advantageous.

---

[8] XY(ϑ) produces a parameterized iSWAP gate, which is not priority for discussion on this chapter. For more details, see (Abrams et al., 2019).

[9] A transpiler is a program that translates a piece of code into another at the same level of abstraction. It is different from a compiler whose output is in a lower level of abstraction than the input.

The amplitude amplification technique is a core component of the so-called Grover's algorithm. Introduced by Grover (1996), this algorithm uses amplitude amplification to search for an element in an unstructured set of N elements. A brute-force classic algorithm would scan all elements in the set until it finds the one that is sought after. In the worst-case scenario, the element in question could have been the last one to be checked, which means that the algorithm would have made N queries to find it. Provided that a given problem can be encoded efficiently in terms of qubits, Grover's algorithm would be able find a solution with $\sqrt{N}$ queries. Thus, Grover's algorithm provides a quadratic speedup. This benchmarking also applies for logic satisfiability problems.

Ignoring for the moment the format of the logical expressions discussed in Sect. 5, let us suppose that a system would need to verify if the values A = True, B = True and C = False satisfy a given logical expression. In this case, the three Boolean variables amount to eight possible combinations for A, B and C; that is, $2^3 = 8$. Therefore, a brute-force algorithm would need to make up to eight checks to get the answer. That is, the system would have to run the algorithm up to eight times. Conversely, Grover's algorithm could solve this with $\sqrt{8} = 2.8$ runs. Indeed, recall that the convincing results shown in Fig. 19 were obtained by running three copies of the circuit.

The difference in performance escalates significantly as the number of logic variables increases. For instance, 20 variables would require up to $2^{20} = 1,048,576$ checks classically, whereas Grover's algorithms would require $\sqrt{2^{20}} = 1,024$ runs.

Brushing aside any thorough comparison between the processing clocks of classical and quantum hardware, a quantum computer running Grover's algorithms would certainly outperform a classical computer running brute-force search. However, the slightly disappointing news is that the number of qubits needed to implement large Grover's-like circuits is prohibitive the present times.

Nevertheless, despite the limited capacity of current quantum hardware technology, the quantum computing research community remains optimistic. The industry is aiming at producing quantum chips with over 1,000 qubits by 2023 (Cho, 2020).

8   A Quantum BCI Musical Instrument

This section presents a practical application of the logic of the mind: a BCI sound synthesiser.

We developed a method to render sounds with the results from checking the satisfiability of the kinds of logical expressions introduced in Sects. 5 and 6 (Fig. 23).

The system generates sounds using an additive synthesiser. The additive synthesis technique is informed by the theory of Fast Fourier Transform, or FFT (Muller, 2015). It is based on the notion that sounds can be characterized as a sum of sine waves.

Additive synthesis works by deploying a number of sine wave sound generators (e.g. digital oscillators) to produce partials, which are added up to produce the final

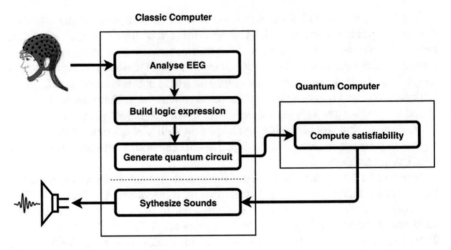

Fig. 23 The example system architecture

result. A sine wave is characterised by a frequency value (i.e. speed of the cycle) and an amplitude value (i.e. strength of the signal).

Our synthesiser comprises eight digital oscillators, each of which requires a frequency value in Hertz (Hz) and an amplitude, whose value varies between 0 (silence) and 1 (loudest). The outputs are summed and a Hanning function is applied to the result to give a smooth fade in and fade out bell shape to the sound.

In Fig. 24, individual partials are represented on the left-hand side of the figure, where a bar on the 'freq' axis (frequency domain) has a certain magnitude on the 'amp' axis (amplitude domain). The spectrum of the resulting sound is represented on the right-hand side; it contains eight partials. A schematic representation of the resulting sound, depicting its bell-like shape is also shown.

For every lapse of time the system checks the satisfiability of the respective logic expression, as discussed earlier, and uses the results to activate the oscillators of the synthesiser. Figure 25 delineates how the qubit measurements plotted in Fig. 18 are associated with oscillators. Notice that there are as many oscillators as the number of different quantum states that the respective circuit can return: in this case, $2^3 = 8$. That is, each possible output is associated with a different partial of the resulting sound. In additive synthesis, changes to the frequencies and amplitudes of the oscillators modify the timbre of the resulting sound. A schematic representation of this is given in Fig. 26.

Each oscillator of the synthesizer is assigned a frequency; for instance: osc 1 = 55.0 Hz, osc 2 = 164.81, osc 3 = 329.63 Hz and so on. This is fully customisable, and the system can be set to change these frequencies algorithmically. The amplitudes are normalized proportionally to the number of times the respective quantum state was observed after a number of pre-specified shots. Thus, in Fig. 25, the first oscillator, whose amplitude is controlled by the number of times the state $|000\rangle$ was observed,

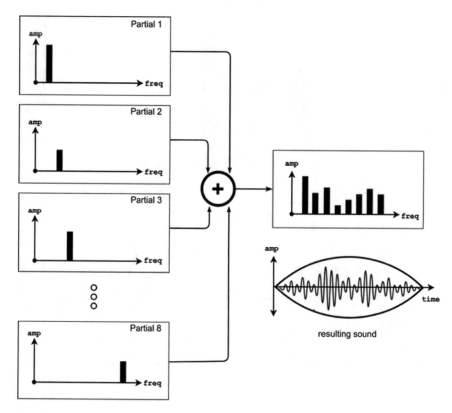

Fig. 24 Additive sound synthesis works by adding up a number of sine waves

will produce a partial that is much quieter than the second oscillator, whose amplitude is given by number of times that state $|001\rangle$ was observed.

The duration of the time lapses is also customisable. However, it should not be shorter than the window of time that is required to read and analyse EEG data. Our experiments suggest that at least 1 s is needed to capture sufficient EEG data to associate with a mental state. But in order to perceive differences clearly in the timbre of the sounds, it is suggested to let it play for no less than 5 s. Of course, time can vary and can be defined algorithmically should one wish to do so.

As a sound is being played, the system processes the EEG for the next time lapse, builds and runs the circuit, and synthesizes a new sound immediately after the current one. And the cycle continues for as long as required.

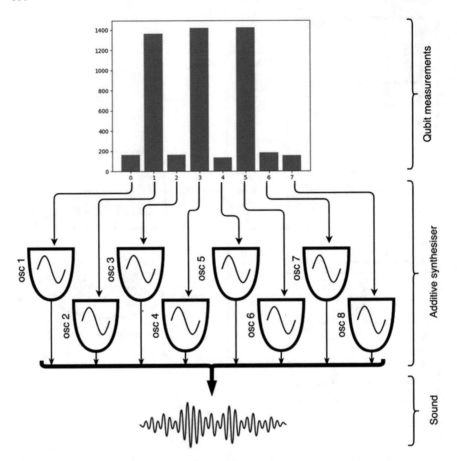

Fig. 25 From qubit measurements to sound synthesis

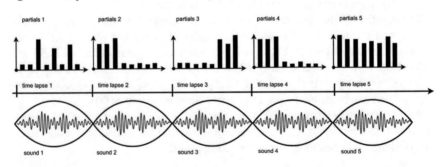

Fig. 26 Schematic representation of a sequence of five sounds and their respective partials shown above the time lapse line

9   Concluding Remarks

This chapter presented an approach to interfacing the brain and a quantum computer. Central to this approach is a method to encode brain activity as logic expressions representing states of mind. These expressions are associated with commands to control systems and/or machines with brain signals.

In addition to building brain–computer interfaces, the chapter put forward the notion of a logic of the mind as a means to study neural correlates of the mind and brain functioning, with potential benefits to medicine; e.g. for diagnosing brain disorder. The EEG is an important physiological signal for diagnostics (Tatum IV, 2008). It is used to detect traces of epilepsy, dementia, cancer, inflammation, sleep disorders and more. For instance, Clarke et al. (2013) reported that persistent excess of beta rhythms at the frontal regions of the brain can indicate hyperactivity disorder. It would be straightforward to encode this as a logic of the mind expression.

The EEG is currently the best signal for BCI systems. However, it is not necessarily the best for research into understanding brain functioning and mental correlates. For instance, functional magnetic resonance imaging (fMRI) measures the minuscule variations in blood flow that occurs with brain activity. It is used to determine which parts of the brain are more active than others when handling particular tasks. It has higher spatial resolution than EEG has, and it can detect information deep inside the brain.

The logic of the mind requires the processing of large Boolean expressions, which are computationally very demanding. Even more so if we were to use fMRI scanning. Today's average classical computer would take hours, if not days, to evaluate a handful of expressions. Hence, the rationale for using quantum computers.

At the time of writing, quantum computing hardware is of limited capacity for realistic applications. But once fully developed, quantum computers promise to be considerably faster than their classical counterparts to run certain types of algorithms, such as the one introduced in this chapter. In the meantime, simulations on virtual quantum machines enable research and development of quantum algorithms, which would eventually run optimally when robust quantum hardware is available.

By way of a practical demonstration, we presented a BCI system that renders the logic of the mind into sounds. We introduced a method to sonify quantum measurements obtained from evaluating logic expressions. In this case, the states of mind do not activate control commands. Instead, they are mapped directly onto sound. Effectively, the system can be thought of as a novel musical instrument.

Incidentally, the sound synthesis technique introduced above is a contribution to the field quantum computing on its own right, in the sense that it provides a method to represent the wavefunction of a quantum system auditorily.

Acknowledgements The author would like to thank Mathew Wilson and James Hefford of the Department of Computer Science at the University of Oxford for reviewing the quantum computing content of this chapter. Also, many thanks to Palaniappan Ramaswamy of the School of Computing at the University of Kent for scrutinizing the BCI content. Their meticulous comments and suggestions contributed significantly to the clarity and rigour of this work.

Appendix 1: Examples from the BCI System

Below are four examples of results taken from a run of the BCI system presented in this chapter, using Rigetti's QVM.

In relation to the BCI musical instrument, the different results produced variations in the timbre of the sounds. Recordings of the respective sounds, and other examples, are available online at SoundClick: https://soundclick.com/LogicoftheMind.

Example 1

This is an example where the logic expression is unsatisfiable (Table 6 and Fig. 27).

Table 6 The EEG analysis and generated expression for Example 1

Expression: $(\neg C \vee B) \wedge (C \vee A) \wedge (\neg C \vee B)$		
Clause 1: $(\neg C \vee B)$	**Clause 2:** $(C \vee A)$	**Clause 3:** $(\neg C \vee B)$
Term 1: Selected electrode = O1 Logic variable = C Frequency = 13.1649 Hz	**Term 1:** Selected electrode = Oz Logic variable = C Frequency = 31.1541 Hz	**Term 1:** Selected electrode = O2 Logic variable = C Frequency = 8.22338 Hz
Term 2: Selected electrode = T3 Logic variable = B Frequency = 24.7721 Hz	**Term 2:** Selected electrode = Fz Logic variable = A Frequency = 31.5992 Hz	**Term 2:** Selected electrode = T4 Logic variable = B Frequency = 27.0611 Hz

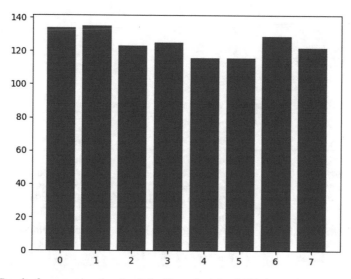

Fig. 27 Results from running the circuit for Example 1 for 1,000 times. Binary numbers were converted to decimals for plotting on the horizontal axis. Vertical axis are the times each of the results were observed

```
H 0                CONTROLLED CONTROLLED Z 5 4 3    H 2
H 1                X 2                              X 0
H 2                X 1                              X 1
X 2                X 5                              X 2
X 2                X 2                              CONTROLLED CONTROLLED Z 0 1 2
X 1                CCNOT 2 1 5                      X 0
CCNOT 2 1 3        X 2                              X 1
X 2                X 2                              X 2
X 1                X 1                              H 0
X 3                X 2                              H 1
X 2                X 0                              H 2
X 2                X 4                              DECLARE ro BIT[3]
X 0                CCNOT 2 0 4                      MEASURE 0 ro[0]
CCNOT 2 0 4        X 2                              MEASURE 1 ro[1]
X 2                X 0                              MEASURE 2 ro[2]
X 0                X 2
X 4                X 1
X 2                X 3
X 2                X 2
X 1                CCNOT 2 1 3
CCNOT 2 1 5        X 2
X 2                X 2
X 1                X 1
X 5                H 0
X 2                H 1
```

Code 4 Generated Quil code for Example 1

Example 2

See Table 7, Fig. 28.

Table 7 The EEG analysis and generated expression for Example 2

Expression: $(B \vee A) \wedge (C \vee A) \wedge ((\neg A \vee \neg C)$		
Clause 1: $(B \vee A)$	**Clause 2:** $(C \vee A)$	**Clause 3:** $(\neg A \vee \neg C)$
Term 1: Selected electrode = T3 Logic variable = B Frequency = 20.6042 Hz	**Term 1:** Selected electrode = Oz Logic variable = C Frequency = 18.7471 Hz	**Term 1:** Selected electrode = Fp2 Logic variable = A Frequency = 8.0119 Hz
Term 2: Selected electrode = Fp1 Logic variable = A Frequency = 21.2267 Hz	**Term 2:** Selected electrode = Fz Logic variable = A Frequency = 32.5744 Hz	**Term 2:** Selected electrode = O2 Logic variable = C Frequency = 10.3202 Hz

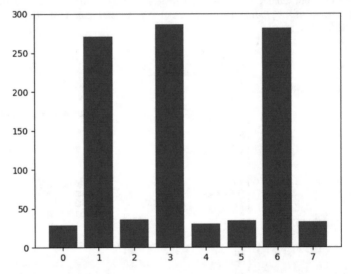

Fig. 28 Results from running the circuit for Example 2 for 1,000 times. Binary numbers were converted to decimals for plotting on the horizontal axis. Vertical axis are the times each of the results were observed

```
H 0
H 1
H 2
X 1
X 0
CCNOT 1 0 3
X 1
X 0
X 3
X 2
X 0
CCNOT 2 0 4
X 2
X 0
X 4
X 0
X 2
X 0
X 2
CCNOT 0 2 5
X 0
X 2
X 5
X 0
X 2
```

```
CONTROLLED CONTROLLED Z 5 4 3
X 0
X 2
X 5
X 0
X 2
CCNOT 0 2 5
X 0
X 2
X 0
X 2
X 2
X 0
X 4
CCNOT 2 0 4
X 2
X 0
X 1
X 0
X 3
CCNOT 1 0 3
X 1
X 0
H 0
```

```
H 1
H 2
X 0
X 1
X 2
CONTROLLED CONTROLLED Z 0 1 2
X 0
X 1
X 2
H 0
H 1
H 2
DECLARE ro BIT[3]
MEASURE 0 ro[0]
MEASURE 1 ro[1]
MEASURE 2 ro[2]
```

Code 5 Generated Quil code for Example 2

Example 3

See Table 8, Fig. 29.

Table 8 The EEG analysis and generated expression for Example 3

Expression: $(\neg C \vee A) \wedge (C \vee A) \wedge (\neg B \vee \neg A)$		
Clause 1: $(\neg C \vee A)$	**Clause 2:** $(C \vee A)$	**Clause 3:** $(\neg B \vee \neg A)$
Term 1:	**Term 1:**	**Term 1:**
Selected electrode = O1	Selected electrode = Oz	Selected electrode = T4
Logic variable = C	Logic variable = C	Logic variable = B
Frequency = 13.7849 Hz	Frequency = 17.5519 Hz	Frequency = 12.6194 Hz
Term 2:	**Term 2:**	**Term 2:**
Selected electrode = Fp1	Selected electrode = Fz	Selected electrode = Fp2
Logic variable = A	Logic variable = A	Logic variable = A
Frequency = 23.9491 Hz	Frequency = 18.6791 Hz	Frequency = 13.1322 Hz

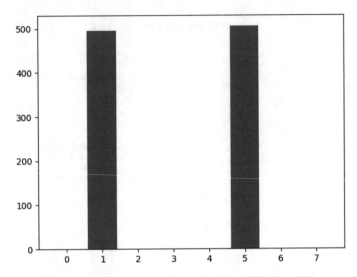

Fig. 29 Results from running the circuit for Example 3 for 1,000 times. Binary numbers were converted to decimals for plotting on the horizontal axis. Vertical axis are the times each of the results were observed

```
H 0                      X 5
H 1                      X 1                      CCNOT 2 0 3
H 2                      X 0                      X 2
X 2                      CONTROLLED CONTROLLED Z 5 4 3    X 2
X 2                      X 1                      X 0
X 0                      X 0                      H 0
CCNOT 2 0 3              X 5                      H 1
X 2                      X 1                      H 2
X 0                      X 0                      X 0
X 3                      CCNOT 1 0 5              X 1
X 2                      X 1                      X 2
X 2                      X 0                      CONTROLLED CONTROLLED Z 0 1 2
X 0                      X 1                      X 0
CCNOT 2 0 4              X 0                      X 1
X 2                      X 2                      X 2
X 0                      X 0                      H 0
X 4                      X 4                      H 1
X 1                      CCNOT 2 0 4              H 2
X 0                      X 2                      DECLARE ro BIT[3]
X 1                      X 0                      MEASURE 0 ro[0]
X 0                      X 2                      MEASURE 1 ro[1]
CCNOT 1 0 5              X 0                      MEASURE 2 ro[2]
X 1                      X 3
X 0                      X 2
```

Code 6 Generated Quil code for Example 3

Table 9 The EEG analysis and generated expression for Example 4

Expression: $(A \lor B) \land (B \lor A) \land (B \lor C)$		
Clause 1: $(A \lor B)$	**Clause 2:** $(B \lor A)$	**Clause 3:** $(B \lor C)$
Term 1: Selected electrode = Fp1 Logic variable = A Frequency = 15.0409 Hz	**Term 1:** Selected electrode = Cz Logic variable = B Frequency = 30.1681 Hz	**Term 1:** Selected electrode = T4 Logic variable = B Frequency = 18.4086 Hz
Term 2: Selected electrode = T3 Logic variable = B Frequency = 18.6357 Hz	**Term 2:** Selected electrode = Fz Logic variable = A Frequency = 32.1824 Hz	**Term 2:** Selected electrode = O2 Logic variable = C Frequency = 19.1024 Hz

Example 4

See Table 9, Fig. 30.

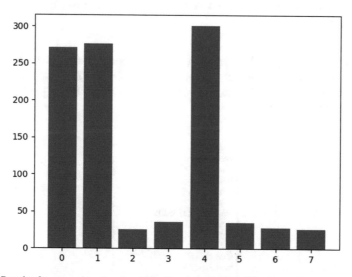

Fig. 30 Results from running the circuit for Example 4 for 1,000 times. Binary numbers were converted to decimals for plotting on the horizontal axis. Vertical axis are the times each of the results were observed

```
H 0                          CONTROLLED CONTROLLED Z 5 4 3    X 2
H 1                          X 1                              X 0
H 2                          X 2                              H 0
X 0                          X 5                              H 1
X 1                          CCNOT 1 2 5                      H 2
CCNOT 0 1 3                  X 1                              X 0
X 0                          X 2                              X 1
X 1                          X 1                              X 2
X 3                          X 0                              CONTROLLED CONTROLLED Z 0 1 2
X 1                          X 4                              X 0
X 0                          CCNOT 1 0 4                      X 1
CCNOT 1 0 4                  X 1                              X 2
X 1                          X 0                              H 0
X 0                          X 0                              H 1
X 4                          X 1                              H 2
X 1                          X 3                              DECLARE ro BIT[3]
X 2                          X 1                              MEASURE 0 ro[0]
CCNOT 1 2 5                  X 2                              MEASURE 1 ro[1]
X 1                          X 5                              MEASURE 2 ro[2]
X 2                          CCNOT 2 0 3
X 5                          X 2
```

Code 7 Generated Quil code for Example 4

Appendix 2: Study Using IBM Quantum Computing Resources

Figure 31 shows the results from running the excerpt corresponding to the logic clause $(A \lor B)$ on an IBM quantum computer simulator, for 2,048 shots. The Quil code shown in Code 2, in Sect. 7.2, was translated into the IBM's OpenQASM language, as shown in Code 8.

```
qreg q[3];
creg c[3];
h q[0];
h q[1];
x q[0];
x q[1];
ccx q[0], q[1], q[2];
x q[0];
x q[1];
x q[2];
measure q[0] -> c[0];
measure q[1] -> c[1];
measure q[2] -> c[2];
```

Code 8 OpenQASM version of the Quil code shown in Code 2

Not surprisingly, the results shown in Fig. 31 are identical to those obtained with Rigetti's QMV, as shown in Fig. 21.

The outcomes from running the exact OpenQASM code on IBM's Santiago processor (*ibmq_santiago* v1.1.1) is shown Fig. 32. Although there were errors, in overall the results are comparable to those produced by the simulator. IBM's Santiago processor proved to be more resilient than Rigetti's Aspen-8 in this case.

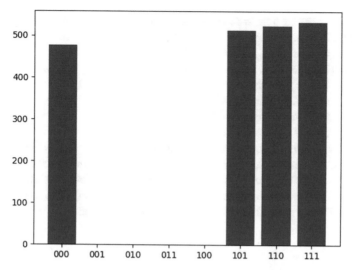

Fig. 31 Results from running the circuit for the logic clause (A ∨ B) for 2,048 trials on IBM's simulator

However, it should be noted the programming for both machines were kept simple, with no error correction and no a priori hardware calibration. Aspen-8 might have performed just as well with appropriate calibration and noise mitigation procedures. A detailed technical discussion on the idiosyncrasies of these machines, calibration, error mitigation strategies and so on, is beyond the scope of this chapter.

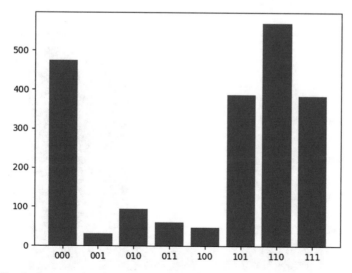

Fig. 32 Results from running the circuit for the logic clause (A ∨ B) for 2,048 trials on IBM's Santiago processor

In order to assess how well IBM's hardware would fare with a deeper circuit, we added a third logic variable and a phase-logic AND operation to implement the expression $((A \lor B) \land C)$, as shown in Fig. 33.

Considering that qubit strings are enumerated from the right end of the string to the left (i.e. $|CBA\rangle$), the expression $((A \lor B) \land C$ is satisfied when:

Case 1, $|101\rangle$: A = True, B = False and C = True.
Case 2, $|110\rangle$: A = False, B = True and C = True.
Case 3, $|111\rangle$: A = True, B = True and C = True.

Figure 34 shows the outcomes from running the circuit on the IBM quantum computer simulator, for 2,048 shots. Not surprisingly, the measurements are fairly accurate. Yet, the outcomes from running exactly the same code on IBM's Santiago processor (*ibmq_santiago* v1.1.1) are not accurate (Fig. 35). The transpiled code comprised 75 gates; the circuit is too deep for this processor.

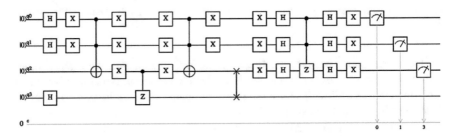

Fig. 33 Quantum circuit for the expression $((A \lor B) \land C)$

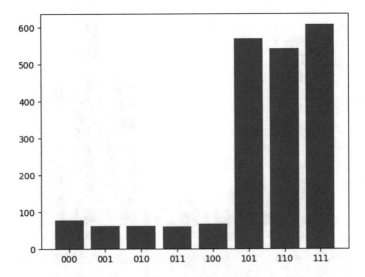

Fig. 34 Results from running the circuit in Fig. 33 for 2,048 trials on IBM's simulator

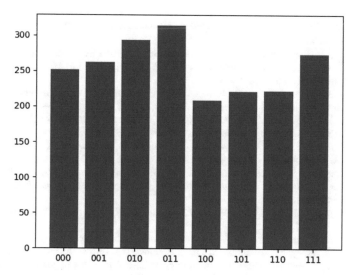

Fig. 35 Results from running the circuit in Fig. 33 for 2048 trials on IBM's Santiago processor

References

Abrams, D. M., Didier, N., Johnson, B. R., da Silva, M. P., & Ryan, C. A. (2019). Implementation of the XY interaction family with calibration of a single pulse. Retrieved October 24, 2020, from https://arxiv.org/abs/1912.04424v1

Aharonov, D. (2003). A simple proof that Toffoli and Hadamard are quantum universal. https://arxiv.org/abs/quant-ph/0301040

Akerkar, R., & Akerkar, R. (2004). *Discrete mathematics*. Person Education India.

Anderson, C., & Sijercic, Z. (1996). Classification of EEG signals from four subjects during five mental tasks. In *Solving Engineering Problems with Neural Networks: Proceedings of the Conference on Engineering Applications in Neural Networks (EANN'96)*, London, UK.

Ball, P. (2014). Questioning quantum speed. *Physics World*, January: 38–41.

Bernhardt, C. (2019). *Quantum computing for everyone*. The MIT Press.

Berta, R., Bellotti, F., De Gloria, A., Pranantha, D., & Schatten, C. (2013). Electroencephalogram and physiological signal analysis for assessing flow in games. *IEEE Transactions on Computational Intelligence and AI in Games, 5*(2), 164–175.

Cho, A. (2020). IBM promises 1000-qubit quantum computer—A milestone—By 2023. *ScienceMag.org* 15 Sep 2020. Retrieved October 30, 2020, from https://www.sciencemag.org/news/2020/09/ibm-promises-1000-qubit-quantum-computer-milestone-2023

Clarke, A. R., Barry, R. J., Dupuy, F. E., McCarthy, R., Selikowitz, M., & Johnstone, S. J. (2013). Excess beta activity in the EEG of children with attention-deficit/hyperactivity disorder: A disorder of arousal? *International Journal of Psychophysiology, 89*(3), 314–319.

Daly, I., Williams, D., Malik, A., Weaver, J., Kirke, A., Hwang, F., Miranda, E. R., & Nasuto, S. J. (2018). Personalised, multi-modal, affective state detection for hybrid brain-computer music interfacing. *IEEE Transactions on Affective Computing, 11*(1), 111–124.

Giannitrapani, D. (1985). *The electrophysiology of intellectual functions*. Karger.

Grover, L. K. (1996). A fast quantum mechanical algorithm for database search. In *Proceedings of 28th Annual ACM Symposium on the Theory of Computing*. https://arxiv.org/abs/quant-ph/9605043

Grumbling, E., & Horowitz, M. (Eds.). (2019). *Quantum computing: Progress and prospects*. National Academies Press. https://doi.org/10.17226/25196

Guenter, F. H.,, & Brumberg, J. S. (2011). Brain-machine interfaces for real-time speech synthesis. In *2011 Annual International Conference of the IEEE Engineering in Medicine and Biology Society*, Boston, MA, pp. 5360–5363.

Hasan, B. A. S., & Gan, J. Q. (2012). Hangman BCI: An unsupervised adaptive self-paced brain-computer interface for playing games. *Computers in Biology and Medicine, 42*(5), 598–606.

Hinterberger, T., Zlabinger, M., & Blasser, K. (2014). Neurophysiological correlates of various mental perspectives. *Frontiers in Human Neuroscience, 8*(637).

Jeunet, C., Tonin, L., Albert, L., Chavarriaga, R., Bideau, B., Argelaguet, F., Millan, J. del. R., Lecuyer, A., & Kulpa, R. (2020). Uncovering EEG correlates of covert attention in soccer goalkeepers: Towards innovative sport training procedures. *Scientific Reports, 10*(1705).

Johnston, E. R., Harrigan, N., & Gimeno-Segovia, M. (2019). *Programming quantum computers*. O'Reilly Media Inc.

Kraft, E., Gulyás, B., & Pöppel, E. (2008). *Neural correlates of thinking*. Spriger Science & Business Media.

Kropotov, J. (2008). *Quantitative EEG, event-related potentials and neurotherapy*. Academic Press.

Ligeza, T., Wyczesany, M., Tymorek, A. D., & Kaminski, M. (2015). Interactions between the prefrontal cortex and attentional systems during volitional affective regulation: An effective connectivity reappraisal study. *Brain Topography, 29*, 253–261.

Marcuse, L. V., Fields, M. C., & Yoo, J. J. (2015). *Rowan's primer of EEG* (2nd ed.). Elsevier.

McFarland, D. J., & Wolpaw, J. R. (2017). EEG-based brain-computer interfaces. *Current Opinion in Biomedical Engineering, 4*, 194–200.

Mermin, N. D. (2007). *Quantum computer science: An introduction*. Cambridge University Press.

Miranda, E. R. (2006). Brain-computer music interface for composition and performance. *International Journal on Disability and Human Development, 5*(2), 119–125.

Miranda, E. R., Magee, W. L., Wilson, J. J., & Palaniappan, R. (2011). Brain-computer music interfacing (BCMI): From basic research to the real world of special needs. *Music and Medicine, 3*(3), 134–140.

Muller, M. (2015). *Fundamentals of music processing: Audio, analysis, algorithms, applications*. Springer.

Nicolas-Alonso, L. F., & Gomez-Gil, J. (2012). Brain computer interfaces, a review. *Sensors, 12*(2), 1211–1279.

Oh, S.-H., Lee, Y.-R., & Kim, H.-N. (2014). A novel EEG feature extraction method using Hjorth parameter. *International Journal of Electronics and Electrical Engineering, 2*(2), 106–110. https://doi.org/10.12720/ijeee.2.2.106-110

Palaniappan, R. (2008). Two-stage biometric authentication method using thought activity brain waves. *International Journal of Neural Systems, 18*(1), 59–66.

Petsche, H., & Etlinger, S. C. (1998). *EEG and thinking: Power and coherence analysis of cognitive processes*. Austrian Academy of Sciences.

Renard, Y., Lotte, F., Gibert, G., Congedo, M., Maby, E., Delannoy, V., Bertrand, O., & Lecuyer, A. (2010). OpenViBE: An open-source software platform to design, test, and use brain-computer interfaces in real and virtual environments. *Presence, 19*(1), 35–53.

Rieffel, E., & Polak, W. (2011). *Quantum computing: A gentle introduction*. The MIT Press.

Rozemond, M. (1988). *Descartes's dualism* (p. 1998). Harvard University Press.

Smith, P. (2020). *An introduction to formal logic. Logic matters*, 2nd edn. Retrieved October 10, 2020, from https://www.logicmatters.net/ifl/

Smith, R., Curtis, M. J., & Zeng, W. (2017). A practical quantum instruction set architecture. Retrieved October 13, 2020, from https://arxiv.org/abs/1608.03355v2

So, W. K. Y., Wong, S. W. H., Mak, J. N., & Chan, R. H. M. (2017). An evaluation of mental workload with frontal EEG. *PLoS One, 12*(4), e0174949.

Squire, L., Berg, D., Bloom, F., du Lac, S., Ghosh, A., & Spitzer, N. (Eds.). (2008). *Fundamental neuroscience* (3rd ed.). Academic Press.

Vartanian, O., Bristol, A. S., & Kaufman, J. C. (Eds.). (2013). *Neuroscience of creativity*. The MIT Press.

Wilson, J. J., & Palaniappan, R. (2011). Analogue mouse pointer control via an online steady state visual evoked potential (SSVEP) brain–computer interface. *Journal of Neural Engineering, 8*(2), 025026.

Yelamanchili, T. (2018). Neural correlates of flow, boredom, and anxiety in gaming: An electroencephalogram study. Missouri University of Science and Technology, Masters Theses 7812. Retrieved October 10, 2020, from https://scholarsmine.mst.edu/masters_theses/7812

The History of Games for (Quantum) Computers

James R. Wootton

Abstract Computer games are not just one application of computers, they are a multitude. A wide variety of computational tasks are combined, all running as fast as possible, to deliver the best possible experience to the player. It is reasonable to expect that, somewhere in this maelstrom of computation, we can find something that quantum computers will excel at. If the quest to combine games and quantum computers was a game itself, it would be one of open-world exploration. Though there may be no well-defined, concrete goals to achieve, there is something that will guide us: this game is a sequel. By looking at how we combined games with computers the first time, we can get some idea of what we can expect in this quantum successor.

Keywords Quantum computing · Computer games · Quantum speedup · Procedural generation · Terrain generation

1 Introduction

Computer games are not just one application of computers, they are a multitude. A wide variety of computational tasks are combined, all running as fast as possible, to deliver the best possible experience to the player. It is reasonable to expect that, somewhere in this maelstrom of computation, we can find something that quantum computers will excel at.

With this we begin the adventure! If the quest to combine games and quantum computers was a game itself, it would be one of open-world exploration. Though there may be no well-defined, concrete goals to achieve, there is something that will guide us: this game is a sequel! By looking at how we combined games with computers the first time, we can get some idea of what we can expect in this quantum successor.

J. R. Wootton (✉)
IBM Quantum, IBM Research – Zurich, Zurich, Switzerland
e-mail: JWO@zurich.ibm.com

© Springer Nature Switzerland AG 2022
E. R. Miranda (ed.), *Quantum Computing in the Arts and Humanities*,
https://doi.org/10.1007/978-3-030-95538-0_10

2   The 1950s: What Can Games Do for Computers?

Since the early history of computing, people have been interested in how it relates to games. This has come in multiple forms: Can computers implement games? Can they play games, and perhaps even beat humans? Could they offer experiences that would be impossible without them? We now know that the answer to all these questions is 'yes' since the last few decades have conclusively shown what computers can do for games.

Back in the 1950s, however, the combination of games and computers got off to a slow start. Early examples were less about what computers could do for games and more about what games could do for computers.

One of the first examples was *Bertie the Brain* exhibited at the Canadian National Exhibition in 1950. It used vacuum tubes and light bulbs and played Tic-tac-toe. Not because it offered a better play experience than using pen and paper, but because it helped to show off the vacuum tubes. The fact that we are writing about it now, over 70 years later, shows that it certainly had some impact. It showed us the first possible reason for making games with computers: to showcase new technology. Unfortunately, in this case, however, the specific technology it was showcasing soon became obsolete.

Then, there was *Nimrod* in 1951. Again, vacuum tubes and light bulbs, but this time the game was *Nim*. It was shown at the Festival of Britain with a stated aim, among other things, to '…illustrate the algorithm and programming principles involved'. This gave us a second reason to make games with computers: for education. Programming would have seemed a strange and arcane art in those days. How better to make it relatable than to show it playing a game?

Next was a research project in 1952 at Cambridge University, now commonly referred to as *OXO*. The idea was to study human–computer interaction, done using another implementation of tic-tac-toe (or 'noughts and crosses' as it is known in Cambridge). It again used vacuum tubes, but this time had cathode ray displays: the first graphical upgrade of the games industry. Nevertheless, the driving principle was research, giving us another reason to make games with computers.

Another project to note was that done at IBM Research in the 1950s: making artificial intelligence (AI) to play *Checkers* and beat a human opponent (Samuel, 1959). This would be something that IBM would repeat decades later, using *Deep Blue* to beat Gary Kasparov at chess, and then *Watson* to beat champions at *Jeopardy*. DeepMind has also been prominent in this area, using *AlphaGo* to beat a champion at *Go*. All of which are more examples of using games to showcase technology.

There were more games in the 1950s, but the ones here show what the decade was all about. Whether it was to showcase the technology, to educate about the technology or to research the technology, games were used to help computers rather than the other way around. An exception was *Tennis for Two*, a game made in 1958 for fun; and it signalled what was coming.

3 The 1960s: What Can Computers Do for Games?

In 1962, a team at MIT decided to put their new PDP-1 computer to good use. They embarked on a project that would:

- Test out the new device (and later, new installations).
- Showcase its capabilities.
- Be fun!

They created a game. But it wasn't just an existing game, with a computer-based implementation tacked on as a gimmick. It was something new: space battles without the need to build your own spaceship. It was called *Spacewar*.

This was the first clear example of computers doing something for games. They were used to make something new, and something fun. Functions made for trigonometric calculations were repurposed to let the player experience flying around a star.

It was not the only game this decade to be set in space. There was also the original *Lunar Lander* as well as the unimaginatively named *Space Travel*. More down to Earth was an economics simulator, *The Sumerian Game*, which was an early ancestor of games like *SimCity*.

In each case, the computer was used to do exactly what computers were made for: crunching numbers and performing simulations. But rather than being used for payroll calculations or planning a real trip to the moon, the computers were used to provide unique play experiences.

4 The 1970s: Commercial Success

Not many people know about *Spacewar!* or *Space Travel*. Any familiarity with *Lunar Lander* is usually due to one of its later incarnations, rather than the 1960s text-based original. This was not due to any secrecy on the part of the developers: they usually freely shared the source code. It was more because few had the room for a room-sized computer, never mind the money.

The 1970s is when it became possible to bring the hardware to the masses. The first arcade game was *Computer Space*, conceived of as a coin-operated *Spacewar!* that ran on hardware dedicated only to the game. The first generation of home consoles soon followed. With these breakthroughs, computer games became something you could sell to the masses, either in an arcade or at home. The games industry had truly begun.

Something else also started in the 1970s, when Paul Benioff began to think about computers that operate according to the principles of quantum mechanics (Benioff, 1980). It has taken a long time since then to fully formulate the idea, explore applications and develop the required hardware. But we are finally close to fruition.

5   The 2010s: What Can Games Do for Quantum Computers?

By 2010 we already had many examples of algorithms that could run on quantum computers, to solve problems that would be intractable for even the biggest supercomputer; see here for a few examples (Jordan, 2021). However, the hardware required to actually run these algorithms was far beyond what was available. Instead of the many thousands of high-quality qubits that are needed, just having two qubits was pretty impressive in 2010. This hardware was also not easy to use, being hidden away in labs around the world and programmed primarily through the manual labour of research students.

The state of affairs did not seem much better in 2015 when this author, a theoretical physicist working on quantum error correction, proposed a quantum computing experiment for 5-qubits (Wootton, 2016). Proposing an experiment was easy. Convincing a research group with the required hardware to actually run it would be much harder. And running it would be harder still.

This all changed in 2016 when IBM released their *Quantum Experience*: A web interface allowing direct access to a 5-qubit device (Quantum Composer and Lab). On one fateful day, just before lunchtime, this author suddenly realized that this provided exactly what was required for his proposed experiment. By the time he left the work, the experiment had been run. A revolution had begun!

Since then, a great many others like this author have used direct and simple access to quantum hardware as part of their scientific research. But that is beyond the scope of this chapter. With access to new computational hardware, there are also other things we can do. It took us until early 2017 to realize that this included making games.

Most probably the first-ever game using a quantum computer is a simple version of *Battleships* (Wootton, 2017). The display was just ASCII in a terminal. The gameplay was no better than you could do with pen and paper. Rather than being made to be played, it was made to be the basis of a blog post.

The idea was to give an example of quantum programming that wasn't a complex algorithm, but instead something more tangible. The states of a qubit were used to encode whether a ship had been sunk or not. Quantum gates were used to implement attacks. Without realizing it at the time, the author repeated the history of *Nimrod*, mentioned above. Just as normal programming seemed like an arcane art then, quantum computing can seem so now. So the game was designed to 'illustrate the algorithm and programming principles involved'. It was a quantum game made for the purpose of education.

The next major quantum game was *Quantum Awesomeness* (Wootton, 2018). This was based on the fact that there are many aspects of a quantum processor that determine how useful it is: number of qubits, which pairs of qubits we are allowed to apply a two-qubit manipulation to, and what imperfections are present. Thus, a puzzle game was designed to help people get an idea of how different devices compare.

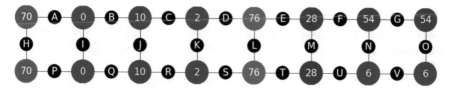

Fig. 1 An example of *Quantum Awesomeness*

The game was designed so that the puzzles were based on the specifications of the device used to run it. The more qubits you had, and the more complex the layout of possible two-qubit gates, the better the puzzles would be. The more imperfections there were, the less it would seem to make any sense. Therefore, to compare devices and see which was best, one only needed to see which was the most fun. It is not as informative as looking at benchmarking data, but it is a lot more accessible to the public.

An example can be seen in Fig. 1. The coloured circles represent qubits on the device. The connections between them show us which pairs of qubits may be directly manipulated through two-qubit gates. And hence the device is itself the puzzle board.

The principle behind the puzzle is simple: a random pairing of the qubits is chosen, and the player has to work out what the pairing is. This is implemented by running a quantum program whose details are defined by the chosen pairing. From the results, a number is calculated for each qubit. This number should be exactly equal for the two qubits of each pair, and so the player's job should be a simple one: they just need to find the pairs of equal numbers.

For an example such as the one above, finding this pairing is indeed a very easy job; it is the pairs labelled H, I, J, K, L, M G and V if the reader is wondering. However, these results come from a simulation of a quantum computer, done without any of the imperfections inherent in real quantum hardware. An example from real hardware (with a different pairing) can be seen in Fig. 2.

In the case shown in Fig. 2, the numbers that should be equal are not as equal as they should be. We see that some pairs of qubits are behaving fairly well. Others qubits, like the one whose number is 14, seem to be giving us total nonsense. This, therefore, allows us to get an idea of how well the qubits work on this device, and also allows us to compare qubits between devices.

This game was designed so that the player didn't need to have direct access to the devices. Instead, puzzles could be made once and played later. This is also

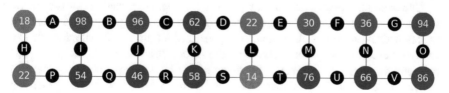

Fig. 2 Running *Quantum Awesomeness* on quantum hardware

demonstrated in the image above, where you can still play the game despite the device it ran on having been decommissioned some time ago. This was an example of making quantum games for the purpose of research, repeating the history of the 1950s.

The next game to use a real quantum device was *Q|Cards⟩*, made in 2019 by a team working at Quantum Wheel, the Fifth Game Jam held in Helsinki, Finland (Cards). This is a card game in which each player has a qubit and each card represents a quantum gate. The quantum computer is used at the very end to judge who played their gates best.

It is a fun game to play, but it also helps to teach people about quantum gates. Instead of getting them to sit through a lecture or read through a textbook, playing the game helps us explain the basics of quantum computing in a friendly atmosphere. Another quantum game made for education.

The motivation behind these games was exactly the same as in the 1950s: education and research. They also served the other big purpose of 1950s games, to showcase the new technology, since their very existence helped to get the word out in the form of blog posts, tweets and more.

The short summaries of the above-mentioned games are intended simply to give a flavour of the quantum games of the 2010s. We will look at one of these examples in detail later in this chapter.

6 Quantum Games Without Quantum Computers

If the history of computer games began in the 1950s, then the 1940s was its prehistory. One relic of this era was *Turochamp*, a chess AI developed by Alan Turing and David Champernowne (Copeland, 2004). It had one big problem: it was too complex for the computer hardware of the time. It was run in 1948, but with a human manually simulating the processes. This too has a parallel in recent experiments with quantum games. It is easier to simulate simple quantum programs on a normal computer than to run them on real quantum hardware. This allows us greater flexibility to explore what qubits can do, within the familiar and fun context of hacking together simple games.

Many games have been made in this way; for instance, *QPong* (a quantum version of Pong) (QPong), *QiskitBlocks* (building quantum programs in a Minecraft-like world) (QiskitBlocks), to cite but two.

One example of our own was not so much a game, but a method for procedural generation. It used just a single qubit (the smallest element of quantum computation) to create a landscape to walk around. The generated terrain was far simpler than can easily be created using non-quantum methods, but that was beside the point. The project was intended to provide a hands-on means by which people could learn all the intricacies of single-qubit manipulations, by creating their own terrain generator. This project will be explained in detail in Sect. 9.

7 The 2020s: What Can Quantum Computers Do for Games?

What can quantum computers actually do for games? Like everything, they bring some benefits but also some design constraints. For one thing, the need to cool the hardware to near absolute zero means that you'll never get one inside a home console. They'll live on the cloud, and so probably aren't something that you want to send jobs to every frame. For any game that might use a quantum computer, the vast majority of it will still run on a normal computer. We just need to find a small corner that quantum computers can excel at.

Though speeding up computation is the big promise of quantum computers, the aim is mostly to take problems that could never otherwise be tackled within a human lifetime and instead solve them within a reasonable timescale. Tasks that can already be done very quickly with conventional hardware, will continue to be done with conventional hardware, with no hope of a quantum speedup.

The most reasonable points in a game to insert quantum results are those where responses are not expected in less than a second. This includes tasks such as setting up the levels, puzzles or worlds that the player will experience. These jobs can be done during a loading screen, or even during the game design process, ensuring that the quantum computer has time to do its thing. The future of quantum computation in games, therefore, looks likely to be dominated by procedural generation.

As with any attempt to find applications for quantum computers, we also need to think about the different eras that the technology will go through. Our main aim is a future era, where all imperfections are removed by error correction, where you'll have more qubits than you'll ever need, and where you'll be able to do any two-qubit gates you want. This is the era of fault-tolerant, scalable quantum computing. With this hardware, we will be able to implement the many algorithms developed over the last few decades. By looking at the kinds of problems that they can solve, we can begin to speculate on how they will be used in procedural generation.

Certain possibilities immediately jump out. Quantum computation has been shown to offer more efficient computational complexity for constraint satisfiability problems, which could be useful in searching the space of different forms of content for those that satisfy desired properties. Also, since many problems in procedural generation can be expressed in the abstract language of nodes connected by a network, quantum speed-ups for the analysis of such networks could offer additional tools. These are motivating examples, but they are admittedly not defined very concretely. This is because quantum computation is a field that has mostly concerned itself with asymptotic properties of a computation's runtime: When thinking about how this time grows with the size of a problem, we know much about what shape the graph would be but little about the exact values. Games, on the other hand, are an application that depends greatly on the exact speed at which calculations can be performed. A factor of 10 in speed can be the difference between something being useful or useless in a game.

The exact runtime of quantum algorithms cannot currently be accurately predicted, nor can the exact size of problems for which conventional hardware must give way to a quantum approach. These quantities will depend on exactly how fault-tolerant quantum computers finally come to fruition, which is still an area of active research. Forecasting exactly how these devices will be useful for procedural generation in games is therefore not possible.

However, quantum computers do exist today in non-fault-tolerant form. They are already pushing the boundary of what can be reproduced with conventional hardware. Though the 'textbook' algorithms might not apply, we can seek to create tailor-made methods for the device we are using and the kind of results we want to get.

Over the next few years, we can seek to discover our own algorithms, to make interesting and useful content for procedural generation. This may not be able to rely on solving complex constraint satisfiability problems and such like, but the devices are already able to generate interesting content. This author has been using them for purposes exactly like these over the past few years, both for games in 'game jams' and papers for research conferences (Wootton 2020a; Wootton 2020b). As the hardware gets ever more sophisticated, so too will the possibilities for what we can do with it. In this way, we can hope to reach some kind of quantum *Spacewar!*, the first example of a unique play experience made possible with the new technology.

8   Battleships with Partial NOT Gates

As discussed in the last section, one reason we might want to combine games with quantum computers is for education. The most obvious method is to make a game such that players will learn about quantum computing through play. But people also enjoy making games, allowing us to also explore a very different possibility: learning about quantum computing by using it to make a game.

This was the idea behind the pioneering *Battleships with partial NOT gates* (Wootton, 2017). It was not made to offer a good player experience but instead was made to form the basis of a blog post. It was used as a simple and relatable example of quantum computers doing something, which would be less imposing than the quantum algorithms of textbooks. It was the *Nimrod* of quantum games, as mentioned earlier, built to '…illustrate the algorithm and programming principles involved'.

The game itself is a variant of the abovementioned *Battleships*. The central idea is that there are ships that get attacked, and after a given number of hits they will sink. Everything else, such as inputs, outputs and the handling of turns, is just bookkeeping.

Like any quantum algorithm, any quantum game will require a hybrid approach. Conventional digital computers will be used for some tasks, and quantum computers will be used for others. In this case we will use quantum computation for the central game mechanic. Thus, it will keep track of how damaged each ship is and implement attacks. Before we do this, let us start with an explanation of what a bit is. Even though they have become a ubiquitous concept in our increasingly digital world, they are not something that people spend much time thinking about.

A bit is simply the smallest possible chunk of information. It can take one of two possible values, which we typically call 0 and 1. We can use bits to store binary values. For example, if we are recording whether a ship has been destroyed or not, we could use 1 for it being destroyed and 0 for it being fully intact. To read out the information, we simply look at the value that was recorded.

The qubits at the heart of quantum computing are the quantum version of a bit. They too can only give outputs of 0 or 1. So we will use exactly the encoding described above, but with a qubit instead of a bit: The qubit state is certain to output 0 to encode the state of a ship that is perfectly undamaged, and that for 1 for a ship that has been destroyed.

A program for qubits takes the form of a so-called 'quantum circuit'. These are like the Boolean circuits with which conventional digital computing can be expressed when considering very low levels of abstraction. We will use the Qiskit framework (Qiskit) created by IBM Quantum and built in Python, to write these circuits. In this *Battleships* game, each player has five ships, so we begin by creating a circuit for five qubits:

```
from qiskit import QuantumCircuit
qc = QuantumCircuit(5)
```

On its own, we cannot do much with this circuit. We cannot even extract an output of 0 or 1 from the qubits. For that we would need a circuit which does not just have a qubit but has a bit on which an output can be written. So, let us modify our circuit to not only have 5 qubits but also 5 bits to serve as their outputs.

```
qc = QuantumCircuit(5,5)
```

To actually extract these outputs, we need to use a so-called measure gate, which specifies exactly when the process of extracting the output should be performed. For convenience we will keep this in a separate circuit, defined for 5 qubits (and bits) as follows.

```
m_z = QuantumCircuit(5,5)
for j in range(5):
    m_z.measure(j,j)
```

Note that qubits and bits are indexed from 0 in Qiskit. So, the five qubits and bits are referred to as 0, 1, 2, 3 and 4. The measure(j,j) command is telling the circuit to extract an output from the jth qubit and write the answer to the jth bit.

After creating a circuit, the final step is to run it. For complex circuits we have no choice but to use actual quantum hardware for this. For 5 qubits, however, it is possible to simulate using a conventional digital computer:

```
from qiskit import Aer
backend = Aer.get_backend('aer_simulator')
job = backend.run(qc.compose(m_z), shots=10, memory=True)
result = job.result()
```

```
outputs = result.get_memory()
```

Let us unpack what is happening here. Firstly, the circuit we ran was specifically the combination of qc and m_z: our original five qubits with all the measurements placed at the end. This combined circuit is created in Qiskit with the command qc.compose(m_z). The circuit was run on a so-called 'backend'. In this case the simulator known as the 'Aer simulator' was used. Using a real quantum device is as easy as changing the backend object to one that refers to a real quantum device so that the 'run' function sends the job to that device over the cloud.

The process is repeated for ten shots, so we can see if there is any randomness in the results. The output of each process will be a 5-bit string, given that our circuit has five output bits. Since we used shots = 10 to ask for ten runs of the same process, the results are returned as a list of the ten outputs. Specifically, we will get the following list:

```
['00000', '00000', '00000', '00000', '00000', '00000', '00000',
'00000', '00000', '00000']
```

Each output is '00,000'. This means that each of the five qubits output a 0 in each case. This is due to the way qubits are initialized: they always start out in the state that is certain to output a 0. Our five ships, therefore, start off fully undamaged.

To implement an attack on a ship we need to apply an operation to its qubits, known as a gate. The simplest example would be a completely fatal attack, flipping the ship directly from undamaged to destroyed. The corresponding qubit must therefore be flipped from the state certain to output 0 to the one certain to output 1. This effect can be achieved by two quantum gates, known as X and Y gates. Both can be regarded as quantum equivalents of the NOT gate for standard bits, which flips between the 0 and 1 states. In the following, we will focus on the X gate to implement the attack.

So, let us get on with it, and launch our most destructive torpedo at the first ship, which is the one corresponding to the qubit labelled 0:

```
qc.x(0)
```

This adds the new operation to the circuit qc: an X gate on qubit 0. If we run the combined circuit now, we will get the result '00,001'. Here the result for the first ship appears at the first bit on the right. It is a 1, denoting that the first ship has been destroyed. Therefore, our X gate had the desired effect.

Now let us load up a new torpedo and take aim for the second ship. But let us suppose that this ship is made of sturdier stuff, so one torpedo will not be enough to sink it. Instead, it will only half destroy its target. In order to implement this attack, we would like to do something like 'half of an X gate'. This means we need to find a gate which, when applied twice, leads to an end result that is exactly the same as an X gate.

If we were using a conventional digital computer, with a conventional bit representing each ship and a NOT gate to implement attacks, this would not be possible. Either you flip the bit value, or you do nothing. There is no in-between. More bits and different types of gate would be required to implement the possibility of a half-damaged ship. Of course, using more conventional bits is exactly the best course of action in this case, but it is not what our game is about. We will doggedly stick with just the one bit for each ship, and happily pay the price of making it quantum.

With quantum bits we can access so-called 'superposition states', which are not quite 0 and not quite 1. That is, when we measure them, the results will appear random, making it easy to think that they are simply the result of an overly engineered coin flip. However, that is not the case. These states are just as definite and well-defined as the ones whose outputs are certain. We could define measurements for which they would give certain outputs instead of random ones, as we will see in the next section. But when we ask the question of whether they are 0 or 1, they do not fit neatly into either box. They cannot use a third option to declare that they are neither: then they would no longer be bits since bits are only allowed two options by definition. Thus, they simply answer randomly. With that in mind, let us implement the following attack on the second ship (the one whose qubit is labelled 1):

```
from math import pi
qc.rx(pi/2, 1)
```

Instead of just an X gate, this is an RX gate. This gate requires two arguments: as before, we specify which qubit it acts on, but we also supply the value of pi/2. If we run the combined circuit now, we will find that the result is randomly either '00001' or '00011'. In both cases, the ships that we have not attacked remain undamaged (represented by their 0 s) and the first ship remains destroyed (represented by its 1). The second ship, however, which is represented by the second qubit on the right, is randomly either 0 or 1. By taking multiple samples we can determine that these two possibilities occur with equal probability.

In order to see that this is not just a fancy coin flip, let us implement the attack again with another `qc.rx(pi/2, 1)`. The entire quantum circuit `qc` is now:

```
qc = QuantumCircuit(5) # attack on the first ship
qc.x(0) # attack on the second ship
qc.rx(pi/2, 1)
qc.rx(pi/2, 1)
```

Running the combined circuit now yields the result '00011' with certainty. The first RX gate took its qubit from the state certain to output 0 to a superposition state between 0 and 1. The second RX gate completed the journey, taking the qubit to the state certain to output 1. Together they had exactly the same effect as a single X gate.

Now let us attack the next ship (the one whose qubit is labelled 2). This is even sturdier than the last and will require three torpedoes to sink. The required effect can be implemented with `rx(pi/3, 2)`. Applying this gate once will give random results, but with a bias towards 0. For two applications the bias will instead favour 1. For three we get 1 with certainty, and the ship is destroyed.

Similarly, we can use pi/4 for a gate that must be repeated four times to form an X gate, and so on. This argument can be thought of as representing an angle, as we will see in the next section. The angle pi is required for an X gate (and so `rx(pi, 0)` would be another way to express the X gate with which we attacked the first ship). For multiple sequential applications of RX gates on the same qubit, these angles effectively add together.

This is basically all there is to the quantum machinery behind *Battleships with partial NOT* gates. It does not show us how or why one might want to build qubits, or why the ability to do partial forms of the gates used in conventional digital computers will confer any computational advantage. The aim that it does achieve, however, is to show them in action. And to show that quantum computers can do a relatable job without the need for an abstract algorithmic approach.

9 Single-Qubit Procedural Generation

Now that the qubit is no longer completely unfamiliar, let us get to know it better. This can again be done in the context of using it to implement something that can be used in a game.

9.1 The Basic Properties of Qubits

One fundamental property of quantum objects is that they must be described by multiple incompatible attributes. This means that there are attributes of the object that cannot simultaneously be well-defined: If one is defined with absolute certainty, the others must be completely ambiguous.

This is not the easiest concept to grasp, since it is not a type of behaviour that we recognize from the large, non-quantum objects of our everyday life. If we know the size of an apple, for example, its colour does not suddenly become undefined. Nevertheless, such things are true of quantum objects. Qubits have been specifically created to be the easiest quantum object for us to access and manipulate. So, by playing with them, we can start to build up some intuition about this quantum behaviour.

We now know enough to write down some rules that any quantum bit must obey.

1. As a type of bit, it can only give us values of 0 and 1.
2. As a quantum object, it has multiple incompatible attributes.
3. As a type of bit, it can store no more than a single binary value.

Put 1 and 2 together, and we find that a qubit must have multiple attributes that can be measured, but all must give values of only 0 or 1. Put 2 and 3 together, and we find that only one of these attributes can be useful to store a binary value at any one time. This is because assigning a definite value for one of the attributes forces the others to be completely random.

9.2 Properties of an Initialized Qubit

As in the last section, we will be considering the concept of quantum circuits. We will also repeat the trick of having a circuit called qc, which contains all the manipulations we want to do before extracting an output:

```
qc = QuantumCircuit(1,1)
```

Next, we will use an additional circuit to extract the output, via the process known as 'measurement'. The standard form of measurement, as we used in the previous section, is implemented simply using the measure gate in Qiskit:

```
m_z = QuantumCircuit(1, 1)
m_z = m_z.measure(0, 0)
```

Whenever one simply refers to 'measurement' in quantum circuits, this is usually what is meant. However, to be more specific, it is known as the 'z measurement'.

The circuit qc defined above is an empty circuit. Since nothing is done to it, it is in the standard state for a freshly initialized qubit. In order to finding out how this behaves, we need to extract an output by combining qc with the measurement circuit and running it:

```
backend = Aer.get_backend('aer_simulator')
job = backend.run(qc.compose(m_z), shots=10, memory=True)
result = job.result()
outputs = result.get_memory()
```

As we might expect from the previous section, the result is 0 with certainty:

```
['0', '0', '0', '0', '0', '0', '0', '0', '0', '0']
```

This is the behaviour we can expect from a freshly initialized qubit whenever it is measured using m_z. However, this is not the only way to measure. Another method is the so-called 'x measurement'. This is done with the following circuit, m_x:

```
m_x = QuantumCircuit(1, 1)
m_x.h(0)
m_x.measure(0, 0)
```

As you might have noticed, this requires an extra element compared to m_z. This is because, at the time of writing, Qiskit doesn't allow us to ask for the x measurement directly, and so we need to use a little hack.

By running the combination of qc and m_x we will get a random mixture of 0s and 1s as our results. To see the probability for each possible output, we need to extract many results. For this it is more convenient to get the results in a less verbose form, which simply lists each type of result along with the number of samples for which it occurred. The syntax for this is as follows, for 1000 samples:

```
backend.run(qc.compose(m_x), shots=1000).result().get_counts()
```

This gets the so-called 'counts' dictionary. Here's an example of the kind of thing you'll see:

```
{'1': 499, '0': 501}
```

Here the two possible results, 0 and 1, both occur roughly 500 times out of the 1000 total samples. From this we can conclude that the results are completely random, giving 0 or 1 with 50/50 probability.

There are actually an infinite number of ways to extract an output from a qubit. But we can create a complete description by using just three. The x and z measurements are two of them. The final one is known as the 'y measurement':

```
m_y = QuantumCircuit(1, 1)
m_y.rx(pi/2, 0)
m_y.measure(0, 0)
```

By running the combination of qc and m_y we will again get a random mixture of 0 s and 1 s as our results. With more samples, we can confirm that it is again 50/50. With statistics on the results of these three types of measurement, we can fully characterize a single qubit state.

Here we characterized the state that a qubit is in when initialized at the beginning of a circuit. The result was that such a qubit is certain to give an output of 0 for m_z, but gives random results for the other two.

Now we can investigate the properties of other possible qubit states. We will do this by applying all of the possible single-qubit operations to a qubit.

9.3   The X Gate

Let us examine the effects of the quantum operator known as the X gate:

```
qc = QuantumCircuit(1)
qc.x(0)
```

Since we are now well acquainted with all the important ways to get an output, we can run them all for this new qc and see what happens. We will find that the x and y measurements still give random results, but the z measurement is now certain to output a 1. The X gate simply flips the bit value that is output for m_z. It serves as the quantum form of a standard NOT logic operator, as discussed above.

If we add another X gate to the circuit, we will see the flip effect again: the output for m_z will go back to being 0.

9.4   The H and RX Gates

Next, let us examine the behaviour of the H gate:

```
qc = QuantumCircuit(1)
qc.h(0)
```

If we run this with our three types of measurement, we will find that it is now m_y and m_z that give random results, and m_x for which the outcome is certain to be 0. This is an effect of what we might call the 'conservation of certainty' in quantum systems. The H gate has unlocked the ability to store a definite bit value in the results of m_x. However, it also takes away the ability to do so with m_z. This prevents us from storing 2-bit values in the same qubit, and so preserves its identity as truly being a type of bits.

A similar effect can be done using the RX gate that we saw earlier. This requires an additional parameter: an angle expressed in radians. For example, with the angle −pi/2:

```
qc = QuantumCircuit(1)
qc.rx(-pi/2, q)
```

The effect of this is to make it m_y that gives the output 0 with certainty, and the other two that are random.

For both H and RX gates with this specific angle, the qubit was certain to output 0 for one of the measurements. With other gates we can make these certain to be 1 instead. The simplest way is to add an x gate at the beginning of each circuit.

With the gates we have so far, we are able to set the output of a qubit to be 0 or 1, and also to change which of the measurements see randomness and which see certainty. These represent the most basic kinds of operations, known as the Clifford gates.

9.5 The Bloch Sphere

The simplest way to move beyond the Clifford gates is to revisit the RX gate, and simply use an angle that is not a multiple of pi/2. Specifically, let us use the following code:

```
qc = QuantumCircuit(1).
qc.rx(-pi/4,q).
```

The results for this will not be as clear cut as we've seen before. Here is an example of what you might see for 1000 samples for each type of measurement:

Results for an x measurement: {'1': 488, '0': 512}.

Results for a y measurement: {'1': 851, '0': 149}.

Results for a z measurement: {'1': 152, '0': 848}.

Here we find that the results from m_x are completely random, but the other two are not. The results from m_z have some randomness, but with a bias towards 0.

The results from m_y are similarly random but with a bias towards 1. The limited certainty of the qubit has therefore been shared between these two possible types of output. Both are mostly, but not completely certain of what output to give.

To fully understand what is happening here, we need a way of visualizing the results. Specifically, we will plot the probability for the outcomes 0 and 1 for each of the three types of measurement on a 3D plot. Since this is well known to be a useful thing to do, it has a name: the Bloch sphere.

Let us first use this visualization on an empty circuit: one that outputs a 0 with certainty for a z measurement, but random results for the others:

```
qc = QuantumCircuit(1)
```

The state of the qubit is represented as a point in 3D space. The x, y and z axes are used to show the probabilities for x, y and z measurements, respectively. If the output is certain to be 0 for each measurement, the qubit state is depicted on one extreme along the corresponding axis. If it is certain to be 1, it is depicted on the other side. For a completely random result, it is in the middle. In this case the certainty of a 0 for the z measurement puts the point at the very top of the image. For the x and y axes, the point lies in the middle. The state corresponds to the point labelled |0⟩ (Fig. 3).

Next, let us plot the state after applying the X gate to the qubit, for which the z measurement is certain to output 1:

```
qc = QuantumCircuit(1)
qc.x(0)
```

Here the point is at the bottom of the Bloch sphere (Fig. 4). The positions of these outcomes are completely opposed to each other. The state in this case has been labelled |1⟩.

Fig. 3 Bloch sphere representing a qubit in state |0⟩

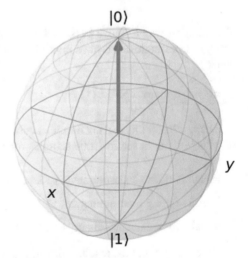

Fig. 4 Bloch sphere
representing a qubit in state
|1⟩

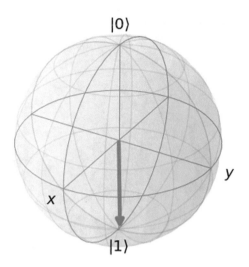

We find similarities for the states whose output is certain for the x measurement. Let us examine again the state where the H gate is used to ensure that the qubit is certain to output 0 for the x measurement:

```
qc = QuantumCircuit(1)
qc.h(0)
```

Although the visualization does not include a specific label for this state, it is usually referred to as the |+⟩ state (Fig. 5).

Let us now construct a circuit that is certain to output 1 for an x measurement:

```
qc = QuantumCircuit(1)
qc.x(0)
```

Fig. 5 Bloch sphere
representing a qubit in state
|+⟩

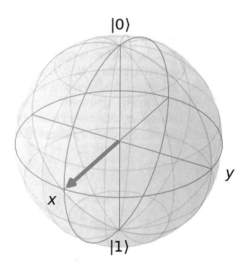

Fig. 6 Bloch sphere
representing a qubit in state
|−⟩

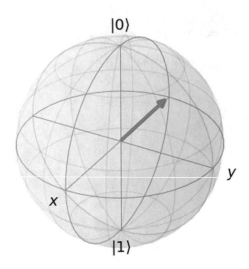

```
qc.h(0)
```

As we should expect, the corresponding point lies on the opposite side to the one above. It is a state known as |−⟩ (Fig. 6).

The states corresponding to the certainty of a 0 or 1 outcome for y measurement similarly correspond to the opposite points on the y-axis. This then gives us six extreme points on the surface of a sphere. Each represents an outcome that is certain for one of the three types of measurement, but completely random for the other two.

With other gates, we can explore the entire surface of the sphere. The RX gate for a given angle corresponds to the rotation of the state around the x-axis by that angle. There are similarly ry and rz gates that rotate around their respective axes.

For example, here is the −pi/4 rotation with RX gate that we considered earlier, with the point corresponding to the state highlighted as a blue point. From its placement, it is easy to see why it is biased towards 0 for a z measurement and 1 for a y measurement: it lies exactly between those points (Fig. 7).

The fact that the qubit state appears bound to the surface of a sphere is a consequence of the conservation of certainty. It is an expression of exactly how the trade-off for certainty between the different measurements works.

Now we have seen all that needs to be known about a single qubit. All single-qubit manipulations can be constructed from the gates discussed above. Everything else, like special names for specific choices of angle and axis, is just administration.

Fig. 7 Bloch sphere representing a qubit after the application of the RX gate with angle –pi/4

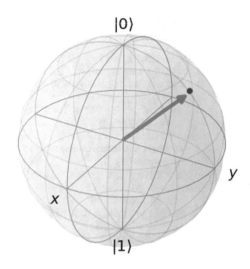

9.6 Single-Qubit Terrain Generation

Using the properties of the single qubit, we will now devise an algorithm for terrain generation. This is based upon a proof-of-principle game created for the *Ludum Dare 45* game jam (Wootton, 2019).

This algorithm for procedural generation is based around a single function: `get_height(i, j)`.

The function above will simply tell us the height of the terrain at any given point, specified by the coordinates i and j. Note that here we use i and j as coordinates, rather than the usual x and y because this section has already been using X and Y as gates, and x and y as types of measurements.

The height in this function is calculated using a single-qubit quantum circuit. Specifically, gates are applied that depend upon the coordinates of the point. The choice of gates presented here was chosen quite arbitrarily. The reader is encouraged to play around with alternatives.

One aspect of the choice of gates that was done very deliberately was to ensure that the circuit created for any given point (i, j) is quite similar to that for its neighbours, to ensure smooth terrain. For a specific example of this, let us consider a circuit that consists of an rx and a ry rotation. For these, we shall slowly change the angle for the RX gate as we move from point to point along the horizontal axis of the map, and similarly change the angle for ry as we move along the vertical axis of the map. Note that this is an arbitrary choice, and does not reflect any deep meaning. There is no connection between the axes on the map and those of the Bloch sphere. It was chosen simply because it seemed like a nice choice. Again, the reader is encouraged to play around with alternatives.

After these gates, a z measurement is performed and the probability of getting the output 1 is calculated. This probability could itself be used as the height. However, it was found that more appealing terrain results when using p as the height. Again, this is a rather arbitrary choice and the reader can try different approaches. Below is the resulting function:

```
def get_height(i, j):
    qc = QuantumCircuit(1, 1)
    # perform rotations, whose angles depend on i and j
    qc.x(0)
    # low frequency rotations to create island shape
    qc.rx((1/32)*i*pi, 0)
    qc.ry((1/32)*j*pi, 0)
    # perform a z measurement
    qc.measure(0, 0)
    # determine the probability of a 1
    counts = backend.run(
        qc,shots=1000
    ).result().get_counts()
    try:
        p = counts['1']/1000
    except:
        p = 0
    # return p² as the height
    return p**2
```

In order to see what kind of terrain this generates, we can run it for a set of points and plot the output. Figure 8 plots the results for a 30×30 set of points around the origin. A terrain colour map is used, which colours low values as blue, and then green, up to high values as brown and then white.

Though this is unmistakably some terrain, it clearly isn't very interesting or realistic. In order to make it nicer, we can add some more gates. As one example, we could break up the single peak by adding gates for which the angles of rotation make larger changes as we move from one point to its neighbours, as follows (Fig. 9):

Fig. 8 Example of a simple terrain generation

Fig. 9 Slightly more
sophisticated example of
terrain generation

```
def get_height(i, j):
    qc = QuantumCircuit(1, 1)
    # perform rotations, whose angles depend on i and j
    qc.x(0)
    # low frequency rotations to create island shape
    qc.rx((1/32)*i*pi, 0)
    qc.ry((1/32)*j*pi, 0)
    qc.rx((1/16)*i*pi, 0)
    qc.ry((1/16)*j*pi, 0)
    qc.rx((1/8)*i*pi, 0)
    qc.ry((1/8)*j*pi, 0)
    # perform a z measurement
    qc.measure(0,0)
    # determine the probability of a 1
    counts = backend.run(
        qc,shots=1000
    ).result().get_counts()
    try:
        p = counts['1']/1000
    except:
        p = 0
    # return p² as the height
    return p**2
```

The example shown in Fig. 9 still is not very realistic. In order to create something
better, one should experiment with various new combinations of gates. We could also
try adding in a randomly generated seed. Note that this would not mean using the
randomness of the quantum computer: our calculation and use of a probability for the
height is an effort to avoid this randomness and make something deterministic from
it instead. Rather it is a random choice of parameters in the circuits we run in order to

make each map unique. There are many ways that the process could be seeded, so we again need to make an arbitrary choice. Since this method for procedural generation was developed for a game jam, the method for seed generation was inspired by the theme of the jam. In this case, the theme of *Ludum Dare 45* was 'Start with nothing'. For this reason, the game starts with no seed, and hence no terrain. As the player explores, their (presumably random) path was used to generate the world as they explore it.

Another way of improving this method is to not just run the circuit with a z measurement, but with x and y measurements too. This additional information could then also be used to determine the characteristics of the terrain at any given point.

There are many further ideas that can be explored, but the limitations of this method must also be remembered. Manipulating a single qubit is, as we saw from the Bloch sphere, much like rotating a ball. As such, we won't get anything uniquely quantum that could not be done using standard tools for 3D rotations.

10   Conclusion

The computer games industry is, by definition, dependent on computational technology to develop ever more engaging experiences. As such, new tools enabled by fault-tolerant quantum computers will certainly be seized upon as soon as they become available.

But there is more to computer games than the industry. It is also an area beloved by hobbyists, used for artistic expression, and a basis for education. These are the approaches that will see the first benefits from quantum computers, as people begin to explore the new technology through the medium of making games.

As with games for conventional computers in the 1950s and 1960s, it is likely that we will see many examples of games being useful for quantum computers before we see quantum computers become truly useful for games. The years ahead will be ones of experimentation, with quantum computing rising from the grassroots of game jams projects and indie developers. It's hard to tell when quantum approaches will reach the big AAA game studios, but we will have a lot of fun on the journey!

References

Benioff, P. (1980). The computer as a physical system: A microscopic quantum mechanical Hamiltonian model of computers as represented by Turing machines. *Journal of Statistical Physics.,* 22(5), 563–591.

Q Cards. https://zhamul.itch.io/qcards

Copeland, B. J. (2004). *The essential Turing* (pp. 563–564). Oxford University Press

Jordan, S. (2021) Quantum algorithm zoo. https://quantumalgorithmzoo.org

Qiskit. https://qiskit.org

QiskitBlocks. https://github.com/JavaFXpert/QiskitBlocks

QPong. https://github.com/HuangJunye/QPong

IBM Quantum Composer and Lab. https://quantum-computing.ibm.com

Samuel, A. L. (1959). Some studies in machine learning using the game of checkers. *IBM Journal of Research and Development, 3*(3), 210–229.

Wootton, J. R. (2016). Demonstrating non-Abelian braiding of surface code defects in a five qubit experiment. *Quantum Science and Technology, 2*(1).

Wootton, J. R. (2017). How to program a quantum computer. Retrieved December 03, 2020, from https://medium.com/qiskit/how-to-program-a-quantum-computer-982a9329ed02

Wootton, J. R. (2018). Benchmarking of quantum processors with random circuits. https://arxiv.org/abs/1806.02736

Wootton, J. R. (2019). Genesis. https://ldjam.com/events/ludum-dare/45/genesis-1

Wootton, J. R. (2020a). A quantum procedure for map generation, *2020 IEEE Conference on Games (CoG)*, 73-80.

Wootton, J. R. (2020b). Procedural generation using quantum computation, *International Conference on the Foundations of Digital Games*, 98.

Printed in the United States
by Baker & Taylor Publisher Services